WEYERHAEUSER
ENVIRONMENTAL
CLASSICS

William Cronon, Editor

WEYERHAEUSER ENVIRONMENTAL CLASSICS are reprinted editions of key works that explore human relationships with natural environments in all their variety and complexity. Drawn from many disciplines, they examine how natural systems affect human communities, how people affect the environments of which they are a part, and how different cultural conceptions of nature powerfully shape our sense of the world around us. These are books about the environment that continue to offer profound insights about the human place in nature.

The Great Columbia Plain: A Historical Geography, 1805–1910 by D. W. Meinig

Mountain Gloom and Mountain Glory: The Development of the Aesthetics of the Infinite by Marjorie Hope Nicolson

Tutira: The Story of a New Zealand Sheep Station by Herbert Guthrie-Smith

A Symbol of Wilderness: Echo Park and the American Conservation Movement by Mark Harvey

Man and Nature: Or, Physical Geography as Modified by Human Action by George Perkins Marsh; edited and annotated by David Lowenthal

Conservation in the Progressive Era: Classic Texts edited by David Stradling

DDT, Silent Spring, *and the Rise of Environmentalism: Classic Texts* edited by Thomas R. Dunlap

Reel Nature: America's Romance with Wildlife on Film by Gregg Mitman

WEYERHAEUSER ENVIRONMENTAL CLASSICS is a subseries within WEYERHAEUSER ENVIRONMENTAL BOOKS, under the general editorship of William Cronon. A complete listing of the series appears at the end of this book.

Gregg Mitman

REEL NATURE

America's Romance with Wildlife on Film

UNIVERSITY OF WASHINGTON PRESS
Seattle and London

Reel Nature: America's Romance with Wildlife on Film is published with the assistance of a grant from the Weyerhaeuser Environmental Books Endowment, established by the Weyerhaeuser Company Foundation, members of the Weyerhaeuser family, and Janet and Jack Creighton.

Library of Congress Cataloging-in-Publication Data

Mitman, Gregg.
Reel nature : America's romance with wildlife on film / Gregg Mitman.
p. cm. — (Weyerhaeuser environmental classics)
Originally published: Cambridge, Mass. : Harvard University Press, 1999.
Includes bibliographical references and index.
ISBN 978-0-295-98886-3 (pbk. : alk. paper)
1. Nature films—History and criticism. I. Title.
PN1995.9.N38M58 2009
791.43'662—dc22 2009003439

The paper used in this publication is acid-free and 90 percent recycled from at least 50 percent post-consumer waste. It meets the minimum requirements of American National Standard for Information Sciences—Permanence of Paper for Printed Library Materials, ANSI Z39.48-1984.

To Debra

ACKNOWLEDGMENTS

Books, like nature films, are the product of many individuals, whose invaluable contributions and efforts remain hidden behind the scenes only to surface in the acknowledgments or credits. This book is no exception. Although my thanks will never repay the debt I owe to those who made this book possible, it at least gives credit to their influence and role in the creative process.

Without the resources of numerous archive and film collections and the help of on-site staff, this research project would never have been possible. These include the American Heritage Center, American Museum of Natural History, Bodleian Library of Oxford University, British Film Institute, Department of Herpetology Archives in the American Museum of Natural History, Department of Rare and Manuscript Collections at Cornell University, Institut für Wissenschaftlichen Film, Library of Congress, Lincoln Park Zoo, Marineland, Museum of Broadcast Communications, Penn State Cinema Register, Rice University, Rockefeller Archive Center, Walt Disney Archives, Western Historical Manuscript Collection at the University of Missouri–St. Louis, Western History/Genealogy Department at Denver Public Library, and the Wildlife Conservation Society. I am especially grateful for the generosity and help Matthew Gies, Madeline Matz, Andrea LaSalla, Tom McKenna, Chuck Myers, Tom Veltre, and Joanne Whaley gave in the use of archive, motion picture, and video collections. Grant support from the National Science Foundation and the University of Oklahoma Research Council provided the necessary funds for research.

My colleagues in the Department of the History of Science at the University of Oklahoma offered the support that enabled me to complete this book in the expected time frame. I am particularly thankful for their graciousness in granting extended research leaves that came at critical points in the writing process and for their intellectual companionship while in Norman. The Department of the History of Science at the University of Wisconsin–Madison welcomes me every summer. The intellectual and social hospitality they extended during a year spent in Madison is warmly appreciated. To the Shelby Cullom Davis Center for Historical Studies at Princeton University and to my comrades there, I extend a special thanks. The Davis Center and its fellows provided an idyllic atmosphere for the final year of writing and reflection.

Many institutions provided forums for productive intellectual exchange that helped in clarifying my ideas and writing. I am grateful to the history of science and science studies programs at Cornell University, Harvard University, Indiana University, Johns Hopkins University, Max Planck Institute, Oregon State University, the University of California–San Diego, the University of Minnesota, the University of Missouri–St. Louis, the University of Pennsylvania, the University of Washington, and Washington University for the opportunity to discuss my work in progress.

Chip Burkhardt, Kevin Dann, Mary Fissell, Betsy Hanson, Karen Merrill, Katherine Pandora, Donald Pisani, François Pouillon, and Zev Trachtenberg, in addition to two anonymous referees, offered invaluable advice and criticism on the manuscript that immensely improved the shape and content of the book. Ari Kelman and David Lobenstine offered helpful comments in the final stages of revision. To graduate students in the traveling road show—Juan Ilerbaig, Gary Kroll, Karin Matchett, Maureen McCormick, and John Snyder—I have a fond appreciation for your willingness to endure such an experiment that was critical to the formulation of many ideas that appear in the book's pages. Thanks also go to John Beatty, Jim Collins, and

Kevin Dann for their willingness to give it a try. Laurel Smith offered valuable research assistance and enthusiasm at the beginning stages of this project. My colleagues in the Interdisciplinary Perspectives on the Environment Program—Rajeev Gowda, Zev Trachtenberg, and Linda Wallace—have been crucial in sustaining an intellectual atmosphere conducive to interdisciplinary thinking on environmental issues central to this project.

No words can quite express the gratitude I feel to the following individuals. Ron Numbers had enough confidence in my scholarship and writing to challenge me to write for a general readership. Joyce Seltzer, my editor, took his recommendation and worked with me patiently and tirelessly. Demanding, yet compassionate, Joyce never failed to bring out the best I could offer, even if at times I couldn't find it. Debra Klebesadel, my wife, never lost faith. She sacrificed family and place, adopting the life of an academic nomad these past few years, to see the project to completion. Intellectual companion, mother of our two children, lover and friend, she is also a talented copy editor who improved my prose considerably. Her work is in these pages as much as mine.

CONTENTS

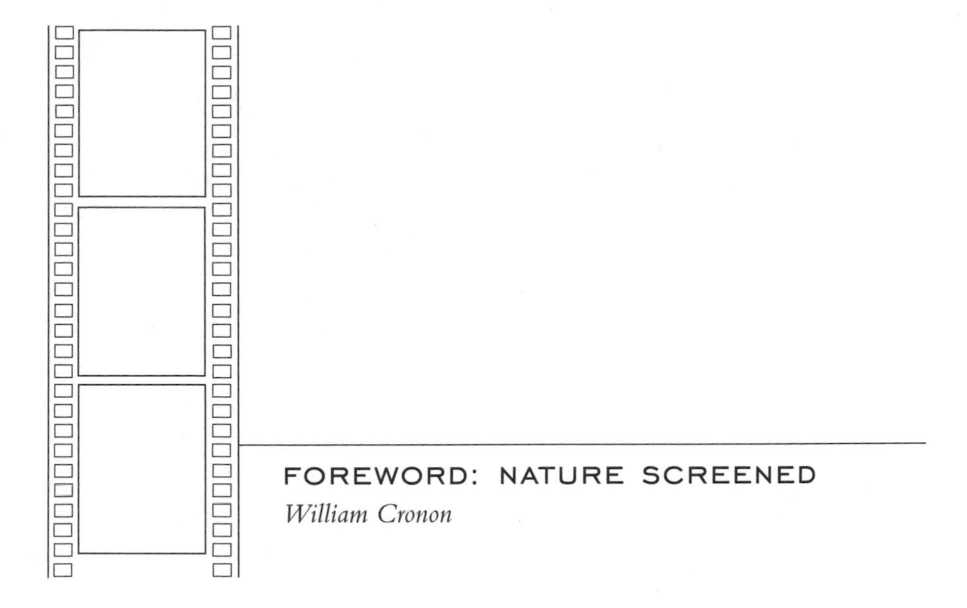

FOREWORD: NATURE SCREENED

William Cronon

In what is surely the most famous passage in Plato's *Republic*, the Greek philosopher Socrates describes a cave in which prisoners, for their entire lives, are shackled with chains so completely that they cannot move their heads and so can only gaze at the rock wall in front of them. A large fire burns behind them, and between the prisoners' backs and that fire is a raised walkway along which various objects—statues and puppets of animals and other figures—can be moved so that shadows of these things are cast upon the wall before the prisoners. In such a situation, Socrates says, the moving shadows are the only reality the prisoners can experience, so that for them the dark outline of a dog or a cat or a bird *is* that creature. The allegory concludes by asking us what it would be like for the prisoners to be released and taken out into the real world to see things in all their three-dimensional reality: not just the puppets whose silhouettes cast shadows on the wall of the cave, but also the creatures those puppets represent, the fire that had cast those shadows, and the sun whose brilliance makes even that fire seem dim indeed. After a period of disorientation and disbelief, the prisoners are forced to rethink their entire past experience and reimagine both themselves and the world.

For Socrates, the allegory of the cave becomes a way of talking about the ideal forms that he and Plato believed constituted the true divine reality, in relation to which the world we inhabit is but a pale reflection. But for those who have lived in the era of photographs and motion pictures, of radio and television, of the video camera and the World Wide Web, this ancient story may suggest quite different meanings. Its deepest insight has always been about the role of representation in human consciousness: the fact that people live as much inside their own heads as they do in the material world, and that the shapes and forms with which we think and experience are never in a one-to-one correspondence with the things and realities we imagine them to depict. Living as we do in what Walter Benjamin called "an age of mechanical reproduction," one is tempted to claim that we now sometimes experience virtual nature with greater intensity and emotional power than we do the non-virtual nature we physically inhabit. Anyone who has watched a modern teenager sitting for hours in rapt attention before a video game—or who has tried to persuade that teenager to leave the game for an activity in the world beyond the screen—will recognize how seductive the pull of the virtual has become in our own time.

Yet this seemingly simple contrast between the virtual and the real does not begin to convey how much the two are entangled in the ways we move through the world. Take, for instance, how we understand wild animals. For most modern city-dwellers, in the United States and elsewhere, observing wild creatures requires long journeys to remote locations, and even then they are generally only seen from afar. Because rats, pigeons, cockroaches, and their ilk are typically assigned to the very different taxonomic category of "vermin," few urban Americans can claim to have regular encounters with wildlife—except, that is, on the screen. One does not have to be a regular viewer of cable networks devoted largely to nonhuman nature—the Discovery Channel, National Geographic Wild, the WildLife Channel, Animal Planet—to see with some frequency images of wild beasts that have far greater visual intimacy and impact than most of us have ever witnessed firsthand. What we now see and know about wild animals comes to us more often than not via television screens, movie theaters, and Web sites. These new virtual relationships with animals have had profound consequences for environmental politics, encouraging people to make intense emotional (and financial) investments in organisms and places with which they might otherwise have no conscious relationship at all.

How this happened—how the virtual nature of wildlife film has come to dominate modern American attitudes toward the creatures with whom we share this planet—is the story Gregg Mitman tells so brilliantly in *Reel Nature: America's Romance with Wildlife on Film*. First published in 1999, the book was quickly recognized as the best and most comprehensive history yet produced of wildlife films and their place in American popular culture.

Mitman approaches this complex subject from diverse and complementary perspectives. He provides, for instance, a fascinating overview of how wildlife films have been produced and marketed since the early twentieth century, when Theodore Roosevelt, only recently departed from the White House, starred in a documentary shot by the British photographer Cherry Kearton depicting one of TR's big-game hunting expeditions to Africa. The film did less well than Roosevelt and Kearton had hoped, mainly because it was visually uninteresting and narratively boring. Inspired by their example, however, an entrepreneur named Colonel William Selig—who tried unsuccessfully to get Roosevelt to reprise his role from the earlier film—shot *Hunting Big Game in Africa* in, of all places, a Chicago studio. Far more melodramatic, with vaudeville-inspired scenes and an actor impersonating Roosevelt stalking a lion in the wilds of Africa, Selig's film made a good deal more money than Kearton's "authentic," but tedious, *Roosevelt in Africa*. Selig's success led him to establish a game farm in Los Angeles where he produced a series of popular jungle adventure films over the next several years.

Here, at the very birth of the wildlife documentary, Mitman argues, were many of the tensions that would typify the genre right down to the present. How much should a filmmaker trade authenticity for entertainment? To what extent are reenactments a legitimate way of obtaining footage that is essential to a story but not easily captured in the field? If a filmmaker intercuts sequences of genuinely wild animals with sequences shot in the easier-to-control conditions of a zoo or a cage or a film studio—or if some of the animals before one's camera are drugged or dead—does this violate an implicit contract with the audience? (And is there an obligation to inform the audience of such tricks?) Can a wildlife film be scripted ahead of time, or is a filmmaker obligated to record "what really happened" on an expedition—however boring that might be to watch? Although techniques for representing and telling stories about animals on film have become ever more intricate over time—having now reached the point that digital technologies can replace animals and actors altogether with computer-generated pixels—the essential dilemmas of wildlife filmmaking have never gone away.

This is undoubtedly because such dilemmas, at their core, aren't mainly about film or cameras or digital technology. They originate in the imperatives of human storytelling, with profound implications not just for box office success and financial viability, but also for the extent to which audiences will empathize and identify with what they see on the screen. Persuading audiences to care about an animal, for instance, requires a filmmaker to discover something in that animal that renders it intriguing or surprising or sympathetic, or some other quality that makes us want to know it better and follow its story. More often than not, this tempts filmmakers to project onto nonhuman creatures attributes that we typically associate with human beings. Do we see animals struggling to make a living? Caring for their young? Enjoying themselves in lighthearted play? Fleeing a terrifying villain? If so, we recognize in them the kinds of stories we tell about ourselves. Filmmakers seeking to attract our attention—and our ticket dollars—will lean into such stories in their efforts to persuade us to watch.

For just this reason, a genre that one might imagine to be more the provenance of wildlife biologists than of Hollywood studios has had an enduring relationship with commercial cinema right from the start. Documentaries narrating African safaris helped lay the groundwork for *King Kong*, and the initially disappointing box office response to *Bambi* led Disney to produce the series of "True-Life Adventure" films—*Seal Island*, *Beaver Valley*, *Nature's Half-Acre*, *The Vanishing Prairie*, and others—that became such a defining feature of wildlife filmmaking in the 1950s. Such films inspired thousands of Americans to want to see for themselves the creatures Disney depicted on the screen. Disneyland itself became one of the destinations designed to meet growing American demand for direct encounters with the well-controlled (and often mechanical) animals that Disney's "imagineers" created for the public, but ever-grander zoos and aquaria and wildlife theme parks were also among the beneficiaries of this cinematically generated demand. Marlin Perkins's *Mutual of Omaha's Wild Kingdom* made carefully scripted wildlife documentary available to

TV audiences each week, while television programs like *Flipper* and *Sea Hunt* created fictionalized dramas for baby boomers fascinated by stories in which people-animal interactions were central to the drama. Such narratives depicted animals behaving in ways that reinforced the gendered family roles idealized in 1950s sitcoms, increasing their appeal to middle-American audiences. When the filmmakers of *Flipper* chose to excise the rambunctious sexual behaviors of dolphins, which might have been more at home on the pages of *Playboy*, perhaps they weren't so different from sitcom producers in that respect either.

Mitman's exploration of the role that films have played in our understanding of wildlife and of the natural world generally is so subtle and wide-ranging that I cannot do it justice in so brief a foreword as this. Suffice it to say that few books offer greater insight into the cultural history of American ideas of nature during the twentieth century than this one. Whether one cares about the history of filmmaking, the evolution of American environmental attitudes, or the ways that science, commerce, and entertainment have shaped each other in the creation of American consumer culture, *Reel Nature* is essential reading. We are delighted to be able to publish it for the first time in paperback as a Weyerhaeuser Environmental Classic. The recent resurgence of interest in environmental film—sparked by the 2005 blockbuster *March of the Penguins* and Al Gore's 2006 documentary *An Inconvenient Truth*—has made understanding the impact of film on environmental attitudes and politics all the more urgent. In an extended new afterword written especially for this edition, Mitman reflects on trends in environmental filmmaking since the book was first published and suggests ways that new linkages among art, science, and activism are being forged to create media that matters in the lives of people and animals throughout the world. If one wishes to comprehend why and how we came to think and feel about nonhuman nature in the cinematic ways we do—experiencing it to a surprising degree via projected images cast like shadows on the walls of Plato's cave—there is no richer or more rewarding guide than this classic study.

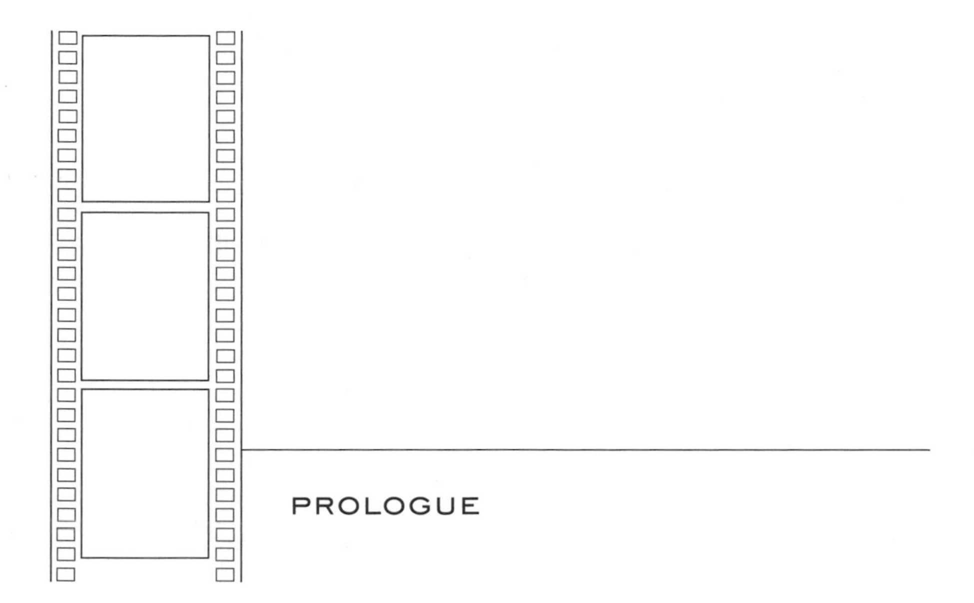

PROLOGUE

In the spring of 1998, the Walt Disney Company opened its newest entertainment attraction, Animal Kingdom. A five-hundred-acre live-animal theme park located on the western edge of Walt Disney World Resort near Orlando, Florida seemed an odd venture for the company associated more with fantasy animals than those of flesh and bone. In the 1950s, Disney's animators, designers, and engineers transformed the company's two-dimensional cartoon fantasy world into Disneyland, a three-dimensional magic kingdom. But Animal Kingdom brought a new challenge to Disney imagineers in the 1990s. It was a challenge that required the advice and expertise of wildlife biologists, conservation organizations, and zoos to recreate an African savanna out of the flat landscape of central Florida cow pasture.[1]

Complete with rolling plains, exotic vegetation, and one thousand wildlife residents, Animal Kingdom is at once an environmental learning center and a tourist's playground. Here nature is scripted—contrived by human hands—to tell a story that is "at once both natural and fantastic," from the thrill ride Countdown to Extinction in Dinoland to the Tree of Life, a "technological marvel" that Disney imagineering's executive designer Joe Rohde describes as "a symbol of the beauty and diversity and the grandeur of animal life on Earth." At the heart of Safari Village, the crossroads where visitors "pass to reach other lands," the synthetic Tree of Life towers fourteen stories high; its massive fifty-foot wide trunk supports a canopy 160 feet in diameter. Some 350 animal carvings adorn its base and sturdy branches, engineered to sway in the wind. Sheltered in the tree's interior is a 430-seat movie theater where spectators can watch a 3D animated adventure from a bug's point of view.[2]

It cost Disney an estimated $800 million to construct Animal Kingdom. The undertaking involved the growth and transplantation of roughly 100,000 trees; the importation of 1,000 animals representing 200 wildlife species; and the inadvertent deaths of at least a dozen animals including cheetahs, rhinoceroses, Asian small-clawed otters, and African crowned cranes as they made the transition from wild jungle to synthetic habitat. In this fantastic creation of authentic nature, immense labor and expense were required.[3]

Intended to lead guests on "dramatic adventures into the mysteries and marvels of the live animal world," Animal Kingdom narrates a story of wildlife and conservation that, in the words of Disney Chairman Michael Eisner, both "informs and entertains." Education and adventure await tourists on the nineteen-minute Kilimanjaro Safari ride. By transporting guests in open-sided, camouflaged trucks, the photographic safari provides close-up encounters with herds of African animals, including elephants, giraffes, and rhinoceroses, on a savanna "so convincingly real that visitors from every corner of Africa itself have said it looks . . . 'just like home.'"[4]

In this packaged tour across the simulated African veldt, the wildlife and conservation messages cannot be avoided. They start while visitors wait in line for the ride within easy view of monitors that show slain elephants harvested for ivory. They continue during the ride as the adventurers participate in a thrilling, fabricated hunt for the ivory poachers responsible. In Animal Kingdom, wildlife poaching is a "social evil." But at forty-five dollars for a one-day adult admission, Disney can equally be viewed as exploiting nature. Animal Kingdom is after all a commercial enterprise. Near Universal Studios Florida and Sea World, the park is hoping to cash in on nature as entertainment in a location that has become an international tourist playground.[5]

Animal Kingdom's nature is a house that Disney built. The park's enchantment grows from the experience of Disney imagineers who created emotion-filled fantasy for over forty years. Yet, everywhere reality intercedes. Constrained by the unpredictable actions of live animals, Animal Kingdom cannot direct this theater of nature on cue. The troops of lowland gorillas on the Gorilla Falls Exploration Trail, for instance, might prefer the privacy of their lush habitat to public display, despite audience expectations. Both animal rights groups and federal regulations prevent the provocation of drama red-in-tooth-and-claw. In Animal Kingdom, nature, except for that produced in the wholly fabricated extinct world of Dinoland, is an innocent and gentle place where animals are "partners in the great web of life."[6]

Designed to tell an "awe-inspiring tale of the interconnected nature of all living things," the park capitalizes on the desire of Americans for intimate contact with wildlife and the natural world. As "a celebration of our emotions about animals and their habitats," says imagineer Rohde, Animal Kingdom is meant to instill an appreciation for the "wonders of Nature." Accordingly, the park has received praise from zoo officials precisely because of its potential to stimulate passionate interest in wildlife conservation. In extolling the park, Sydney Butler, executive director of the American Zoo and Aquarium Association, remarked that "before you teach conservation, you have to fascinate. When you go to this park, you will immediately be fascinated."[7]

Inspired by Disney's True-Life Adventures, a genre of sugar-coated educational nature films initiated by Walt Disney in the late 1940s, the theme park is the latest and most dramatic manifestation of America's ambiguous relationship with nature. Nature films, like naturalistic displays found in animal theme parks, museums, and zoos, have sought to capture and recreate an experience of unspoiled nature. They have blended scientific research and vernacular knowledge, education and entertainment, authenticity and artifice. As art and science, nature films seek to reproduce the aesthetic qualities of pristine wilderness and to preserve the wildlife that is fast vanishing from the face of the earth. As entertainment, they promise enlightenment and thrills simultaneously. In the early twentieth century, these films were patronized by an expanding urban middle and upper class who were drawn to depictions of distant nature as well as its conservation. In the motion picture houses, these audiences could flee the artificiality and complexity of modern life and immerse themselves in a wildlife landscape where instincts and behaviors were seemingly more natural, more authentic. From these early days, to engage and sustain audiences and to insure commercial success, emotional drama had to be made a part of filmed nature. Such fabrication made the line separating artifice from authenticity difficult to discern and even to maintain. Whether crafted to elicit thrills or to preserve and educate audiences about the real-life drama of threatened wildlife, nature films then and now reveal much about the yearnings of Americans to both be close to nature and yet distinctly apart.

"Wilderness," Aldo Leopold wrote, "is the raw material out of which man has hammered the artifact called civilization." Longing for the authentic, nostalgic for an innocent past, we are drawn to the spectacle of wildlife untainted by human intervention and will. Yet, we cannot observe this world of nature without such intervention. The camera lens must impose itself, select its subject, and frame its vision. The history of nature film reverses Leopold's claim. Cultural values, technology, and nature itself have supplied the raw materials from which wilderness as artifact has been forged.[8]

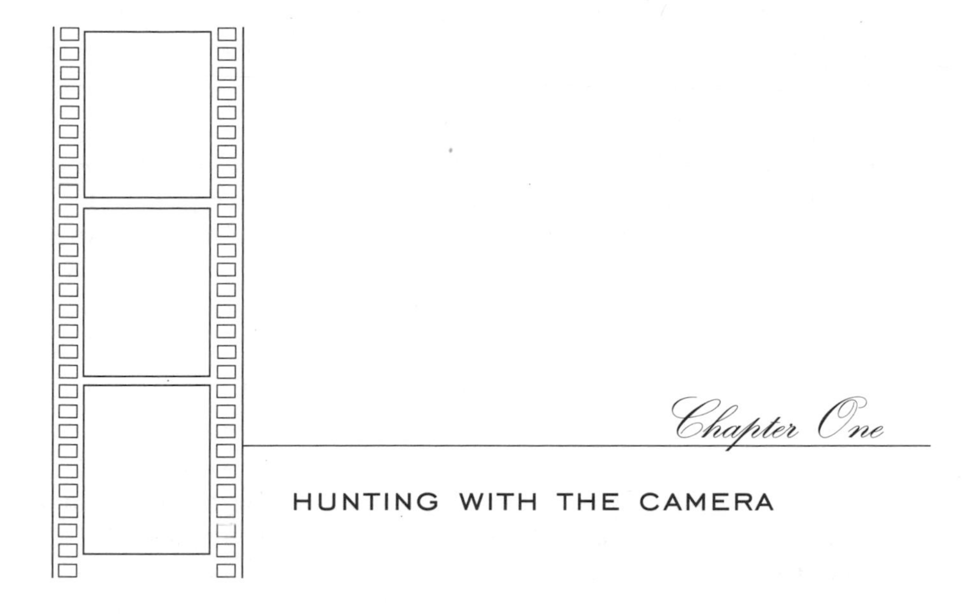

Chapter One

HUNTING WITH THE CAMERA

From the town of Gondokoro in the Congo region of British East Africa, a cable dispatch sent to the *New York Times* on the last day of February in 1910 announced the end of an expedition that had been making front-page copy in American newspapers. Theodore Roosevelt, former president, amateur naturalist, conservationist, and great white hunter, had set sail aboard a steamer headed north on the White Nile for Khartoum. His departure from the Congo marked the end of a year-long expedition undertaken on behalf of the Smithsonian Institution and sponsored by the wealthy industrialist Andrew Carnegie to collect African fauna and flora, particularly big game, for the National Museum in Washington, D.C. A barge in tow contained the final shipment of more than eleven thousand vertebrate specimens; expedition members had killed an unprecedented average of forty animals a day for an entire year. Even without colonial holdings in the "Dark Continent," the United States, through Roosevelt's expedition, brought a vast quantity of African nature to the American metropolis. Transformed through the taxidermist's craft, the Roosevelt collection of lion, zebra, white rhinoceros, Coke hartebeest, and oryx on display in the National Museum represented the technological forefront of animal exhibition—the museum diorama, a realistic portrayal of wild animals in their natural surroundings meant to spellbind the public and serve as material for scientific study.[1]

As a small army of artisans, craftsmen, and naturalists worked to transform the limp, putrefying animal skins from Roosevelt's expedition into vibrant representations of wildlife, audiences in New York City viewed herds of hartebeests and giraffes roaming across the veldt and hippopotami at play in the Tana River in East Africa, shot on Roosevelt's African expedition by the camera, not the gun. Developed in the 1880s to investigate the physiology of animal motion, the motion picture camera had by the early 1900s become associated with entertainment as much as science. Both the film and museum diorama were mechanical reproductions of nature that through their respective technologies sought to capture and preserve the wild animals, which turn-of-the-century naturalists feared were "being rapidly civilized from the face of the earth." Of the two, time would prove the camera to be the vanguard technology. Poised at the intersection of art, science, and entertainment, natural history film would transform American perceptions of and interactions with wildlife over the course of the twentieth century.[2]

When the Motion Pictures Patent Company released *Roosevelt in Africa* to theaters on April 18, 1910, the prospects of natural history film securing a place in American culture seemed anything but assured. The motion pictures of Roosevelt in Africa had been shot by Cherry Kearton, a famous naturalist-photographer from London on safari in British East Africa, with the assistance of James L. Clark, a sculptor and taxidermist on a collecting expedition for the American Museum of Natural History in New York City. Kearton and Clark met the Roosevelt party at their camp in Nyeri in August of 1909. The footage shown to American audiences contained thirty-six scenes in all, and included Roosevelt planting a tree in front of the Bomba Trading Company's Office in Nairobi, a native ceremonial war dance performed in Roosevelt's honor at the request of Governor Jackson, and herds of giraffe at a distance of 150 yards. Some of the most striking scenes were of the courtship dance performed by the Jackson dancing bird, clear shots of hippopotami at play, and a close study of a young serval cat.

None of the shots, however, showed Roosevelt bringing down game, nor was Kearton successful in his attempt to secure the first motion picture footage of lions in the wild. In place of lion footage, Kearton spliced in a flash-picture still of a lion that one reviewer described as "flatter than a pancake. It looks like a dead lion, or a poor wash drawing."[3]

Lacking drama, *Roosevelt in Africa* proved a great disappointment to motion picture audiences. One movie-house exhibitor complained that there wasn't a "picture in the 2,000 feet that is fit to be called a picture. . . . Anybody could take a .22 rifle and go out in the sagebrush in Idaho and get more excitement hunting jack rabbits." Although the majority of audiences shared in this exhibitor's appraisal, a journalist from *The Moving Picture World* noted that the audience response to *Roosevelt in Africa* varied along class lines. At theaters located in working-class neighborhoods like Elizabeth, New Jersey, *Roosevelt in Africa* had little appeal. "The class of people who came there," wrote H. F. Hoffman, "were expecting to see Teddy slaughtering lions and tigers and wallowing in their gore." On Broadway, however, Hoffman found an audience made up of a more "intelligent class of people," who remained in their seats during the entire picture and seemed to enjoy the show. "It would seem," Hoffman reasoned, "that the Roosevelt pictures are going to take better at houses that cater to the middle class than at those who cater to the shopgirl trade."[4]

Roosevelt in Africa appeared at a time when the middle and upper class of American society were struggling to assert control over the motion picture industry. Only two years earlier, Mayor George B. McClellan had shut down every motion picture theater in New York City during the week of Christmas by revoking the licenses of the 550 nickelodeons and movie houses found predominately in the tenement districts and immigrant neighborhoods of the city. While vaudeville or stage theaters paid five hundred dollars in annual license fees to operate in the city, a storefront movie theater paid only twenty-five dollars. For the price of a nickel, three hundred to four hundred thousand working-class men and women a day in New York City found a readily accessible, brief, and cheap form of amusement. It was a welcome diversion from the long hours spent in the workshop and factory. But the dark lighting and dank air of the theater, the crowds of foreigners, unaccompanied young women and children, and, above all, the proclivity toward lewd, violent, and carnivalesque subjects on screen did not readily conform to what established Anglo-Saxon Protestant Americans deemed acceptable public amusement.

The closing of New York City nickelodeons in 1908 was not meant to eliminate the new form of working-class entertainment but rather to clean it up. Although exhibitors were successful in getting a court injunction against the mayor's action, McClellan's move warned producers and exhibitors that some form of regulation, either voluntary or enforced, was in order. In 1909, the People's Institute, a famed settlement house founded by Columbia professor Charles Sprague Smith, took the lead in helping establish the movies as a progressive form of mass entertainment. The Institute brought together prominent New York civic organizations, such as the Federal Council of Churches, the YMCA, and the Society for the Prevention of Crime, and wealthy New York Protestants, such as Andrew Carnegie, to serve on a National Board of Review of Motion Pictures that would screen motion pictures and sanction acceptable films with a seal of approval, indicated by an open scissors overlaid upon a four-pointed star. By working in cooperation with the National Board of Review, producers gained some respectability among the city's charitable, educational, and religious leaders, and thereby opened up wider markets for the showing and promotion of their films. Civic leaders had feared the degenerative influence of motion pictures on the tastes and morals of the American public. But by 1913, respected members of society were more likely to agree with Harry Downer, chairman of Davenport, Iowa's motion picture review committee, that the motion pictures represented a "new social force which reaches more people annually" than any other instrument for "social betterment."[5]

Downer suggested that "science [had] built a delightful means of recreation, a graphic influence in education, [and] a splendid force for moral and wholesome life."[6] His remarks drew attention to the instrumental role biological science played in the technological development of cinema. In 1882, the French physiologist Etienne-Jules Marey invented the chronophotographic gun to record the motion of birds in flight: on a single plate he could take twelve photographs in one second to reveal living processes and movements unobservable to the human eye. Similarly, in 1877, the American photographer Eadweard Muybridge, with an apparatus of twelve cameras operated by trip wires and electronic shutters, exposed the true nature of a horse's gallop wherein all four feet lift off the ground. If Marey, Muybridge, and others contributed to a technology that could control time and motion in the interests of science, by the early 1900s inventors and showmen had made cinema as much a technology of entertainment as of education. Emphasizing the scientific and educational dimensions of this new form of mass entertainment helped counter its image as morally and socially corrupting. When Roosevelt publicly endorsed Cherry Kearton's motion pictures of Buffalo Jones lassoing wild animals in Africa at a preview before the New York Press Club in September of 1910, *The Moving Picture World* took the occasion to salute Roosevelt for adding "enormous prestige" to the important role of motion picture photography in the science of natural history. "When vested interests, when the press, when the pulpit, when the law and learning come out, as they frequently do, to malign and defame the picture, we have the greatest man of his time present at a moving picture, and saying kind things about the scientific value of kinematography."[7]

Educational and religious leaders regarded natural history film as an important venue in the reform of motion pictures. Not only did animal subjects appeal to a wide audience, but they were "entertaining in the best sense of the word and at the same time rich in educational value." Nature films made ideal billings for a Sunday program catering to the family or to more "refined" motion picture audiences. Pictures of animal life furnished a means for reinforcing moral values. When a film on the life of the stickleback fish appeared in 1913, one reviewer praised the "singular lessons of special importance [revealed] to the classes in natural history," particularly the "unselfishness and devotion to offspring . . . shared by the male stickleback," which resembled the "dove in his home-building and family-raising characteristics."[8]

Natural history films might help uplift the masses, but only if such films could draw crowds to the box-office. As the Roosevelt film proved, audiences craved drama over authenticity. Chicago film entrepreneur Colonel Selig, unable to convince Roosevelt to film his adventures for the Selig Polyscope Company, staged his own production of Roosevelt's hunting exploits in his Chicago studio. Selig capitalized on his experiences as a minstrel showman and those of a vaudeville actor talented in Roosevelt impersonations to recreate Roosevelt's tropical expedition in the sub-freezing temperatures of the windy city. Much more popular than Kearton's authorized film, Selig's *Hunting Big Game in Africa* faked a scene of a lion being shot and carried away by native porters. Kearton had tried for months to photograph a similar scene in the wild without success. Selig's melodramatic reenactment of a lion stalked in the jungle by the great white hunter attracted great attention. *The Moving Picture World* found the scenes convincing enough to remark that "there is no doubt about this lion; he stalks majestically about the picture, thus enabling an audience to realize how a lion would look, not on the war path, but peaceably ambling about among natural surroundings. Your captive lion in a zoological park does not do much prowling about except in a small cage."[9]

The financial success of *Hunting Big Game in Africa* convinced Selig of the market for thrilling jungle wildlife adventures, and he established a game farm in Los Angeles where over the next five years he embarked on the production of a series of immensely popular animal films, including *Alone in the Jungle, In Tune with the Wild,* and *The Leopard's Foundling.* Selig's novelty lay in using actresses Bessie Eyton and Kathleen Williams as heroines of his stories. A Selig film was a "blood curdling romance of the dangerous animal infested jungleland of Africa," where in every thicket there may lurk a ravenous or a savage foe. Although Selig's films drew large crowds, his use of sensationalism to appeal to more "base" human emotions and his sacrifice of authenticity for melodrama were elements closely tied to what reformers feared contributed to the degeneration of morals and tastes among the lower class. Some reformers like Downer remained skeptical of the ability of the wholesome, educational picture to commercially compete with pictures of the "base-sort" without philanthropic support.[10]

The difficulty in distinguishing between wholesome education and bawdy entertainment was a problem faced by educators, scientists, and philanthropists who wished to cultivate natural history film for a more serious-minded audience. One could not just dispense with the dramatic, since this would largely eliminate the attendance of the popular theater-going public. Emotional drama was necessary, but the question of whether such drama had been authentically captured in the wild or had been created through artifice in order to elicit thrills and generate mass appeal increasingly became a subject of inquiry and concern.

Two years prior to the release of Selig's faked film, Roosevelt had been engaged in a highly publicized dispute that involved similar issues about the authentic representation of nature. This dispute concerned not the motion picture screen, but the written word. During the opening decade of the twentieth century, a large market appeared for the realistic wild animal story made famous by writers such as Jack London, Ernest Thompson Seton, Charles D. Roberts, and William J. Long. The popularity of animal stories such as Seton's *Wild Animals I Have Known,* which went through sixteen printings in four years, and London's best-selling novel, *The Call of the Wild,* was intimately connected to a back-to-nature movement in full force during Roosevelt's presidential years (1901–1909). The emergence of environmental preservation organizations such as the Audubon Society and the Sierra Club, the federal government's preoccupation with conserving the nation's natural resources, the push for nature study in the public schools, and the growth of landscape architecture initiated by Frederick Law Olmsted's work in the late nineteenth century all reflected an interest among the nation's urban middle class for greater contact with nature. This public fascination with nature was not based on nostalgia for a rural, agrarian past. To the city dweller with a weekend country retreat, the farmer whose livelihood depended on the land had little appreciation for the aesthetic qualities of untouched woodland or meadow. Instead, nature offered a place of regeneration and renewal for a growing managerial middle class with increased leisure and money to pursue the luxuries of country life previously confined to the rich. Literary naturalists such as Seton found in the animal story a genre that corresponded to the artistic, educational, and recreational values prized by the urban middle class in their reverence of nature.[11]

In March of 1903, the acclaimed Catskill nature writer, John Burroughs, used the pages of the *Atlantic Monthly* to lambaste the growing literary genre that he dubbed "mock natural history." In Burroughs's evaluation, literary naturalists such as London, Long, and Seton had simply seized on the public's fascination with nature and turned it "into pecuniary profit." Burroughs disdained the sentimentalism and anthropomorphism found in books such as Seton's *Wild Animals I Have Known* and Long's *School of the Woods* that claimed to be faithful and accurate representations of nature. Stories of a fox that enticed chasing hounds into the path of a train or a porcupine rolling down the hill for fun had, in Burroughs's words, crossed "the line between fact and fiction." Any nature writer faces the danger, Burroughs wrote, "of making too much of what we see and describe, of putting in too much sentiment, too much literature, in short, of valuing these things more for the literary effects we can get out of them than for themselves." The boundary between fact and fiction and between scientist and artist was crossed when the nature writer imparted more drama to nature than was found. In the opinion of Burroughs, these "sham" naturalists had overstepped these bounds in pursuit of monetary gain.[12]

Over the next four years, the battle unfolded between popular nature writers and naturalists over whether science could adequately understand and capture the life of the individual animal. Roosevelt, so impressed by

Burroughs' critique that he invited the writer on a two-month trip to Yellowstone in the spring of 1903, watched from the sidelines. Finally, in the spring of 1907, in an interview published in the June issue of *Everybody's Magazine,* the president questioned the veracity of stories like those of Long that portrayed the life of Wayeeses the wolf, said to have killed a caribou by a single bite into the heart. He feared that Long's books, which were particularly popular in the public schools, would ultimately hinder public appreciation, study, and preservation of nature. "If the child mind is fed with stories that are false to nature," Roosevelt argued, "the children will go to the haunts of the animal only to meet with disappointment, . . . disbelief, and the death of interest." Firmly believing that ruthless competition, survival of the fittest, and instinct were the authentic features that defined life within the animal kingdom, not individuality and sentimentalism, Roosevelt championed the knowledge of the hunter over those whom he regarded as armchair naturalists. Roosevelt found "real" nature through the touch of a steel trigger and the sight down a gun barrel, rather than through the poet's pen.[13]

Three months later, Roosevelt, along with prominent naturalists such as William Hornaday, J. A. Allen, and C. Hart Merriam, appealed to the authenticity of science to squelch the debate. Roosevelt suggested that Long's stories bore the "same relation to real natural history that Barnum's famous artificial mermaid bore to real fish and real mammals." By invoking Barnum, Roosevelt associated nature fiction with fakery and deprecated the manipulation of nature by some in the interests of crass commercialism. Embedded within his criticism was the implication that only those who confronted the struggle for existence in nature could claim to truly understand it. Yet, Roosevelt's social Darwinist vision of savage nature was no less fanciful than the largely benevolent vision offered by Long. It too was crafted to meet the needs and expectations of a specific audience—in this case, wealthy sportsmen and naturalists found in the upper echelons of American society.[14]

The quest for authentic nature evident in the response of Roosevelt and other naturalists to the nature-faking controversy was as critical to natural history film as it was to the realistic wild animal story. Just as Burroughs cautioned writers against taking undue license with the facts of nature for literary effect, so the naturalist-photographer in developing a faithful picture walked a thin line between capturing the authentic drama and beauty of nature and artificially fabricating drama to create a box-office sensation.

For the upper class of American society, particularly its male population, much was at stake in trying to maintain a distinction between the authentic and artifice in nature on screen. Amidst the material comfort and moral complacency found in modern, civilized life, those men who profited most from urban-industrial capitalism and ushered in the machine age were precisely those who most feared the corrupting influences of modern, commercial society. For individuals like Roosevelt, nature's wilderness was the therapeutic balm that healed the debilitating effects of modern urban life. It was the last place of grace—a landscape of authenticity where the trappings of civilization might be shed, the purity of God's hand felt, and the real self found.[15]

The camera embodied this inherent tension between authenticity and artifice within early twentieth-century American culture. Would this machine, which offered a cheap, readily accessible, mechanical reproduction lead to a wider and more democratic appreciation of authentic nature? Or would it turn nature into artifice, yet another imitation among the many that flooded the marketplace to entice an emerging consumer culture at the end of the nineteenth century? If nature on screen was a sham, simply another money-making product crafted by savvy writers or motion picture men, then its power to soothe the antimodernist anxieties of the urban elite by offering a brief respite in a more innocent place would be lost.[16]

In their quest for authenticity, the integrity of the naturalist-photographer became central to establishing the value of a film among the urban gentry and upper middle class. The boundaries between the real and the fake were not always easy to discern; the eye alone could not be trusted to distinguish the authentic from the artificial. Virtue and experience separated the scientist from the showman. In his introduction to Cherry Kearton's *Wild Life Across the World,* for example, Roosevelt attested to the "absolute trustworthiness" of Kearton's photographs. "Any photograph presented by him as of a wild animal," vowed Roosevelt, "can at once be put down as having been taken under precisely the circumstances which he describes. His work, therefore, is of first-rate scientific importance. It should appeal to every hunter and lover of outdoor life, and it should be studied by every Naturalist." Along with Kearton's trustworthiness, Roosevelt highlighted the virtues of "hardihood, daring, resourcefulness, patient endurance and unflinching resolution" which made Kearton a successful and reliable wildlife photographer. Authenticity embodied more than just a faithful representation of wildlife, as if that in itself was unproblematic. It also depended upon whether the photographer had experienced the thrill and danger of the hunt while capturing wildlife on film. To test one's mettle against the harsh conditions of wilderness brought the photographer closer to Roosevelt's vision of the struggle for existence reenacted in the noble hunter's pursuit of challenging game.[17]

The promotion of an early sequel to *Roosevelt in Africa* illustrates the importance of individual integrity and experience in affirming the genuineness of natural history film. Colonel C. J. Buffalo Jones, a sixty-seven-year-old Nebraskan rancher, embarked on an expedition to East Africa shortly after Roosevelt's return "to show the world how easy it would be for American cowboys to rope and subdue the fiercest and biggest game." Jones had devoted his life to the preservation of the buffalo and the roping of big game from the Arctic Circle to the Gulf of Mexico. In Africa, two cowpunchers—Marshall Loveless and Ambrose Means—and Guy H. Scull, a writer and former soldier in Roosevelt's cavalry regiment, joined the expedition. Famed naturalist photographer Cherry Kearton documented the escapades on film. Buffalo Jones and his entourage were successful in their efforts both to rope and capture on film eland, warthog, rhino, serval cat, cheetah, and lion in their native African veldt.[18] They returned with nine thousand feet of film that was then edited into two thousand feet in which "we see the animals captured right out in the open in good sunlight, and there is absolutely no chance of a fake." The film's authenticity was vouched for by Roosevelt at the New York Press Club and in an introduction to Guy H. Scull's *Lassoing Wild Animals in Africa*. In Roosevelt's opinion, Kearton held an equal place of status to Jones and his two cowpunchers. "For cool courage and proficiency in his art," Roosevelt asserted, "there is nothing to choose between Mr. Kearton and any one of the men actively engaged in the work of roping the dangerous wild beasts hunted by the party." In invoking moral and physical attributes such as courage, prowess, and gallantry, Roosevelt ennobled all of these men with the chivalrous qualities that he regarded as essential though threatened by the degenerative influences of modern life.[19]

The character traits Roosevelt praised in Kearton were many of the same virtues used to distinguish the gentleman sportsman from the pot-shot hunter by the patrician class of American society. Roosevelt was a founding member of the Boone and Crockett Club, established in 1887 by a prominent group of wealthy New York sportsmen that feared the commercialization and extermination of game animals, particularly in the West.

Founded "to promote manly sport with the rifle," the Boone and Crockett Club actively worked for the preservation of big game through legislation and the establishment of gentlemanly codes of sport hunting. Limited to a membership of one hundred men, the Boone and Crockett Club elected only individuals who had "killed with the rifle in fair chase" at least three different kinds of American large game. In their emphasis on "fair chase" and preservation of wild game, members of the Boone and Crockett Club meant to dissociate themselves from the game butcher, who killed large quantities of game for profit and, from their view, desecrated nature's magnificent creations by turning them into products of commerce. In his sojourns across Africa, Roosevelt emphasized time and again that "neither he nor the other members of his party shot any animal not needed for food, which could not properly be called a 'specimen' and considered worthy of preservation."[20]

Roosevelt's reverence of the hunter was reserved for the mythic heroes in America's past who had, like Boone and Crockett, overcome their common origins by hunting only to satisfy basic needs of food and clothing and displaying chivalry in their role as Indian fighters on America's frontier. Roosevelt felt that men of his class were threatened by the physical and moral effeminacy of modern times. By combating wilderness, and living the strenuous life that such a struggle entailed, they could reinvigorate themselves with the prowess and republican virtues of their pioneer ancestors, making them once again worthy leaders in public affairs. Fearful that man overly civilized by industrial society would lose the "great, fighting masterful virtues," Roosevelt championed wilderness as the place to restore "vigorous manliness for the lack of which in a nation, as in an individual, the possession of no other qualities can possibly atone." The creation of fifty-three wildlife refuges, the expansion of national parks, national monuments, and forest reserves systems, and the reclamation of arid lands under Roosevelt's administration were all part and parcel of his effort to preserve for future generations a remnant of America's frontier wilderness, which he believed helped forge the essential character of America as a nation.[21]

Despite Roosevelt's claim that his African expedition only killed game for food or science, others found little to distinguish Roosevelt, the great white hunter, from that of the game butcher. The Reverend William Long suggested that Roosevelt knew little about nature, for "every time [he] gets near the heart of a wild thing he invariably puts a bullet through it." Long was not alone. Increased antihunting sentiment at the turn of the century cast doubt upon the claim embodied in Roosevelt's exploits and in the museum collection that death itself was an act in the preservation of life. Shunning the melodrama of fiction and the trophy of the hunter, the camera captured authentic nature, some thought, while embracing a more humanitarian sentiment in the preservation of vanishing wildlife. William Hornaday, director of the New York Zoological Park and ardent wildlife conservationist, for example, argued that "any duffer with a good checkbook, a professional guide, and a high-power repeating rifle can kill big game, but it takes good woodcraft, skill and endurance of a high order, to enable man or woman to secure a really fine photograph of a fine wild animal in its unfenced haunts . . . Sometimes the dangers involved are a hundredfold greater than those encountered by a well-armed man in hunting." Hornaday reinforced the esteem for the naturalist-photographer by resolving in the 1908 code of ethics for the Camp-Fire Club of America to give "a particular fine photograph of a large wild animal in its haunt . . . more credit than the dead trophy of a similar animal," a statement also affirmed by the Lewis and Clark Club, the North American Fish and Game Protective Association, and the Society for the Preservation of the Wild Fauna of the British Empire. Naturalist photographers, like Cherry Kearton, often proclaimed that the great risks taken and the skill and patience required for securing a good photograph far surpassed those of the sport hunter. "It is much easier, much less dangerous, to shoot a lion, after your boys have led you up to him, and your white hunter—who is by your side, ready for emergencies—has given you explicit instructions," wrote Kearton, "than it is to creep close to that lion and take a moving picture of him." Even Roosevelt acknowledged that while "it is not a very hard thing to go off into the wilderness and kill an elephant . . . , it is a very hard thing to get good photographs of them." Still, the camera could lie and it was thus indispensable to audiences in the early years of nature film for the truthful character of the naturalist-photographer to be guaranteed, as in Roosevelt's endorsement of Kearton. Armed with the camera, the scrupulous naturalist-photographer promised to benefit greatly the study of wildlife and its conservation and to instill in the public an appreciation of nature.[22]

While Roosevelt traveled to the steamy jungles of the "Dark Continent" to purify and restore masculine virtues corrupted by the artificial surroundings of the industrialized metropolis, Paul Rainey, "big game hunter, naturalist, millionaire, and sportsman-extraordinaire," as *Outing Magazine* described him, found similar restorative powers in the harsh, frigid environmental conditions of the Arctic wilderness. Born to a prominent Cleveland industrialist, Rainey managed the family's estate from offices in New York City, where he came to associate with leading patrons of education and science through such institutions as the American Museum of Natural History and the New York Zoological Society as well as sporting clubs such as the New York Yacht Club and the Racquet and Tennis Club. In 1910, he set out upon an expedition to the Arctic to capture and secure photographs of arctic wildlife for the New York Zoological Park.[23]

Rainey returned to New York in September of 1910 with the two largest live polar bears ever captured, Silver King and Queenie, in addition to six musk oxen and two walrus. The animals were promptly put on display at the New York Zoological Park. Under the directorship of William Hornaday and the financial backing of wealthy patrons such as Madison Grant, Henry Fairfield Osborn, and Theodore Roosevelt, the New York Zoological Park sought to establish itself as a place for the educational and moral uplift of New York City's immigrant and working-class population. For the city's more genteel culture, the zoological park also served as a place of respite and rejuvenation from the hustle and bustle of city life. The proliferation of zoological parks in American cities during the early twentieth century was, like the development of natural history film, partially an effort in moral and social reform. Just as religious leaders and the wealthy elite found in natural history film a more respectable form of entertainment than that offered in the nickelodeons catering to a working-class clientele, they hoped the zoological park would be a place of more refined education and entertainment than the bawdy, carnivalesque atmosphere associated with the nineteenth-century circus and fair. When plans for creating the New York Zoological Society and Park were debated within city and state government during the 1890s, for example, wealthy residents along upper Fifth Avenue, opposite Central Park, offered strong support. Their hope was that the Central Park menagerie, a popular attraction among residents of the slums and ghettos of Manhattan's Lower East Side, would be abolished. Hornaday strongly opposed "the Bronx Zoo" nickname for the New York Zoological Park, since he believed that a zoo implied "a small, cheap and usually smelly affair, with little scientific standing, or none at all." A zoological park, in contrast, stood for "a public institution of large size and more or less dignity."[24]

In linking his Arctic exploration with a collecting expedition for the New York Zoological Society, Rainey added an official, scientific, and hence authentic dimension to his voyage. The popular press described Rainey's capture of the ferocious polar bear, Silver King, as a heroic venture, in which Rainey risked life and limb in the name of science. When Silver King had shaken off the noose of his captors in the icy arctic waters, Rainey placed, rather than threw, the rope around the "infuriated" animal's neck. "Those who were with him on these occasions say that he does not know the meaning of fear," noted one reporter.[25]

Only four months after his Arctic expedition, Rainey set sail for Mombasa on a privately financed $250,000 hunting and semi-official scientific collecting expedition for the New York Zoological Society and the Smithsonian Institution. One year later, the moving pictures of Rainey's African expedition arrived in the United States. Destined "to go down in the archives of many governments as an authentic record in the study of natural history," *Paul Rainey's African Hunt* was also destined to become the largest money-making wildlife film of the decade.[26]

Before embarking on his expedition to "darkest Africa," Rainey informed a correspondent for *The New York Times* that his interests as a gentleman-sportsman had advanced beyond the stage "where numbers of dead animals count most." Instead, Rainey expressed what he clearly regarded as a more enlightened attitude. His "principal desire" was "to trap wild animals and bring them back alive." The photographing of animals came next in importance, to give Americans a realistic sense of what life in the jungle was like. To help in his task, Rainey shipped fifteen of his best Southern foxhounds to Africa in the hopes of training them to hunt lion. Rainey's dogs proved themselves worthy of the challenge. They helped kill, not capture, twenty-seven lions in less than thirty-five days and wounded a great many more. An editorial in *The New York Times* questioned how sportsmanlike Rainey's attitude toward hunting was, equating it with "butcher work." But for motion picture audiences, watching the dogs tree and kill a leopard and duel to the death with a lioness was thrilling entertainment. A visit to the New York Zoological Park could assure audiences of the film's authenticity. There in a cage was the hyena, whose capture in the wild they had witnessed on film as a gruesome struggle by the animal to escape from a leg trap.[27]

Paul Rainey's African Hunt, however, did give the viewer something more than thrilling chase and gore. One of the reels that reviewers remarked upon most comprised particularly long and clear-focused takes of zebras, rhinoceros, giraffes, gazelles, oryx, and wild boar gathered at a water hole. Such scenes, where "all species are at peace," demonstrated that "darkest Africa" did not teem with dangerous and savage beasts in the way that audiences had come to expect from the tales of hunters and novelists such as Joseph Conrad and Edgard Rice Burroughs. Here was "nature in all its innocence," a sight that reminded one reviewer of Noah's Ark. The water hole became a fitting allegory for the motion picture theater, a place where exhibitors hoped people from different economic, national, and social backgrounds would gather for an evening's entertainment. It was an allegory put to great advantage by the advertisers for *Paul Rainey's African Hunt*:

> One touch of nature makes the whole world kin. That is the secret of the extraordinary success of this picture. It has that one vital appeal to all classes which makes them forget for the time their stations in life in their absorbing interest in the domestic joys and sorrows of god's obscure four-footed kingdom, which, after all, are much the same as their own. There is something in it that appeals to every mother, that appeals to every father, and, best of all, it appeals to every child.

Heralded by Henry Fairfield Osborn, president of the American Museum of Natural History, as the "greatest contribution to natural science of the decade," *Paul Rainey's African Hunt* offered what prominent educational, religious, and business leaders had urged in their appeals for motion picture reform: a picture that was simultaneously entertaining, educationally and morally uplifting, and, last but not least, profitable.[28]

The film opened on April 15, 1912 at Daniel Frohman's Lyceum Theater, one of New York City's exclusive playhouses. Billed as a motion picture that attracted "wealthy people at top prices," the film extended its reach beyond New York City's high society to include a "better class," one identified as a "morally conservative, church-oriented culture tied to the older middle class." A magazine advertisement in November of 1913 tells of the marketing efforts to bridge morals and profit. A minister in the upper left-hand corner espouses the need for motion pictures of entertainment and educational value to benefit the moral welfare of his congregation, as a well-dressed businessman in the lower corner speaks of the need for a motion picture to make money. The two are united in the center of the ad with the title *Paul Rainey's African Hunt*. Although the film's unparalleled financial success in 1912 may have alleviated initial reluctance to book travelogue-expedition films, the largest patrons in the early years of natural history film continued to be the urban middle and upper classes. Advertisements for *Paul Rainey's African Hunt* underscored nature's alleged power to unify people of different socioeconomic backgrounds, but the development of a mass market for natural history film would not come until after the Second World War. Rainey's success was itself short-lived. Two years later when the second series of his African pictures was released, it met with little enthusiasm. American audiences had become familiar enough with baboons and giraffes, zebras and lions that the stakes for thrilling, educational entertainment had been raised.[29]

The integrity of the naturalist-photographer and the harsh conditions endured photographing animals in the wild were important elements if a nature film was to win the patronage of more respected members of society. Educational links to a scientific institution also enhanced the status of a natural history film. The gift of a captured lioness by Colonel Buffalo Jones and of polar bears and a hyena by Paul Rainey to the New York Zoological Park added to the public treasure and assured audiences that the motives of the naturalist-photographer were noble and pure. As a science, natural history offered a path of adventure and a secular, yet ascetic pursuit of truth, far removed from the commercial excesses of modern society.[30] Hunting with the camera and gun in the quest of scientific discovery became a common test of manhood and avocation for the sons of New York's prominent industrialists and financiers in the early twentieth century. Leisure and wealth afforded them the opportunity of adventure in a period when America's male role models, like Charles Lindbergh, proved their character through heroic feats that pitted human ingenuity and stamina against nature, and Hollywood leading men, like Douglas Fairbanks, upheld sport as the "antidote to too much civilization."[31]

William Douglas Burden was one such son of a prominent New York industrialist. His family made a fortune in iron and steel and owned a posh country estate on Long Island that was used by dignitaries, including the Prince of Wales. At nine years of age, Burden was introduced by his father to the northern wilderness of Quebec, where he learned, during the summers of adolescence, the ways of the woods from the famed Indian hunter Archie Miller. Upon his graduation from Harvard in 1922, he traveled to distant lands—Alaska, Indochina, and Central and South America—in search of big game, lost silver mines, and his own, authentic, self.[32]

During his Far East adventures hunting ram on the Mongolian frontier, Burden crossed paths in Peking with Roy Chapman Andrews, director of the American Museum of Natural History's Central Asiatic Expedition, a five-year scientific and film expedition to the Gobi desert from 1922 to 1927 in search of the evolutionary birthplace of mammals and humans. Andrews, a Midwesterner by birth, started out in the museum in 1907 working for forty dollars a month as an apprentice to the taxidermist James L. Clark. As an employee of the museum, Andrews crafted himself a life of exploration and adventure.[33] Despite their different socioeconomic backgrounds, Burden found in Andrews a kindred soul, a man who epitomized the life of exploration and adventure, and did not languish under the corrupting influences of wealth and fame. When oil and mining companies interested in mineral resources of the Gobi offered to sponsor the Central Asiatic Expedition, Andrews was adamant that the "expedition should be strictly scientific and its objectives be . . . without taint of commercialism." Like Roosevelt, Andrews championed the harsh environment of nature where the body, threatened by the softness of modern corporate life, could be fortified and reinvigorated and manly character restored. "When one is at grips with crude nature, when the struggle to maintain oneself and do one's work against physical odds is really on," wrote Andrews, "the gloss of civilization quickly fades. It leaves character bare for all to see." In the introduction to Burden's book, *Look to the Wilderness,* Andrews bestowed upon Burden the highest accolade: "indifferent to what the city dweller considers to be the essentials of comfort," Burden was, like Andrews, "a primitive at heart."[34]

Burden returned from his crusade to remote lands determined to follow a path in scientific natural history, a field that seemed to offer a corrective to the modern, synthetic, corporate world. "With a strong taste for the wild places of the world, and an equally strong distaste for crowds and pavements," he entered Columbia University in 1925, pursuing interests in physiography, geology, and paleontology—natural historical sciences that would continue to satisfy his passion for adventure.[35] Sparked by a course on paleontology offered by the herpetologist and animal behaviorist Gladwyn Kingsley Noble and intrigued by reports of a new species of giant lizard discovered in Java, Burden, with Noble's enthusiasm, encouragement, and support, journeyed to the Dutch East Indies in 1926 to film and capture *Varanus komodensis* for scientific and popular exhibition in the American Museum of Natural History's new Hall of Reptiles, scheduled to open to the public in 1927. Burden's donation of $15,000 toward the expedition secured him a place among the select members of the museum's Board of

Trustees, an expensive honor. Burden not only contributed most of the capital for the expedition, he also invested his labor. He was accompanied on the expedition by his first wife Katherine White Burden; the Smith College herpetologist Emmet Reid Dunn; a professional hunter from Indochina, J. M. Defosse; a cameraman hired in Singapore, Lee Fai; and numerous porters, cooks, and hunters hired in the region who remained nameless throughout Burden's stories. The expedition was both a scientific and public success. Burden returned with two live specimens and twelve dead. The captured dragons were donated to the Bronx Zoo, where attendance increased by 30,000 people a day, and the dead specimens were used to make up the Komodo lizard group at the Museum's new Hall of Reptiles.[36]

Unfortunately, the live specimens did not live long at the zoo and the "lethargic, deflated captive" failed to give "any impression of his aggressive, alert appearance in his wild home."[37] The dragon in captivity bore little resemblance to its true self, precisely because the zoological park was purposefully designed in the early twentieth century to be a mixture of nature and culture, a middle landscape that offered a more tamed, civilized nature.[38] In seeking an escape from civilization, Burden found the zoo too much a reminder of all that he purposefully left behind in his mythic quest; it stifled not only the dragon's true nature, but his own. Only when the expedition had reached Bali did Burden feel as though he had finally broken with civilization. In the jungles of this tropical paradise, Burden "had come to life again." Burden could not restore the life in the captive beasts, but he did bring back from the island of Komodo thousands of feet of film. Through film, Burden believed he could offer the urban populace a look at animals in their natural habitat more captivating and real than found in the Bronx Zoo. He used the raw footage of nature to recreate for the metropolitan spectator a primeval scene of the Komodo dragon in its free and wild state, a scene more vivid than if the spectator had been in Java with the real thing.[39]

The ending sequence of Burden's Komodo dragon film reveals much of the behind-the-scenes operations that went into Burden's recreation of the event. The scene begins with an unidentified porter, Burden, and his wife Katherine entering a blind or "boma." The shot establishes the racial and gender hierarchies within the expedition: the unnamed Indonesian porter bears the heavy camera equipment; Burden, in true white hunter fashion, bears the gun; Katherine follows behind. Burden closes the blind and cranks the camera, while Katherine holds the gun. The drama unfolds. A large

Komodo dragon enters the screen from the left. Center-screen is the carcass of a wild boar. A smaller dragon is on the right. As the large dragon goes to feed on the boar, the scene cuts to a close-up of the dragon's jaws enveloping the bait. A medium shot follows of the large Komodo with head and body raised high upon its muscular forelegs. With our gaze upon the blind, the climax builds. Now Katherine runs the camera while Burden takes aim and fires. Only Burden takes the shots, from the camera and gun, that will, through the miracles of cinema and taxidermy, transform the dragon into a scientific and entertainment spectacle in the Hall of Reptiles at the American Museum of Natural History. The Komodo dragon diorama was the most prominent and spectacular display within the Hall of Reptiles. Taxidermy preserved the Komodo dragon for posterity, but motion picture technology reanimated the beasts for visitors' viewing pleasure next to the diorama display.[40]

In writing popular accounts of the expedition, Burden came to realize that authenticity entailed much more than faithful attendance to factual detail. His "Komodo diary though absolutely exact" as to the chronological order of events "was utterly useless," for "it would have bored anyone to extinction." In preparing an article for *Natural History,* Burden related the live capture of the largest Komodo dragon they had seen on their travels, a beast "so large and so villainous of aspect that I [Burden] trembled with instinctive revulsion." Once snared, lassoed by Defosse, and hog-tied, the dragon was put into a special cage with steel netting placed over a large air hole at the top. The next morning, Burden wrote, all were dismayed to find that the beast, with its prodigious strength, had ripped the steel apart and escaped.[41]

But the herpetologist Dunn complained that the event that Burden described was "witnessed by no white man." Burden was irritated by Dunn's reaction. Their dispute epitomized a fundamental disagreement over what constituted the boundary line between authentic and faked nature. Although the escape of the large Komodo dragon occurred early in the expedition, and although Burden had not actually seen its capture, his retelling of the story—based upon the descriptions of Defosse and Dunn, who had witnessed the event—was written to "make the reader feel as one felt at the time." Burden prided himself in "adhering to a high standard of truthfulness in everything [he had] written," and, if anything, thought his writing suffered from "too much thought of accuracy and too little imagination." In order to capture an emotional truth, Burden felt it legitimate to alter the precise sequence of events. An authentic factual record did not necessarily correspond to authentic emotional experience. And it was the latter that would attract and interest the public.[42]

Like Burden's diary of the expedition, the film footage initially contained lengthy segments of boring activity that, unedited, conveyed little of the emotional drama Burden felt upon encountering these majestic creatures. By the 1920s, drama had become an essential element of natural history film, especially if such films were to have any chance of Hollywood distribution. The question for wealthy sportsman and naturalists like Burden was whether the drama was fabricated to elicit cheap thrills or to authentically capture a natural distillation of the emotional experiences of hunting and collecting wildlife in the remote regions of the globe. Although the scenes of Burden and Katherine in the blind were added later in the hopes of obtaining a theatrical release, the scenes of the dragons were neither staged nor faked. But Burden chose the particular shots of dragon postures for precise reasons. The close shot of the large Komodo enveloping its jaws around the wild boar resembled, according to Burden, "*Tyrannosaurus* as restored in modern paintings" and gave a "fairly accurate picture of the way in which carnivorous dinosaurs devoured their prey." The dragon's raised head posture was reminiscent of another Komodo pose witnessed only rarely when the dragon sat back on its hind legs and tail, a "striking though superficial resemblance to certain dinosaur restorations." These two dramatic postures from the film were juxtaposed in the museum diorama to represent the essential nature of the Komodo dragon. Both the raised head posture and the enveloping of the jaws around the wild boar occupied very short pieces of film footage and they never occurred together. Yet the museum-goer takes these postures—both meant to convey an emotional impression of experiencing a primeval monster in a primeval time—as representative of the dragon's habits.[43]

In drawing parallels between the Komodo dragon's behaviors and those of *Tyrannosaurus*, Burden was clearly hoping to draw upon the success of the 1925 Hollywood blockbuster, *The Lost World,* a film version of Sir Arthur Conan Doyle's romantic adventure in South America. Through the special effects work of Willis H. O'Brien, *The Lost World* brought to the screen dinosaurs "in the flesh and as big as life," and, as one commentator remarked, "about 1000 times as big as your imagination or your visits to the Natural History Museums would have led you to believe them to be." Realistic fights between *Tyrannosaurus* and *Triceratops,* coupled with thrills, comedy, and suspense, made *The Lost World* the most sensational, engrossing, and smashing entertainment picture of the year. Even with the suspense of the hunt, a Komodo dragon devouring a wild boar could not rival a brontosaurus ravaging the streets of London. In the talented hands of special effects artist O'Brien and Burden's good friend, film director Merian Cooper, the plot of Burden's Komodo dragon story would be refashioned into a major theatrical release: *King Kong*. But in 1927, the Komodo dragon picture was, in the words of one producer, "too educational and without sufficient dramatic or adventure interest to make a good theatrical release." Burden's dragon picture did find an audience among a select social network of sportsmen and conservationists. The movie played to members of the Wilderness Club and the Boone and Crockett Club; at dinner meetings of the museum trustees; to explorer-traveler organizations like the National Geographic Society; and at private parties in the homes of New York City's social elite, such as the publisher George Putnam. By turning down a lucrative offer to make a dramatic picture that would in the end, Burden feared, misrepresent the Komodo lizards, he believed he acted honorably by sacrificing drama for authenticity, commercialism for scientific truth. A noble attitude, perhaps, but one that Burden would gradually revise.[44]

Burden's preoccupation with authenticity in nature on the silent screen signified a deep ambivalence within early twentieth-century America, particularly among the upper class, toward modern life. Framed as the antipode of civilization, wilderness was esteemed to hold curative powers that could soothe the antimodernist anxieties found within the industrialized metropolis. Wilderness was a landscape of authenticity, a place where masculine virtues were honed and restored, a place where individual character was laid bare. The camera was a means of recreating this wilderness experience, but only if truthfulness could be assured. But the eye, Roosevelt, Burden, and others knew, was easily deceived. Elements that helped distinguish authenticity from artifice for a wealthy New York social elite included the gentlemanly codes of the hunt, the austere conditions faced in the wild, the links to scientific institutions such as the zoo and natural history museum, and the eschewal of individual financial gain. But members of New York high society were not the only patrons of nature on screen. By the 1920s, Hollywood also offered naturalist-photographers a means of support, as well as a route to fame and fortune. In Hollywood's commercial exploitation of nature for entertainment, some, like Burden, found the distinction between artifice and authenticity more and more difficult to maintain.

Chapter Two

SCIENCE VERSUS SHOWMANSHIP ON THE SILENT SCREEN

On May 20, 1923, *Trailing African Wild Animals,* billed as the first "purely commercial animal picture . . . endorsed as really 'natural,'" opened at New York City's Capitol Theater, a 4,500-seat, lavish, first-run picture palace built in 1919. Filmed by the famed adventurer-photographers Martin and Osa Johnson, *Trailing African Wild Animals* was financed through the Johnsons' personal investments, proceeds from the sale of a Midwest jewelry store owned by Martin's father, and the patronage of wealthy members of New York City's Explorer's Club and the American Museum of Natural History. Trustees of the American Museum hoped that their patronage and endorsement of Martin and Osa Johnson would bring publicity to the world of natural history and the halls of the museum. But their patronage alone was no guarantee that the film would reach a wide audience. By the mid-1920s, the commercial success of natural history film rested largely in the hands of an oligopoly, formed by major Hollywood studios including MGM, Warner Brothers, Twentieth-Century Fox, Paramount, and RKO, that monopolized control of the commercial entertainment film industry through vertical integration of production, distribution, and exhibition. Metro Pictures, for example, which promoted, distributed, and exhibited *Trailing African Wild Animals,* began as a modest Hollywood studio with a national distribution network that in 1924 merged with Goldwyn Pictures and Louis B. Mayer Productions to become one of the most powerful corporations in Hollywood. Throughout the 1920s, the Johnsons successfully straddled the worlds of Hollywood and science, but it was an alliance fraught with tension. In the eyes of wealthy patrons at the American Museum of Natural History, to exploit wildlife on camera in the interests of crass commercialism threatened to degrade the sanctity of nature and its importance as a therapeutic retreat from the profane influences of modern civilization. For them, the preservation of wildlife on film served as a lasting record for future generations of a natural heritage that was being erased by the modernizing forces of civilization. In contrast, the Johnsons had far fewer scruples about the preservation of authentic nature when commercial success depended upon their ability to make nature conform to the conventions of Hollywood entertainment. Increasingly, the influence of Hollywood played an instrumental role in determining the conventions and market through which nature films might reach a popular audience.[1]

Born in Lincoln, Kansas to a self-made Swedish businessman and a mother who was a Christian Scientist, Martin Johnson, unlike Burden, had few reservations about turning nature into a commercial commodity, particularly when the packaging of nature as entertainment offered a means to achieve wealth and standing in American society. In 1907, Martin Johnson accompanied Jack and Charmian London as a cook and photographer on a highly publicized two-year voyage aboard the *Snark* through the South Seas. Upon his return home, Johnson, spellbound by the glamour of the public limelight, capitalized on his ventures with the famous writer by opening the *Snark* theater in Independence, Kansas. Johnson successfully lured audiences with his tales, slides, and motion pictures of adventure through the Pacific, which included male-only shows containing pictures of nude women and details of the sexual mores of New Hebrides men. In 1910 Johnson took the show on the road with his sixteen-year-old bride, Osa Leighty, whose soprano voice became a featured act in his travelogue lectures. Through the purchase of slides and motion picture films of places and indigenous peoples that he had neither visited nor seen, Johnson crafted a travelogue-lecture show of the South Seas that earned him a six-year booking on the Orpheum Vaudeville Circuit. In 1917, backed by a group of Boston investors, the couple traveled to the Solomon Islands where they obtained footage for their first travelogue-adventure film, *Among the Cannibal Isles of the South Pacific.* The film's commercial success, attributed in part to the striking contrast of a photogenic, petite, yet "plucky" American woman amidst cannibals and headhunters and to titles replete with racial slurs and gags, attracted additional investors into the Martin Johnson Film Company. A second trip to Melanesia in 1919 provided material for their popular film, *Jungle Adventures,* which was successfully marketed through tie-in advertisements for adventure clothing at several New York City department stores.[2]

In 1921, Johnson's chance meeting at New York City's Explorer's Club with Carl Akeley, a leading taxidermist, sculptor, inventor, and foremost authority on African wildlife, changed the course of Martin Johnson's career and helped rocket him from flim-flam showman to respected naturalist photographer. At the time, Johnson was negotiating with *Asia* magazine, one of the sponsors of Roy Chapman Andrews's Central Asiatic Expedition, to become the expedition's official cinematographer. Akeley, who had other plans for Johnson's talents, intervened on Johnson's behalf and recommended that the American Museum of Natural History instead hire J. B. Schackelford as the expedition photographer. Andrews, who expressed reservations that Martin Johnson would make his Central Asiatic Expedition nothing but a "motion picture show," favored the new arrangement. In its stead, Akeley offered Johnson a photographic safari to a continent filled with adventure and wildlife: Africa.[3]

Akeley met with Johnson at the Explorer's Club a decade after his famous 1909–1910 collecting expedition to British East Africa for the American Museum of Natural History, where he hunted elephants with Teddy Roosevelt. On that expedition Akeley's idea for a monument to Africa's vanishing wildlife in the American Museum of Natural History first took form. Akeley envisioned a hall that was to include dioramas of over forty animal groups: an illusion such that a visitor would "think for a moment that he has stepped five thousand miles across the sea into Africa itself." By the early 1920s, he had completed the first of his artistic and scientific masterpieces for the African Hall. Lifelike and majestic African elephants to this day occupy a place of central prominence within the museum's Theodore Roosevelt Memorial. In 1921, Akeley had just returned from yet another East African expedition, this time to secure the most noble of quarry—the male silverback gorilla. At the forefront of the taxidermic profession and expert in the techniques of museum display, Akeley created a habitat diorama of stunning realism in which the male silverback, "the Giant of Karisimbi," is prominently displayed towering above other female and juvenile gorillas in the family group. The "Giant of Karisimbi" stood in the foreground of a vast panorama of edenic Africa to which Akeley longed to return.[4]

Lacking the financial resources and animal specimens to bring his vision of Africa to completion, Akeley believed that Martin Johnson alone had the talent to take the subject of African wildlife and turn it into a commercial success. Through such publicity, Akeley hoped he would be able to arouse public interest and philanthropic support for his African Hall. Although Akeley had invented a lightweight motion picture camera that revolutionized wildlife photography in the field, his own motion picture *Meandering in Africa* showed that he lacked Johnson's eye for composition, drama, and lighting.[5]

Having relieved the Johnsons of their obligations to *Asia* magazine, in 1921 Akeley introduced them to Kenya's chief game warden, Arthur Blayney Percival. Although he idolized Paul Rainey and made a point of tracing the footsteps of his expedition, including a stop at the famous peaceful water hole, Martin Johnson went to Africa with little experience photographing wildlife. The couple returned to the United States after two years with 100,000 feet of film, which they edited down to a seven-reel motion-picture release, *Trailing African Wild Animals*. A testimonial, signed by both Akeley and Henry Fairfield Osborn, assured the public that *Trailing African Wild Animals* was "free from misleading titles, staging, misinterpretation, or any form of faking or sensationalism." Johnson's skills as a wildlife cinematographer were unsurpassed, according to Akeley and Osborn, and they praised his ingenuity, resourcefulness, physical strength, and unlimited energy. "In short," they concluded, "we believe in Martin Johnson."[6] The endorsement by the American Museum of Natural History (AMNH) raised the reputation of Martin Johnson from a vaudeville performer to a gentleman naturalist, who conducted himself, in the words of Helen Bullit Lowry of the *New York Times Magazine*, "according to the standards of this new world of science into which his achievement has precipitated him."[7]

Trailing African Wild Animals is contrived around a story in which the couple set out in search of a lost lake teeming with wildlife, which Johnson claimed to have learned of through the memoirs of an eighteenth-century

Scotch missionary. In her feature article, "New Adam and Eve Among the Gentle Wild Beasts," Lowry portrayed the Johnsons' search as nothing less than a quest for the Garden of Eden. Their eventual discovery of Lake Paradise revealed a place of jungle peace that knew "no original sin." This was the vision of Africa that Akeley had hoped the Johnsons would capture, a place not of jungle horrors and impenetrable forests, but, in Lowry's words, an "idyllic oasis, where nature is adjusted and where God is in His Heaven instead of in the First Baptist Church." Lowry praised the Johnsons for their "conscientious truthfulness" and for refusing to pander to public demand for thrilling spectacle and dangerous adventure. She chided "Mr. Babbitt and family who take their adventure canned" and demand "action and adventure . . . even if reinforced by a leopard on a chain filmed out in Hollywood," and lauded the Johnsons' willingness to risk commercial success for the sake of scientific accuracy. The museum's endorsement added further assurance, Lowry concluded, that *Trailing African Wild Animals* was an authentic picture and "not a patchwork of elephants taken in Africa in 1921 and men taken in khaki out in Hollywood in 1923."[8]

While the museum endorsement moved Johnson into the world of science, his showmanship also brought the Museum into the world of Hollywood and advertising, where artifice and hype were instrumental in developing a mass market.[9] Despite publicity and testimonials to their innocence, purity, and faithfulness to real nature, the Johnsons' story about the lost lake was cleverly fabricated; the lake was well known to Samburu pastoralists. The Boma Trading Company had established an outpost on a ridge overlooking the lake in 1907. Scenes of danger and suspense in which rhinos charge and lions threaten to pounce betrayed the vision of Africa as a peaceful paradise and were also conveniently staged. The Johnsons and their staff often provoked wildlife into action. In a carefully crafted climactic scene, the heroine, Osa, drops the camera for her Winchester rifle and rescues Martin from the charge of an elephant without a moment to spare. Osa's thrilling escapades were reminiscent of earlier commercial films of African adventure, staged on Colonel Selig's southern California game farm, with "beauty and the beast" themes. Thus, the cinematic conventions that Martin and Osa Johnson drew upon to produce *Trailing African Wild Animals* owed much more to Hollywood artifice than museum officials publicly acknowledged. Akeley and Osborn had been knowingly duped, but it was a small price to pay for advancing their educational and conservationist cause.[10]

Trailing African Wild Animals moved the Johnsons closer to the social circles of wealthy patronage operating behind the exhibit halls of the American Museum. With the film behind him, Akeley managed to secure financial backing for a four-year photographic expedition to East Africa from Daniel Pomeroy, director and vice-president of Banker's Trust Company and a prominent AMNH trustee, and George Eastman, president of the Eastman Kodak Company. The four-year Pomeroy-Eastman-Akeley expedition enabled the Johnsons to establish a permanent camp at Lake Paradise with the intent of producing three feature-length wildlife films. The museum's board of trustees, despite initial reluctance to become involved in a commercial venture, hoped the expedition and motion pictures to follow would help popularize the museum and generate revenue for construction of the African Hall and a memorial befitting the nation's former president, Theodore Roosevelt. But the complex financial agreement eventually made indicates the caution with which the museum's board of trustees approached this combination of commercialism, mass entertainment, and nature. Akeley may have believed in Johnson; others were less certain. While the board agreed to scientifically endorse the expedition, they were unwilling to finance it directly. Instead, a separate company, the Martin Johnson African Expedition Corporation, was established with Pomeroy as president. This company in turn entered into a trust agreement with the museum, which agreed to endorse any commercial films produced while refusing any legal liability. Once the theatrical release of the films had ended, the corporation was to donate all of the expedition's photographs, films, and specimens to the museum. In addition, seeking collateral for a $37,000 advance, the corporation required Martin Johnson to donate all negatives of still and motion pictures from previous films to the museum. Johnson was himself ecstatic with the final arrangement, exclaiming that his "showmanship, mixed with scientific knowledge" would result in the "biggest money maker ever placed on the market."[11]

Simba, the celluloid result of the couple's four years in African paradise, opened at the Earl Carroll theater in New York City in January of 1928 with a top seat price of $1.65 and proved to be the moneymaker Johnson envisioned, earning $2 million in box-office revenues. In a series of articles in the *New York Times Magazine* intended to publicize the Johnsons' adventures and the film's release, Johnson portrayed their Lake Paradise home as a place far removed from civilization, with only "elephants and natives for [their] nearest neighbors." Highlighting the image of nature as a landscape of authenticity, Johnson emphasized that Lake Paradise was a place where "simple pleasures stand out in their true values, unsullied by the artificial entertainments of civilization." Through their ventures into the African veldt, where they displayed "tireless patience and endless courage" and endured "privation, perils, thirst, and fevers," the Johnsons linked themselves in the opening titles of *Simba* to other explorers such as Livingstone, Stanley, Akeley, Roosevelt, and Rainey who similarly shunned civilization in search of "true Africa." *Simba* proclaimed itself to be an "authentic record" of the lion as it lived "in the unspoiled freedom of his native wild," not as the public knew him in the artificial environments of civilization, "caged in circuses and zoological gardens." Authenticity was promoted not only by the endorsement of a major scientific institution, but also by the depicted simplicity of Martin and Osa Johnsons' life on Lake Paradise.[12]

The intimacy the Johnsons achieved with their wild neighbors, revealed through superb close-up photographs, enabled motion picture audiences to identify with wildlife as individuals on screen. In *Simba,* the influence of Jack London's craft is readily apparent. Johnson incorporated conventions from the realistic animal story, relying upon anthropomorphism, character development, and distinct personality traits to highlight individual animals. In the elephant sequence in *Simba,* for example, footage of an elephant walking aimlessly is identified in the title as "Mama Tembo . . . walking in her sleep." Shots of supposedly the same "Mama Tembo" scratching herself against a tree are accompanied by a comic remark: "Dam [sic] that tick." "Willie Tembo," Mama Tembo's eldest, is next introduced to the audience, followed by scenes of the elephant herd in migration. In the migration footage, a newborn baby elephant is featured tagging along with the herd, a character that one reviewer singled out as contributing to the most lighthearted scenes in the picture. Naming animals became one device in helping audiences identify with wildlife on screen. It also created associations between wildlife and domestic pets within American culture, an association accentuated by the real-life pets that were often acquired on photographic expeditions, such as the Johnsons' pet gibbon from Borneo, Kalowatt, who accompanied them throughout their travels.

While the development of individual characters added a familial and intimate sense of wildlife on the African veldt, *Simba* also utilized the typecasting of species to add further comic fare and basic natural history information. The Grant zebra (or common zebra) was introduced at a watering hole as "just a snappy jackass trimmed with awning stripes."

Footage of the Grant zebra juxtaposed with that of the Grevy zebra, said to "wear a more conservative suit of neat narrow stripes," illustrated the different coloration patterns and size in these two species. The "timid and dubious" nature of the giraffe was attributed to its great height, which gave it good cause to be suspicious, since it continually saw "a lot to be alarmed at." And scenes of hyenas feeding at a carcass offered visual testimony to the Johnsons' impressions of the hyena as a "cruel, sneaking scoundrel."

In *Simba,* stereotypes of both animal species and human races abound. Until well into the 1950s, natural history films of Africa often portrayed the continent's indigenous people with less dignity and integrity than that reserved for the majestic species of African wildlife. In *Simba,* native Africans are stripped of their culture and identity. In the movie's taxonomic natural history, native Africans represent an ancient and lowly "half-savage" stage in the evolutionary advancement of the human race. The taxonomic hierarchy of human races evident in *Simba* found support in Henry Fairfield Osborn's exhibit, the Hall of Man. The exhibit prominently displayed within the American Museum of Natural History the evolutionary advancement and superiority of the Nordic race and, thereby, gave scientific sanction to Osborn's own eugenic, anti-immigrationist, and racist sentiments. In *Simba,* the incessant degrading situations into which the Johnsons cast individual Africans and tribal groups reinforced a notion of the racial superiority of white American audiences and the inferiority of blacks. Typecast as the coon, the most degrading of all black stereotypes in film, Africans in *Simba* were nothing more than objects of amusement for white audiences. Racial sight gags were rampant throughout the film and relied upon placing Africans in 1920s American society backdrops. A shot of a young Samburu woman with short hair and bare breasts is followed by a title that labels her as "just a little black flapper." Repeated cuts to scenes of an old African man, identified as the "village soak," trying to open a beer bottle throughout one section of the film further reinforced racial stereotypes of blacks as lazy and inferior. At the same time, natural history films that included nude shots of young women offered a form of titillation that evaded the gaze of movie censorship boards, since such films were deemed of scientific and educational value. In *Simba,* for example, a profile shot of an African woman dancing with her bare breasts in rhythmic motion is followed by the title: "All the scenery looked better after the rain." An immediate cut to a shot of another young bare-breasted African woman reinforces the impression that, like wildlife, the native human inhabitants of Africa were nothing more than a spectacle to gratify American audiences.[13]

Replete with comedy and titillation, two essential elements for commercial success, *Simba* also ended with a thrilling climactic adventure. Lions purportedly threaten the livestock of Lumbwa herdsmen one thousand miles south of the Johnsons' Lake Paradise home. Footage of Lumbwa villagers dancing, supposedly working themselves into "a frenzy of courage," is followed by shots of Lumbwa warriors dressed in lion-mane headdresses, carrying their shields and spears in search of lions on the African veldt. Lumbwa warriors then circle a lion and spear it as the lion thrashes out and injures one man. Another lion is found, but gets away, and then charges out of the bush in rage "now with his eye on the camera." A series of close-up shots of the lion lurking in the grass and intercuts of Osa looking on with her gun build suspense. When the lion charges, Osa brings him down. In a jarring contrast, *Simba* ends with Osa preparing an apple pie for dinner at camp that night. Her self-reliance and courage were portrayed not in opposition to her traditional role as homemaker, but as that which renewed vitality and excitement in her life with Martin and their surrogate child, the pet gibbon Kalowatt. Depictions of the emancipated American woman, such as those of the Hollywood film star Mary Pickford, often showed how the heroine's newly realized freedom revitalized life within the domestic sphere. It is a telling example of the ways in which *Simba* played off narrative conventions established in Hollywood fiction films.[14]

While the realistic lion-hunting scenes invoked the thrill and danger of safari and kept American audiences on the edge of their seats, a closer look at the lion-spearing footage reveals something amiss. The native tribesmen in this footage are clearly from different tribes. The first segment consists of Lumbwa (Kipsigis) spearing a lion; later in the sequence men are wearing the headdresses of Maasai warriors. The Johnsons, Akeley, Eastman, and Pomeroy had, in fact, hired forty Lumbwa with spears to stage the lion-killing scenes. Automobiles were used to prevent the lions from escaping and to drive them into an area suitable for filming and spearing. To ensure footage of sufficient length, Johnson purchased Alfred Klein's film, *Equatorial Africa: Roosevelt's Hunting Grounds,* which included scenes from an actual Maasai lion hunt. Johnson spliced the purchased film with his own staged Lumbwa lion-spearing scenes. Only Eastman, in a privately published diary, revealed what really took place behind the scenes. Any public knowledge of the event would have jeopardized the film's claim to authenticity and undermined the American Museum of Natural History's endorsement of Martin Johnson.[15]

Simba illustrated the immense investment, labor, and talent required to bridge the gulf separating science from showmanship within American culture. In the film, Johnson successfully hid artifice behind the mask of authenticity, and thereby ensured support among museum members and patrons. Simultaneously, by providing canned thrills, adventures, and stereotypic comic fare, he catered to the tastes of a wider moviegoing public. In the end, sensationalism, not authenticity, propelled the Johnsons into celebrity status during America's roaring twenties. Hollywood was a route by which the aspiring classes might achieve fame and fortune. And the Johnsons were proof that the camera could be a means to experience the life of travel and adventure afforded by the wealth and leisure of the upper class. When the Fox Film Corporation agreed to finance their next photographic adventure to produce a sound film of the Mbuti pygmies and gorillas of the Congo region, the Johnsons—now with the backing of Hollywood—gave up any pretensions to science. In turn, members of the American Committee of International Wild Life Protection, which included representatives from the Boone & Crockett Club, the American Museum of Natural History, the New York Zoological Society, and Harvard's Museum of Comparative Zoology, began to doubt Martin Johnson's character. While Martin was in Africa filming the Fox feature *Congorilla,* Harold Coolidge of Harvard's Museum of Comparative Zoology wrote to William Douglas Burden to express concern that Johnson, whose expedition was undertaken on behalf of a Hollywood motion picture studio, had falsely used the name of the American Museum to secure a gorilla-collecting permit from the Belgian government. Coolidge also feared that Johnson's procurement of three live gorillas, without official confirmation that a single collecting permit had been granted, did serious damage to the international goodwill gained by the American Museum and "cast reflections on the veracity of Martin Johnson." In the eyes of museum naturalists and their wealthy patrons, nature could be exploited as commercial entertainment only if it also promoted science, not merely to secure individual financial gain.[16]

If the Johnsons found in nature a commercial commodity that capitalized on the entertainment industry and enabled them to participate in the fame, luxury, and leisure of Hollywood society, Burden remained steadfast in his belief that authenticity need not be sacrificed in generating audience appeal for nature subjects on the motion picture screen. During the winter months of 1928, while *Simba* played to packed theaters in New York City, Burden was at work on the story outline for a natural history film that would in his words "combine true drama with authenticity." Perhaps reflecting a nostalgia for summers spent as a youth in the wilderness of northern Quebec, Burden set out to make a "picture of the North American Woods Indian before the arrival of the white man." By combining a dramatic story with a historically accurate ethnographic representation of the North American Indian, Burden hoped to produce a picture that would simultaneously appeal to the moviegoing public and be of educational value in the classroom. Optimistic about the incorporation of nontheatrical film into the educational curriculum, Burden believed that educational films, unlike theatrical releases, had the potential to generate a permanent income since they could be shown year after year. In utilizing Indian artifacts and ethnographic information housed at the American Museum of Natural History, Burden hoped to demonstrate to the museum's board of trustees that the educational moving picture field offered a "method of obtaining funds that are forever necessary for research." Burden made certain that his interest in a permanent income was not misconstrued as a desire, like the desire of Martin and Osa Johnson, to achieve wealth and fame. In a letter to Jole B. Shackelford, the cinematographer for Roy Chapman Andrews' Central Asiatic Expedition, Burden insisted that he was "not primarily interested in making a great financial success of the enterprise." "First and foremost," Burden remarked, "it must be a great picture, an artistic picture, and a truthful one insofar as these elements are compatible."[17]

The inspiration for Burden's film *The Silent Enemy* came from Merian Cooper and Ernest Schoedsack's production, *Chang,* a 1927 Paramount release that critics hailed as the "most magnificent wild animal picture ever brought to the screen." Cooper and Schoedsack had experienced in real life the drama that Hollywood sought to craft on screen. A newspaper reporter before he joined the Army Aviation Service, Cooper was shot down during the First World War and taken by the Germans as prisoner of war. In association with Major Cedric Faunt-le-Roy, Cooper formed in 1919 the Kosciusko Flying Squadron for the Polish Army. Shot down by Budenny's Cossacks, put in a Moscow prison, and sentenced to death, he miraculously escaped with the aid of a professional smuggler. Schoedsack also got his taste for the life of adventure during the First World War. A cameraman in 1915 for Mack Sennett's Keystone Film Company, a fun factory devoted to the production of short comedies, Schoedsack was recruited into the newly established Photographic Section of the Army's Signal Corp. There he turned his camera to capturing the horror, thrill, and danger experienced on the front line. He continued to satisfy his passion for adventure as a newsreel photographer during the Polish-Russian War and journeyed to the Balkans to capture on film the dramatic realities of the 1920–1922 Greco-Turkish war.[18]

The two adventurers first crossed paths at a train station in Vienna, both en route to Warsaw to aid in the relief of Polish refugees during the Bolshevik invasion. In 1923, while on board Captain Edward Salisbury's yachting expedition through Ceylon, the South Seas, and Arabia, Cooper secured Schoedsack as the expedition's photographer. The partnership, through films such as *Grass* and *Chang,* helped establish what would later be characterized as the "naturalist (romantic) tradition" of nonfiction film. Praised as "dramatists of reality" by the editor of *Asia* magazine, Cooper and Schoedsack, like Robert Flaherty in *Nanook of the North,* sought to immerse themselves in the lives of their subjects and thereby discover a dramatic story latent within the real.[19] In the naturalist tradition, the story emerged from the location. Adventure and melodrama were not imposed on the material for entertainment appeal. Rather, the task of the photographer was to bring forth the essential drama found in natural surroundings observed and discovered through firsthand experience.

Cooper criticized those who were quick "to decry the motion picture as a thing of fakes and sham." In 1925 he envisioned a time "not far off, when the motion picture will play a very real and vital role in education. With the flexible means of expression given by the film," he suggested, "it is possible to record the great natural geographic dramas which go on all over the world, wherever Man contends against Nature in the struggle for existence. And, given the proper technical experts and equipment, there is no reason why the screening of such a drama may not have almost universal interest. When man fights for his life, all the world looks on. And where does man have to fight harder," Cooper ended, "than when he finds his opponent the unrelenting and stern forces of Nature?" Convinced that reality had sufficient dramatic interest in and of itself to attract audiences to the motion picture screen, Cooper and Schoedsack endeavored to capture "natural drama" of heroic proportions acted out in remote regions of the world where the artificial rhythms of a modern, technological civilization could be only dimly heard.[20]

The search for an epic natural drama led Cooper and Schoedsack, along with Marguerite E. Harrison, to Turkey in 1924. Harrison, a reporter for the Baltimore *Sun,* worked for American intelligence in Germany and Russia and had come to Cooper's aid while he languished in a Moscow prison. With $10,000 raised by Harrison, the three headed to Constantinople with a rough scenario for a film in which Harrison, "[w]eary of modern life," sets out on a "quest for a forgotten people living somewhere in the heart of Asia . . . to relive the existence of . . . remote ancestors." Unable to secure permits from the Turkish foreign ministry to film on location in Kurdistan, the three headed south to Persia. There they accompanied the Baba Ahmedi tribe on an annual forty-six-day migration. Fifty thousand Bakhtiari people and half a million animals from the sun-scorched plains near the Persian Gulf crossed the treacherous ice-cold waters of the Karun River and braved the spectacular snow-covered Zardeh Kuh mountains to reach their summer pasturelands in the central Persian plateau. In living with the Bakhtiari, Cooper and Schoedsack had gone beyond the superficial "gossip and kodaks of an onlooking tourist" to experience, in the words of fellow explorer William Beebe, the "direct touch of one of the very tribesmen." By enduring the same physical hardships faced by the tribespeople in their native environment, Cooper and Schoedsack believed they had partially shed the trappings of modern civilization to discover the authentic drama lived by the Bakhtiari in their struggle for existence. Footage in the 1925 film *Grass* of the perilous crossing of the frigid, torrential waters of the Karun River and the climb through precipitous snow-covered mountain passes with bare, frozen, and bleeding feet provided both drama and visual documentary evidence of the hardships endured by the Bhaktiari nomads and their filmmaker companions. To live as the Bhaktiari lived became an important element in establishing the authenticity of the drama on screen. In the romantic vision that both Cooper and Schoedsack shared, the body served as the link to some more innocent, primitive relationship with nature. "You risk your skin," Cooper wrote, "and in the moment when life balances with death, no matter how afraid you may be, you get a touch of the animal value of existence. . . . These are the moments when conscience and memory alike are drowned in the fine physical or spiritual beauty of life."[21]

Although *Grass* captured through stunning photography the natural sublime, its lack of individual characters and limited use of comedy failed to captivate the interest of the moviegoing public. Voted the worst picture of the year by Princeton students, *Grass* was distributed by Paramount and earned the three a modest return on their initial $10,000 investment. When Cooper and Schoedsack turned their attention and cameras to the jungles of southeast Asia, they combined their interest in "natural drama" with a scenario that included character interest and lighthearted comedy. As in *Grass,* they arrived in Asia only with the most general idea for a script, confident that drama would emerge out of the material gathered during their eighteen-month stay among the Laos in the Nan district of northern Siam. In *Chang: A Drama of the Wilderness,* the romantic theme of man's struggle against nature is again at the fore, but it is told through the labors of an individual Lao family, who carve out an innocent, simple, life amidst the threatening forces of the jungle.

Chang begins with an idyllic scene that portrays the "peaceful rhythm" of a Lao family—Kru, his wife, Chantui, and their children, Nah and Ladah—felling trees, tending their animals, and hulling rice in a small plot of land surrounded by dense jungle. The drama unfolds when a leopard and tiger kill the family's goat and cherished water buffalo. Kru, who "bolder than the rest of his tribe [had] ventured deeper into the Jungle to make his home," returns to his village for help in trapping the wild jungle animals that threaten his family and domestic livestock. The villagers use pit traps, snares, deadfalls, and nets to clear the jungle of predatory beasts. Having slain the tiger, Kru and his friend Tahn head for the village boastful and triumphant when another tiger unexpectedly appears. Fearful for his life, Kru scrambles up a tree. As the tiger leaps at Kru, now perched on a limb, its ravenous, snarling face fills the entire motion picture screen, a thrilling close-up that Schoedsack achieved by building a tree platform that turned out to be within six inches of the tiger's reach. But the most stirring scene, one that every reviewer commented upon, was the film's climax: a great elephant herd stampedes through the village, destroying everything in its wake. Nestled in a pit covered with logs, Schoedsack shot spectacular footage of the stampede from the ground level. Aerial shots from high in the jungle canopy added to the dramatic footage. With their homes destroyed, Kru and the villagers embark upon an elephant drive in which they round up the elephants into a large corral or *kraal.* The wild spirits of the magnificent beasts are broken, and the captured elephants are put into service rebuilding the village and Kru's jungle home. Peace and contentment return to the Lao family's pastoral life.

Throughout the film, Cooper and Schoedsack, like the Johnsons, utilized the personality of individual animals, both wild and domestic, to elicit an emotional identification with animals on screen. In particular, the mischievous antics of the family pet, a white gibbon named Bimbo, cute scenes of an adorable mother honey bear and her cub, and an obstinate baby elephant offered comic fare. The character development of individual animals, which borrowed from the craft of the realistic animal story, would become an important convention in generating mass audience appeal for natural history subjects. It was a theatrical convention, however, that existed in tension with the narrative forms within the professional world of science where individuality and emotion were eschewed in accounting for the life of animals.[22]

One of the top five Paramount pictures of 1927, *Chang* cleared $500,000 in the first few months of its opening. Nominated for an Academy Award, the movie played to packed houses in Boston, Chicago, Detroit, Los Angeles, New York, and other major cities. Urban audiences may have found in *Chang*'s wilderness drama an escape and respite from the strain of city life; small-town moviegoers were less enthusiastic. One small-town motion picture exhibitor in Colorado, for example, failed to recoup his rental costs and suggested that "farmers have too many animals to gaze upon daily" to make wildlife films such as *Chang* of interest to rural residents. In the 1920s, the market for natural history film continued to be tied to urban areas, where the back-to-nature movement flourished largely among city dwellers in search of arcadia.[23]

Critics credited the success of *Chang* to a combination of superb wildlife photography, animal comedy, thrilling drama, and authenticity. Not only were the animals "the real thing," *The Moving Picture World* noted, but the native actors were also "wholly natural . . . with a sincerity all their own." In fact, the Lao family in *Chang* had been contrived; the native actors who played the parts were cast by Cooper from the locals he employed in the Nan district for the making of the film. Like Flaherty, who when accused of having staged scenes argued that "one often has to distort a thing to catch its true spirit," Cooper and Schoedsack used art to present a dramatization of reality. It was a dramatization they believed captured the essence of the Lao's struggle with nature, a struggle that the filmmakers had come to know through an intimate association with the Lao in their native environment. Some reviewers, however, were skeptical about the ability to mix drama with authenticity. Richard Watts of the *New York Herald Tribune,* for example, praised Cooper and Schoedsack not for their naturalist vision but their "shrewd" showmanship and suggested that although "*Grass* . . . fell . . . considerably short of the marvelous show provided by [*Chang*], it had a stark, heart-breaking sincerity that must of necessity be lacking from a production in which comedy and drama are mingled with a showman's conscious skill."[24]

Watts may have expressed doubt about the compatibility of drama and authenticity, but *Chang* convinced Burden that it was possible to make a picture with both entertainment interest and educational value. While at work on his script for *The Silent Enemy* in the spring of 1928, he wrote to Cooper explaining the difficulty that lay before him. "We are trying our level best to combine true drama with authenticity so that we will have when we are finished a record of lasting value. It is this very combination which makes it so difficult," Burden lamented. "Either one alone would be a cinch by comparison."[25]

Like *Chang,* Burden's *The Silent Enemy* had a plot that revolved around the human struggle for existence against nature. Hoping to produce a picture that would stir the public's imagination, Burden sought an idea for a dramatic situation comparable to "the scene of the elephant charge in *Chang.*" He read numerous accounts of travel in the far north, including the seventy-two volume *Jesuit Relations* written by Jesuit missionaries in New France during the seventeenth and eighteenth centuries, all in vain. A query to Cooper in search of ideas led nowhere. Then, in late April of 1928 Burden received a letter from Captain Thierry Mallet that furnished the clue he had been looking for. Mallet, president of the Reveillon Frères company, had spent twenty years of his life in the northern regions of Canada between the eastern Atlantic seaboard and the Yukon interior. He had worked with Robert Flaherty in filming *Nanook of the North:* the "only truthful depiction," Mallet argued, "that was ever made of that part of the North." But *Nanook of the North* lacked drama, according to Mallet, because the drama that he and Flaherty knew was "too terrible to show on the screen. Censorship would have interfered." Deeply taken by Burden's resolve to not "sacrifice truth for the sake of drama," Mallet suggested that the "torrent of the caribou fording and swimming a river in the Barren Lands" was just the climax that Burden sought to end his film. Flaherty had wanted to include footage of the great caribou herds in *Nanook of the North* but was furnished with the wrong location; Mallet would tell Burden precisely the place to capture the magnificent, natural drama on film.[26]

The story of *The Silent Enemy* follows the Ojibwa's race against time, starvation, and the harsh Arctic winter to reach the great caribou migration.

It opens during the haze of Indian summer in the Hudson Bay region of northern Ontario and Quebec. Chief Chetoga's son Cheeka, armed with the ancient charm from his father's once-mighty bow, spears fish in nearby waters and aspires to become a man like the tribe's great hunter, Baluk. The hero of *The Silent Enemy,* Baluk, is a symbol of masculine virtue; his muscular frame and prowess attest to his noble and pure character. When Chief Chetoga's daughter, the tender-hearted Neewa, becomes trapped on a cliff with a black bear and her two cubs, Baluk rescues her. He slays the mother bear with his arrows and gives the two captured cubs to Cheeka as pets. Although Baluk and Neewa appear destined for one another, the evil Medicine Man Dagwan uses trickery and false magic to try and disgrace Baluk's reputation as a great hunter and claim Neewa for his own. When Baluk's hunting party returns from the south with no game to tide over the tribe in the coming winter months, Chief Chetoga reminds his council that they are in the seventh year, a year of famine, and heeds the advice of Baluk to take the tribe north where they will find the great caribou herds. But the trek takes the life of Chief Chetoga before they reach the open, wind-swept barrens. And still there are no signs of caribou. Dagwan invokes his magic and tells the tribe that Baluk, the cause of all their suffering, must be sacrificed to appease the Great Spirit. Bare-chested in the snow, Baluk chooses the death of a chief and climbs his funeral pyre. As he recites his death chant, engulfed in flames, the scouts send a signal fire to alert the tribe of a coming caribou herd. Baluk is rescued from the flames and leads his men in the slaughter of abundant game. In a harrowing scene, Cheeka narrowly escapes being trampled to death in the caribou stampede. With stomachs full and plenty of food for the coming months, happiness and joy return to the tribe. Dagwan, having "made false medicine in the name of the Great Spirit," is sent into the wilderness without weapons or food, to die the slow death of a coward.[27]

Script in hand, Burden solicited Jesse Lasky of Paramount Pictures for a distribution agreement. In the late 1920s, since the studios owned and controlled bookings in the large theater chains, this meant independent producers had little hope of getting their films shown to a wide public without the distribution and marketing of a Hollywood studio. Enthusiastic about the box-office success of *Chang,* for which he had negotiated distribution rights, Lasky was optimistic about the prospects of *The Silent Enemy* and offered a release contract. To finance the film's production costs, Burden, along with his coproducer William Chanler, a former Harvard classmate and junior partner in a New York law firm, had raised $100,000, much of it from personal investments and the investments of close friends and family. Burden had hoped to recruit the Central Asiatic Expedition photographer, Jole B. Schackelford, as the cinematographer for *The Silent Enemy.* But Shackelford expressed skepticism about the film's box-office appeal. He believed that "the American Indian [was] too commonplace an element in the everyday life of the average American" to make the picture a success. "His battle for existance [sic]," Shackelford mused, "is not fraught with the menace as is the primitive who is further removed from civilization." "It is the strange and unknown that thrill and hold the blasé public today," Shackelford insisted. It was a prescient warning. Unable to hire Shackelford, Burden and Chanler chose Hollywood cameraman Marcel le Picard. H. P. Carver, whose previous credits included *Grass,* served as the film's director.[28]

Suffering from an ulcer that was diagnosed later in life as amoebic dysentery, Burden longed to escape the nervous energy of the city and recuperate in the north woods of his youth. Filming *The Silent Enemy* afforded such an opportunity. In the summer of 1928, Burden and Chanler traveled to northeastern Ontario and northwestern Quebec to find a location and a supporting cast. The two were determined to find "pure blood families" of Ojibwa whose memories of traditional tribal ways had not been completely erased by modern life. With this goal in mind, they accompanied Father Evain, a plump and charming Roman Catholic priest with a forked and flowing white beard on visits to his congregation. A missionary for thirty years, Evain journeyed by canoe in summer and dog sled in winter to the scattered villages of Ojibwa along Lake Abitibi, Temagami, Temiskaming, and Kippewa. Evain helped Burden and Chanler win the support and cooperation of local Ojibwa, whose economic livelihood normally depended upon fur-trapping in the winter and tourist demand for cooks and guides in the summer. But the expansion of the Canadian transcontinental railway system during the First World War brought an influx of white trappers into the area, resulting in depressed fur prices, depleted game, and a sharp economic downturn for the Ojibwa. Burden's offer of sixty dollars a month thus appeared as welcome economic relief to many Ojibwa in the area.[29]

In the fall months on location at the mouth of the Kippewa River on the Quebec side of Lake Temiskaming, the crew began work on the preparation of props and costumes for the film. While the American Museum of Natural History lent an assortment of original Ojibwa clothing and objects, and some Ojibwa brought family heirlooms, many of the props had to be made, including wigwams, canoes, bows, arrows and quivers, fur clothing, snow shoes, and sleighs. To ensure the authenticity of objects, Burden relied upon the craft knowledge of elder Ojibwa in the reconstruction of their traditional material culture.

Described as "the first time that a race, realizing it is about to die, has itself acted out its own story, on its original stage and with the original settings, as its final 'beau geste' to the race that destroyed it," *The Silent Enemy* endeavored to preserve for future generations the story of "America that used to be." Picard used light and shadows in *The Silent Enemy* to good effect. It was an effect reminiscent of that Edward S. Curtis achieved in his still photography and films, for example, *In the Land of the War Canoes,* to convey the impression of an innocent and pure race whose time had passed. In the romantic naturalist tradition, the story of that vanished past is inextricably linked to the "natural environment," to the "primeval forests" once roamed by native Americans, "and the wild animals [they] used to hunt." This link is one reason Burden chose Rabbit Chutes in the Temagami Forest Reserve as the location for filming the Indian village and winter scenes. Rabbit Chutes was an area where virgin red pine still grew. As one of the few remnants of pristine wilderness, Rabbit Chutes offered a vision of nature unadulterated by the arrival of white civilization.[30]

Traditional artifacts and untouched wilderness figured prominently in Burden's notion of authenticity. So did racial identity. Although Hollywood producers often used white actors in lead Indian roles, Burden insisted on the need for racial purity in his cast. The preservationist mission underlying Burden's film project was one in which nature and race were closely linked. Throughout the early twentieth century, eugenic concerns for racial purity went hand in hand with environmental concerns for the preservation of nature. Many conservationists, including Madison Grant, Henry Fairfield Osborn, and Theodore Roosevelt, were also prominent eugenicists. Both movements, eugenics and conservation, embraced antimodernist anxieties about the degenerative influences associated with urban life.[31]

Two of the lead actors in *The Silent Enemy*, Paul Benoit, who played the part of Dagwan, the evil medicine man, and a thirteen-year-old who played Cheeka, were native Ojibwas discovered by Burden on his travels through the north woods. The part of Chief Chetoga was played by Chauncey Yellow Robe, whom H. P. Carver had spotted one day viewing exhibits at the American Museum of Natural History. Yellow Robe, the son of a hereditary chief of the Sioux tribe and great nephew of Sitting Bull, had been shipped East at the age of fifteen to General Pratt's Carlisle Indian School, where "[his] dreams of glory in the Indian world vanished from [his] vision." Stripped of his native clothes, his long hair shorn, Yellow Robe became indoctrinated into "the process of civilization." An advisor to the Federal Indian Residential School in Rapids City, South Dakota, Chauncey Yellow Robe did not completely embrace assimilation; he spoke out harshly against commercial entertainments such as wild west shows that deprived Indians of their "high manhood and individuality" by "permitting hundreds of them to leave their homes for fraudulent savage demonstrations before the world." Wary at first of Burden's intentions, Yellow Robe became convinced of his sincerity and agreed to play the part. Throughout the harsh winter months of filming, he shared a teepee with Burden and later dined at the Burden family estate on Long Island.[32]

Burden and Chauncey Yellow Robe may have shared living quarters, but the return to a romantic, nostalgic past reenacted in the wilderness of the north woods had far different consequences for native Americans than it did for white middle- and upper-class urbanites. In search of an actor to play the lead role of Neewa, Burden visited one of Texas Guinan's speakeasy clubs frequented by the socialites of New York City where a popular dancer, Molly Spotted Elk, performed. A Penobscot from Maine, Molly Spotted Elk began her stage career performing native American dances on the vaudeville circuit and at the famous Miller Brothers' 101 Ranch in Marland, Oklahoma. Disheartened by the phony quality and carnivalesque atmosphere of circus-type road shows, she longed for performances more "authentic" and "true." She made her way to New York City, where she began to get bookings at more upscale theaters and cabarets. But the city wore upon her. "God never intended the Indian to live always in the city," she wrote, "and see nothing but man-made things. Even the trees in the park are trained to grow to suit the mind of men."[33] When Burden offered her the role of Neewa, she seized the opportunity of "being in the woods" and "living and feeling the part of an Indian girl of long ago." On location, however, Molly Spotted Elk identified herself much more closely with the modern world than with the "primitive attitude" of the "bush" Indians hired as extras whom she viewed as "unchanged after centuries."[34] She may have found in the north Canadian forest a romantic escape from city life, but to New York City socialites, the wilderness unveiled the noble savage that she was presumed to be, regardless of her assimilation into modern, industrialized society.

At the exclusive five-dollar-seat premiere of *The Silent Enemy* at Broadway's Criterion Theater on May 19, 1930, Molly Spotted Elk, surrounded by people of wealth and social standing, asked of herself: "Am I civilized only on the surface and still just a savage underneath?"[35] The awkwardness she felt was no doubt reinforced by an article that recounted the making of *The Silent Enemy* in a souvenir pamphlet available for purchase by movie audiences. "The Indian is still at heart what he was 400 years ago," William Laurence observed in "The Death Chant of Red Gods and Man," "and after a few months of scratching his memory he just shed his outer layers of white man's civilization and became himself again." Pristine nature may have freed Burden from "frayed nerves," but for native Americans like Molly Spotted Elk, it shackled them to a frozen past. Like the animals of the forest, their history was deemed by European Americans to be forever tied to the seasonal cycles and rhythms of nature, not to changes wrought by civilization and culture. The attraction of this alleged tie to nature explains the popularity of ethnographic exhibits in the country's leading museums of natural history. It also explains why ethnographic film and natural history film were so closely linked during the first part of the century in the imagination of the moviegoing public.[36]

To critics, the wilderness location, the fight between the mountain lion and bear, the attack on the moose by timber wolves, and the migration of the great caribou herd in *The Silent Enemy* offered a "sincerity and authenticity" unparalleled in any other film of its kind. Madison Grant, a prominent lawyer, influential naturalist in conservation organizations such as the Boone and Crockett Club and the Save the Redwoods League, and author of the racist polemic, *The Passing of the Great Race,* gave his scientific stamp of approval to the film after a private screening. As president of the New York Zoological Society, Grant congratulated Burden and Chanler for their "great service to science and to zoology in placing on record the life in the northern woods." "I am thoroughly familiar with the country from the upper Ottawa River to the Hudson Bay region," wrote Grant, "and I can testify that your pictures are accurate in every detail." In all his years of experience in the north woods, Grant "never had the opportunity of seeing a wolverine on such intimate terms." Some scenes, such as a wolf attack on a bull moose, were so striking that Grant at first had a hard time believing they were real. He suspected that the wolves were actually huskies with their tails wired down, until Chanler convinced him otherwise. The caribou migration Grant judged to be the most amazing and authentic wildlife footage he had seen.[37]

Grant was right to be suspicious, but he erred significantly in judging what was authentic and what was artifice. In filming wildlife, Burden faced the real difficulty of combining true drama with authenticity. Mass appeal depended upon dramatic interest, but true drama in nature could not always be found readily. Count Ilia Tolstoy, grandson of the Russian writer Leo Tolstoy, led a special expedition to obtain footage of the great caribou migration for the film's climax. A one-thousand-mile, six-month journey to the location supplied by Captain Theirry Mallet in the Windy Lake Region of the Northwest Territories ended in failure. Like Flaherty before them, they missed the crossing by several hundred miles and were only able to secure photographs of small herds. Burden needed drama on a much grander scale than nature supplied. To create stirring scenes that paralleled the elephant stampede in *Chang,* he obtained the assistance of Carl Lomen, head of the Lomen Reindeer Corporation, based in Nome, Alaska. Lomen supplied a herd of reindeer to the north slope near Point Barrow, where a small crew staged Cheeka's near-death encounter and filmed much of the spectacular great caribou hunt. Although considered to be of the same species, reindeer are a semi-domesticated Eurasian equivalent of wild caribou. Nervous about this departure from authenticity, Burden asked Lomen to keep "information in regard to the caribou pictures" secret, since interest in the picture depended "to a very considerable degree on the sense of reality that it gives."[38]

Given that other wildlife scenes throughout the film were staged, Burden's caution about the caribou footage seems at first glance peculiar. By keeping animals in a natural enclosure large enough to not impede authentic movements and behaviors, yet small enough to take action shots with relative ease, Burden staged animal scenes that he argued were "absolutely truthful from a natural history point of view." The technique was first used to great effect by Merian Cooper in *Chang.* A mountain lion and bear fight in *The Silent Enemy,* for example, occurred after the two animals, kept in the same enclosure, were starved for several days and then provided with a deer carcass. Only in this way was it possible, Burden reasoned, to "secure most interesting pictures of animals . . . which are impossible to photograph in nature."[39] Unlike the mountain lion, however, which was captured in the Kaibab wilderness of the American southwest, the caribou were not pure animals of the wild, untainted by human touch. The Lomen Reindeer Corporation was a large commercial enterprise that harvested reindeer for meat. The reindeer were as much products of modern industrialized society as they were of nature. As such, they defied the sharply defined boundaries erected in early twentieth-century America between authentic nature and synthetic civilization. But they were not the only subjects of the film whose impurity Burden found disconcerting.

Chief Buffalo Child Long Lance cut a striking figure in his part as Baluk in *The Silent Enemy*. Clothed only in a loin cloth, his athletic build, hardened by days spent training with Jim Thorpe at the Carlisle Indian School and by active combat in the Canadian army during the First World War, impressed reviewers and audiences alike. The famed sports writer Grantland Rice compared Long Lance's battle with the moose in *The Silent Enemy* to the first round of the Dempsey-Firpo fight, while Regina Crewe in the *American* delighted in his "superb figure" and "utterly fearless" character. In the opinion of *Variety,* Long Lance's "full-blooded" stature as a Blackfoot Indian made the picture all the more authentic.[40]

When Burden and Chanler chose Long Lance for the role of Baluk, he was already a celebrated figure in the gossip columns of New York society. In his 1928 autobiography, Long Lance recalled the nomadic existence of his childhood, traversing the plains as a member of the Blackfoot tribe. A glimpse of the vanished life of the northern Plains Indian, *Long Lance* became a bestseller and made its author New York City's "social lion of the season." His distinguished good looks, accented by his copper skin and straight black hair, coupled with a gallant manner and flair for the dramatic won Long Lance the title of "the Beau Brummell of Broadway." He appeared the perfect choice to play Baluk. During the filming of *The Silent Enemy,* however, suspicions about Long Lance's true identity mounted. His punctuality at meals, his boisterous and cheery manner, his poor knowledge of Indian sign language, and his unconventional dancing all struck Chauncey Yellow Robe as unusual for a Plains Indian.[41]

In the spring of 1930, prompted by the doubts of Yellow Robe, Willie Chanler launched an investigation into Long Lance's ancestral past. A letter from the Bureau of Indian Affairs confirmed Yellow Robe's suspicions. Chief Buffalo Child Long Lance was an impostor. Born Sylvester Long in North Carolina, Long Lance was, according to Bureau records, Croatan, a turn-of-the-century designation for the Lumbee, a federally recognized tribe of mixed Indian, English, and African descent. Chanler also learned from a teacher at the Carlisle Indian School that Long Lance had claimed Cherokee heritage in his original application, but among Cherokees at Carlisle he was considered an outcast, "more of negro blood than Indian."[42]

The child of Joe and Sallie Long, Sylvester grew up in Winston-Salem in a culture where there was no place for individuals who transgressed racial boundaries. His father, a slave in childhood, was likely of mixed African-American, Catawba Indian, and white parentage, although he believed his parents to be of white and Cherokee background. His mother was the daughter of a prominent plantation owner and a slave of mixed Lumbee and white ancestry. To the white residents of Winston-Salem, Sylvester Long was colored. Suffocated by the oppression of a Jim Crow society, Sylvester Long fled at the age of thirteen and joined a traveling wild west show. Among whites, his physical features enabled him to pass as Indian. When he returned to North Carolina in 1909, he applied to the Carlisle Indian school with the assistance of his father, who listed his son's heritage as one-quarter Cherokee. After the war, while working as a freelance writer for a series of Western Canadian newspapers, he was adopted by the Blood Indians of Alberta and given the name Buffalo Child. Thereafter, Long Lance assumed the identity of a Blackfoot tribal member and fabricated a childhood among the Plains Indians in northern Montana.

It didn't take long for Chanler to discover the location of Long Lance's real parents. Dispatched to Winston-Salem, Ilia Tolstoy befriended Joe and Sallie Long and alleviated Burden and Chanler's worst fears. The Longs assured Tolstoy that they were of Cherokee and white background. Their lineage was confirmed in a sworn affidavit that Tolstoy procured from Colonel William Blair, a Harvard graduate and prominent banker of Winston Salem who had known Joe Long for forty-five years. Tolstoy telegraphed Chanler one month before the scheduled premiere of *The Silent Enemy:* "Have full information and affidavit of him being Indian and white." To Chanler that was enough to assuage any accusations of fraud. "Being Indian is all we care about," he told Tolstoy. "What tribe he belongs to is entirely secondary." Long Lance, however, could no longer live the lie. Two years later, amidst renewed rumors of deception, he put a forty-five caliber pistol to his head and committed suicide.[43]

In making *The Silent Enemy,* Burden hoped to produce a film in which dramatic interest remained faithful to scientific truth. The reconstruction of traditional Ojibwa artifacts and clothing, the hiring of native Americans for the cast, the shooting on location in northern Canada during sub-zero winter temperatures, and the scripting of scenes based on the historical accounts of Jesuit missionaries and stories told by elder Ojibwa all contributed to what Burden believed to be an authentic and lasting record of the native American's struggle for existence before the arrival of Columbus. He had lived in the forest of the Ojibwa, endured their hardships, listened to their old men around the campfire. "That is why," Chauncy Yellow Robe asserted, "the picture is real." Artifice, however, threatened to undermine the film's integrity at every turn. Facts were not necessarily more compelling than fiction. The incident with Long Lance made that clear. And nature could not always be counted upon for spectacle, as the staged wildlife scenes revealed. Immersed in the wilderness, Burden felt he touched an essential nature that eluded the mass of tourists, who "buzz past blurred landscapes, photograph a mountain peak, their wives, their children, the bears, and pass on with scarcely a glimmering of understanding or appreciation of what they have seen." Yet, *The Silent Enemy* depended upon those masses for commercial success.[44]

Critics heralded the majestic scenery and striking realism of *The Silent Enemy*. Box-office revenues failed to match the artistic acclaim the film received. Theater owners found the picture "a good educational subject but not entertaining enough to be satisfying to the average theatregoer." The lack of sound also limited the film's appeal to a public that, sparked by the release of *The Jazz Singer* in 1927, clamored for the latest in talking pictures. Richard Dana Skinner, film critic for *The Commonweal,* thought the silent format, supplemented by a musical score based upon traditional Ojibwa melodies, strengthened *The Silent Enemy*'s "beauty and impressiveness." Audiences did not share in Skinner's enthusiasm. Distribution executives at Paramount considered the film a problem picture and limited its release primarily to the third-run, small theaters located outside the downtown metropolitan centers. Given that natural history film appealed to a more urban middle- and highbrow clientele, Paramount's promotional strategy proved disastrous. "In the small movie houses in country districts," audience reaction to *The Silent Enemy* was described as "apathetic." One year after *The Silent Enemy* premiere, gross revenues from the film fell considerably short of the total amount owed Paramount and private investors. With $95,000 in outstanding notes to individuals and the Guaranty Trust Company of New York, Burden Pictures filed for bankruptcy.[45]

The Silent Enemy's failure to attract a mass audience revealed the extent to which the Hollywood studio system, from production through exhibition, determined the commercial success or failure of natural history film in the 1920s. The successful travelogue–expedition film fed the public's fascination for the exotic, the danger of the lurid jungle, the caricatures of animals and of tribes and races of people deemed primitive by white society. Shackelford was right. The North woods and its inhabitants were too familiar for the American public. The drama that Burden sought to wrest from the Canadian wilderness was perhaps too close to home; it lacked the sensational qualities found in faraway lands.

Yet limits to showmanship in the name of science existed, even in Hollywood. One month before the opening of *The Silent Enemy,* a film of an alleged expedition into the Belgian Congo by Sir Hubert Winstead of London created a public splash. Although reviewers doubted the film's authenticity, sensational advertising coupled with footage of beasts being slain and the "suggestiveness of gorilla scenes" guaranteed *Ingagi,* in the opinion of *Variety,* a "sure box-office." RKO bought national rights from Congo Pictures, Ltd. after the film grossed $23,000 in one week at San Francisco's Orpheum theater. In Seattle, *Ingagi* broke existing house records. "Has the Man Ape Been Found? Gorillas! Wild Women! Apparently Half Ape! Half Human!" ran the ads in Chicago, where the film played continuously at the Garrick theater for six weeks. Crowds came especially for the last ten minutes of the film, which purported to show a group of fully naked "ape women," shadowed by thickets, cohabiting with gorillas and a tribal sacrifice of a woman by native Africans to a 600-pound gorilla. A graphic film poster depicting a gorilla groping a bare-breasted black woman titillated audiences with its implied scenes of bestiality. Sex and violence were sure box-office draws and *Ingagi* had enough to attract not only the public but the attention of the scientific community, the Better Business Bureau, and the Will Hays office of the Motion Picture Producers and Distributors of America.[46]

In May of 1930, coincident with *The Silent Enemy*'s Broadway opening and a few weeks after *Ingagi*'s showing on the East Coast, the American Society of Mammalogists gathered in uptown Manhattan at the American Museum of Natural History for their annual meeting. At the society's traditional smoker, the museum treated naturalists to a private showing of motion pictures of African mammals selected from their collection of Martin

Johnson films. The selection was an ironic choice for the evening's entertainment. Johnson was in Africa filming *Congorilla* for Fox. And just that afternoon, at the request of William K. Gregory, curator of the museum's Department of Comparative Anatomy, members unanimously passed a resolution that publicly condemned the film *Ingagi* for its gross misrepresentation of "the natural history of Africa, while pretending to be a truthful record of a scientific expedition."[47]

Naturalists feared *Ingagi* might do serious harm by offering moral grounds for the gorilla's extermination precisely at a time when conservation efforts to save the species from extinction were imperative. The 1930 annual meeting of mammalogists featured for the first time a day-long symposium on the gorilla, a species that had eluded study by scientists in its natural habitat. But this was not their only concern. At a time when naturalist-photographers were just beginning to show "the possibility of bringing wild life from all parts of the globe within the sight of the cinema patron," declared *Nature Magazine,* a popular monthly published by the American Nature Association, "one single 'faked' picture will cast an irremovable stigma on the whole field, and raise doubts in the minds of the public regarding all Nature films. If anybody with a sense of the dramatic and a motion picture camera may go to Africa—or Hollywood—and pass off on the public, without censure, fakes purporting to reveal new scientific truths, the motion picture as a Nature educator will have utterly lost its value."[48]

Prompted by the appeals of the Better Business Bureau and the scientific community, the Hays office in June of 1930 barred exhibition of *Ingagi* not on the grounds of the film's nudity or innuendo of bestiality but for its nature faking. Experts judged the gorilla scenes in *Ingagi* to be a compilation of orangutan stock footage and a man in a gorilla suit. A newly discovered animal, "so 'venomous' that it could not be brought to America," was considered a hoax, fabricated by shellacking bird wings onto the hide of a scaly anteater that was then cloaked around the shell of a leopard tortoise. An investigation by the Hays office determined the film to be faked at the Selig zoo in California using captive animals and actresses in blackface.[49]

Although *Ingagi* deluded some individuals like *Chicago Tribune* film critic Mae Tinée, who considered the poor cinematography convincing evidence that it was "the real thing and not a staged adventure," the real cause for concern among the scientific community was less the obvious fake than the cunning forgery. In the fall of 1930, Burden became incensed by the release of *Africa Speaks,* a picture staff and board members from the American

Museum of Natural History regarded "as much more dangerous than *Ingagi* because the fake [was] so much better done." A record of the Colorado African Expedition's twelve-thousand-mile journey by motor truck across equatorial Africa, *Africa Speaks* advertised itself as the first picture to authentically record and reproduce the sounds of African animals in the wild.[50]

Led by Paul Hoeffler, a photographer-journalist for the *Denver Post,* the Colorado African Expedition initially set out in search of an alleged Bantu tribe of seven-foot giants in the heart of the Belgian Congo. Hoeffler first learned of this mysterious tribe during a 1925 expedition, backed by a group of prominent Denver businessmen, to South West Africa in search of the bushmen of the Kalahari. Undertaken to bring worldwide attention to the Colorado cow town, the 1925 expedition publicized its mission as an attempt to "demonstrate that [the African bushmen] form the so-called 'missing link' between man and the lower animals" and to "bring them back to civilization for study and scientific research." Hoeffler's 1927 film version of the Kalahari expedition, *The Bushman,* failed to attract public interest or a national distributor, despite the widespread press coverage the expedition received. Upstaged by the thrill and drama of *Chang,* Hoeffler returned from his 1928–1929 motor safari much more savvy about how to exploit African wildlife and people in the interest of commercial success.[51]

The official Colorado African Expedition traveled from Mombasa in British East Africa to Lagos on the west coast of Nigeria. But with few lions and no warriors of the same heroic stature as the Maasai, the Gold Coast lacked drama sufficient to arouse the interest of motion picture audiences or a commercial distributor. Neither was the novelty of animal sounds and a Movietone lecture enough to capture the attention of a public weary of the standard travelogue-expedition fare. To obtain a distribution contract from Columbia Pictures, Hoeffler had to fabricate nature to meet the conventions of Hollywood entertainment. On film, the path of the explorer-naturalist charts eastward in order to build drama that culminates with a Maasai lion hunt. Back projection enabled Hoeffler and his Hollywood companion Harold Austin to appear in scenes of wildlife, such as those of white rhinoceri. In reality they had acted out the scene in a Hollywood studio against footage of wildlife shot on location. Further questions about the film's claim to authenticity were raised by the picture's most gruesome scene, in which a Maasai boy, alleged to be the son of a tribal chief, is killed by a lion as he runs across the open savannah upon orders from Hoeffler to retrieve rifles from the truck. *Variety* thought the whole lion section "fishy" and suggested the morbid scenes were not fitting for younger children or women.[52]

In a letter to Arthur James of the *Exhibitor's Daily Review,* Burden lashed out against James's public defense of *Africa Speaks.* The lion scenes, according to Burden, had been staged in Hollywood using tame lions; the animal sounds he believed were dubbed in a Hollywood studio. "The picture is a fundamental lie," insisted Burden, "and if this kind of thing is to be tolerated in the industry those of us who are interested in the real thing might as well pack up and quit. It is surely impossible that our understanding of Truth should be so different!" To Burden's moral tirade, James responded: "I don't believe the morals of the world are injured by the use of artifice to arrive at a correct impression, nor do I believe that for other than purely scientific purposes the depiction of wild life, if the impression is faithful to the fact, is bettered by leaving out dramatic moments that are difficult to arrive at except by artifice. Certainly the public must have its drama or it remains at home. The scientist is in a different position. He wants the exact facts and finds his drama in the depiction of many things which would be boring as hell to the laymen."[53]

In their effort to combine science with showmanship, naturalist-photographers found themselves on the horns of a dilemma not easily resolved. Julian Johnson, a Hollywood writer who composed the titles for *The Silent Enemy,* expressed the dilemma well. As the motion picture "depends for its life upon mass production, a picture equally depends for its life upon mass favor," wrote Johnson. Audiences had "to be catered to, to be fostered, to be cajoled." In the interests of mass appeal and economic profit, nature had to conform to the conventions of Hollywood entertainment. But in this mix of showmanship and science, where could the line between artifice and authenticity be confidently drawn?[54]

Dismayed by the financial failure of *The Silent Enemy* and by the popularity of sensational films like *Ingagi* and *Africa Speaks,* Burden searched for a niche for natural history film outside the commercial mass-production system of Hollywood. In Britain, John Grierson used the Empire Marketing Board to create a successful system of distribution to educational, industrial, and governmental organizations for films of "the living fact," but the market for nonfiction film in the United States during the early 1930s remained fragmented and undeveloped. Those controlling the economic interests of Hollywood and the institutional structures in place for showing and promoting films did not believe a market existed for educational films, despite the enthusiasm among educational theorists for this new medium. Only with the extensive use of 16-mm film to disseminate information and propaganda to the public during the Second World War did the prospects for documentary and instruction films in the United States expand. Between 1937 and 1947, for example, the number of university and college film libraries in the United States more than doubled, from twenty-seven to sixty-five. Similarly, the estimated number of 16-mm sound projectors jumped from 6,500 in 1936 to 100,000 in 1948.[55]

In the early 1930s, Burden faced the harsh economic realities of the market for nontheatrical film. He established, with an advisory board of prominent relatives and friends, including C. V. Whitney, Marshall Field, Kermit Roosevelt, and Willie Chanler, a short-lived nontheatrical motion picture service, Beacon Films, Inc., that was also a distributor for the Religious Motion Picture Foundation. Inspired by the modest returns *The Silent Enemy* received during its nontheatrical distribution to public schools, churches, lodges, and other civic organizations, Burden looked optimistically on the potential market beyond Hollywood for informational and entertaining motion pictures that contained more of what he described as "the real and wholesome material that the world has to offer."[56]

The existence of Beacon Films amounted to little more than a letterhead on business stationery, but it forged a relationship between Burden and Merian Cooper, who both served as company vice-presidents. Burden's cousin, Cornelius Vanderbilt Whitney, a member of the advisory board, was also an important link in the acquaintanceship of these two adventurers, an acquaintanceship that set in motion the development of RKO's classic blockbuster of 1933, *King Kong*.[57]

In the spring of 1931, Burden and Cooper initiated plans to secure a photographic record of the mountain gorilla in Africa that could serve as the basis for a motion picture along the lines of *Chang* and *The Silent Enemy*. Both expressed interest in adapting the narrative film for educational purposes, although they were unsuccessful in convincing the Rockefeller Foundation to finance an experimental venture in educational film. The first three months of the expedition would be spent in the Lake Kivu district of the Belgian Congo studying the habits of the gorilla by following families of them day after day. Then, an enclosure would be built in a thinned-out section of the jungle, to afford perfect photographic opportunities, and arranged in such a way that the enclosure would not appear on film. Gorillas would be captured and placed in the confined area. Their earlier observations would ensure that the captive gorillas were fed their natural diet and formed the normal social groupings found in the wild. In this way, Burden and Cooper believed they could "get the real thing," since they presumed the gorillas "could be kept healthy and contented" and "would act . . . just about the same as they do in a wild state."[58]

Harold Coolidge, a leading authority on African primates at Harvard University's Museum of Comparative Zoology, believed that if any man could accomplish the photographic expedition proposed, it was Merian Cooper. But Coolidge thought the difficulties faced in capturing gorillas posed challenges that neither Burden nor Cooper had imagined. To capture the gorillas alive, Coolidge recommended the use of some type of asphyxiating gas. Even with this technique, gorillas would undoubtedly be killed, which necessitated a collecting permit from the Belgian government. As secretary of the recently founded American Committee of International Wildlife Protection, which oversaw applications by American citizens to European governments for special hunting and collecting permits, Coolidge had influence in international conservation circles that neither Burden nor Cooper could ignore. Given the recent controversy over *Ingagi* and questions raised about Martin Johnson's *Congorilla* expedition, Coolidge advised Burden that the American Committee was "very doubtful as to the ethics of any picture proposition which involves the capture of animals especially when the proposition is primarily a commercial one." Within the scientific community, the mix of science and showmanship that proliferated in Hollywood travelogue-expedition films of the 1920s had become increasingly suspect. Shunning Hollywood, Burden favored the establishment of an independent production company that would be financed by wealthy New York businessmen such as Sonny Whitney and Marshall Field. But Cooper, unwilling to be indebted to personal friends, preferred to establish an independent company secretly financed through Paramount. Such an arrangement would effectively camouflage Hollywood interests, distance the expedition from the taint of commercialism, and hopefully make it easier to acquire the necessary scientific collecting permits.[59]

In the fall of 1931, Cooper accepted an offer by David Selznick to join RKO as an associate producer and executive assistant. Selznick had just left Paramount to rescue the failing studio hard hit by the Depression. He promised Cooper full reign at RKO in developing the gorilla picture. Fascinated by Burden's expedition to capture and film the dragon lizards of

Komodo, Cooper imagined a setting for the gorilla film identical to the remote volcanic island described by Burden. Based upon his successful use of animals as film characters, such as Bimbo, the pet gibbon in *Chang,* Cooper began to outline a plot in which a giant gorilla played the feature role. Footage of gorillas shot in Africa could be enlarged in the studio to make the animal appear gigantic in size. The gorilla scenes would then be intercut with magnified footage of Komodo dragons to create fights between the gigantic gorilla and prehistoric beasts.

In Hollywood, Cooper found the need to film authentic nature obsolete. Through the triumphs of special effects artist Willis H. O'Brien, he and Schoedsack produced and directed the entire film on the studio lot of RKO. Nevertheless, much of *King Kong* was patterned on Burden's real-life adventure. Carl Denham, the explorer-naturalist-photographer who sets out in search of the prehistoric island, was "a deliberate combination," Cooper wrote Burden, "of you, Schoedsack, and me." Katherine White Burden, who accompanied her husband on his search for the Komodo dragon, furnished the idea for the female lead played by Fay Wray. Coolidge's suggestion of asphyxiating gas to capture gorillas in Africa provided the basis for the gas bombs used to capture Kong. Even the film's climax had its basis in real life, which Cooper, in the interests of Hollywood, sensationalized into mythic proportions. "When you told me that the two Komodo Dragons you brought back to the Bronx Zoo, where they drew great crowds, were eventually killed by civilization, I immediately thought of doing the same thing with my Giant Gorilla," Cooper confessed. "I now thought of having him destroyed by the most sophisticated thing I could think of in civilization, and in the most fantastic way." Icons of technological triumph and industrial efficiency, the Empire State Building and the airplanes that bring about the death of Kong were fitting symbols for the modern machine age.[60]

Cooper and Schoedsack drew upon their experience as dramatists of reality to create through artifice a Hollywood sensation that hyperbolized many of the popular elements that had made travelogue-expedition films of the 1920s successful: the mystery, intrigue, and danger of unknown lands; the violence and raw sexuality of nature waiting to explode; the thrill and adventure of capturing and taming savage beasts. *King Kong* was a stunning climax to the brief reign of the travelogue-expedition film. Frank Buck and Clyde Beatty might continue to attract Great Depression audiences with their thrilling, death-defying capture of animals in films like *Bring 'Em Back*

Alive, Wild Cargo, and *The Lost Jungle,* but only by creating on a Hollywood set the same sensationalism and melodrama manufactured in *King Kong.* In Hollywood, artifice had surpassed nature, taking the spectacle of death and the exotic beyond anything the drama of "real" nature could offer.

Killing game and predators killing prey were staples of the travelogue-expedition film of the 1920s, scenes that audiences had come to expect. When W. S. Van Dyke returned after a seven-month photographic safari to film *Trader Horn,* a 1931 screen adaptation of *The Life and Works of Alfred Aloysius Horn* shot on location in equatorial Africa, MGM executives expressed disappointment over the lack of carnage in the wildlife footage. Van Dyke's expedition was the largest safari on record: ninety tons of equipment, two hundred native porters, and a Hollywood crew of thirty-five. Still, Hollywood spared no expense to procure the macabre. MGM sent a second crew to Mexico, beyond the watchful eye of the Society for the Prevention of Cruelty to Animals (SPCA), to stage gory scenes of an antelope mangled by lions and a hyena whelp mauled by a leopard.[61]

The violent drama of nature's struggle for existence that Hollywood sensationalized in an effort to make travelogue-expedition films a commercial success existed in tension with the scenes of jungle peace that Akeley had sought to portray in the African Hall of the American Museum of Natural History. Developments in film as an instrument of scientific research and education, coupled with the importance of documentary expression in 1930s America, enabled a more dramatic story to be told about the interdependence of life. As a research tool in the study of animal behavior, the camera offered a close-up, intimate look into the daily and seasonal life of animals. In the growing discipline of ecology, the camera offered a wide-angle panoramic vision that revealed the intricate relationships among life. Science and showmanship would come together once again to bring a drama of nature to the screen. Facilitated by Walt Disney and the markets opened by television, and conforming to changing environmental, moral, and family values in postwar America, the private lives of animals united through the ecological web of life would come into focus as a more violent, savage nature faded into the background.

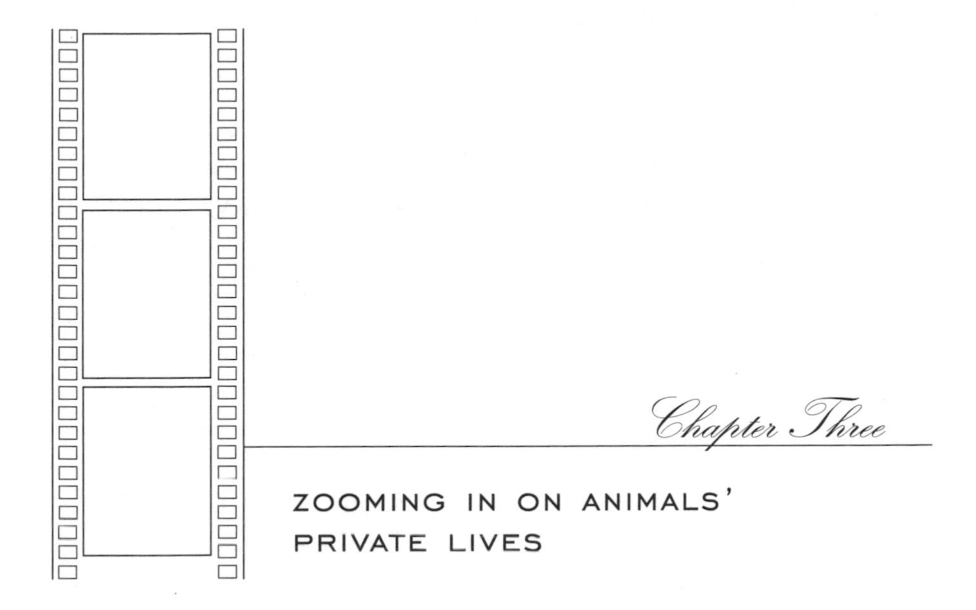

Chapter Three

ZOOMING IN ON ANIMALS' PRIVATE LIVES

In an autobiographical essay written near the end of his life, the Nobel laureate Niko Tinbergen reflected on why he chose to spend so many hours of his life observing the behavior of animals in the field. Tinbergen traced his "deep-rooted love of natural beauty" to a "largely innate, typically masculine love of the hunting range."[1] Thirty years earlier, in the conclusion to his famous 1953 study of the herring gull's world, Tinbergen described bird study as "sublimated hunting, as is bird photography. All aspects of hunting: habitat selection, stalking our quarry, trying to outwit it, and finally experiencing intense satisfaction in getting what we want, are present in both bird watching and bird photography." "I know through introspection," Tinbergen continued, "that scientific bird study and bird photography give me exactly the kind of experiences and satisfaction as I once found in hunting seal on the arctic ice."[2]

Tinbergen was not the only ethologist of his generation to comment on the similarities found in the study of animal behavior, photography, and the hunt. The watchfulness and patience by which the hunter came to know an animal on intimate terms were observational qualities essential to wildlife photography and to the biological study of behavior. All primitive hunters, Tinbergen suggested, possessed a craft knowledge that made them "exceptionally good students of animal behavior."[3] Konrad Lorenz, who shared the 1973 Nobel Prize in Medicine and Physiology with Tinbergen and Karl von Frisch for his contributions to the study of animal behavior, also spoke admiringly of the remarkable patience acquired by wildlife photographers, whom Lorenz traced to stone-age men that had developed a photographic eye for the animals they hunted.[4] But the hunt differed from ethology as a natural historical science and scientific photography in at least one significant respect. Whereas the gun was an indispensable tool of the nineteenth-century naturalist preoccupied with collection, classification, and description, it ultimately precluded investigation of the underlying mechanisms of life, which was a focus of biological science in the twentieth century. Analysis in the scientific study of animal behavior began with the observation and recording of animal motion, a task for which still and motion pictures were ideally suited.[5]

The development of film as a research and educational tool within science coincided with film's rise as a Hollywood technology of mass entertainment. Since film proved an ideal method for teaching surgical technique, medical cinematography flourished. But other fields soon capitalized on this new technology: in particular, anatomy, embryology, psychology, anthropology, and animal behavior. By the 1920s, film demonstrations were an integral part of such scientific meetings as the Association of American Anatomists, the American Society of Zoologists, the American Ornithologists' Union, and the American Society of Mammalogists. Spurred by the advent of 16-mm film and lighter equipment such as the Akeley camera, the application of film to the study of behavior in both laboratory and natural settings had become commonplace by the late 1930s.[6]

In their use of the camera, biologists who studied animal behavior under natural conditions sought a close-up picture of animal life, one that revealed an intimate understanding of the daily activities and social interactions among individuals of a species in the wild. American and European ethologists such as G. K. Noble, Julian Huxley, Konrad Lorenz, and Niko Tinbergen who employed film as an instrument of research developed an experimental and theoretical direction within the study of animal behavior oriented around visual communication within the animal world.

As a research tool, film captured an intimate look at animal life. It also, as a media of mass communication and entertainment, presented a concentrated view of the private lives of animals that the public would rarely, if ever, encounter in nature. Even within the professional world of science, a sense of the dramatic prevailed over the mundane. The experiential knowledge that united scientist and photographer—the endless hours spent in the field, immersed in the senses, waiting and watching for that rare moment to capture on film—was left on the editing room floor. Unlike other technologies created and developed specifically within the scientific laboratory, film was a technology of art and entertainment as well as scientific research. Scientists could never escape its entertainment role. When Tinbergen polled students about the merits of educational nature films, 91 percent described such films as both instructive *and* entertaining. In adopting the camera as an instrument of research and education, ethologists had inadvertently acceded the spectacle of animal life.[7]

In June of 1937, the American Museum of Natural History treated guests attending the annual members' day reception to a prototype display envisioned as part of the natural history museum of the future. Located in the basement of the Theodore Roosevelt Memorial building, "Some Suggestions for a Future Hall of Animal Behavior" contained a series of five displays that depicted the way the world appears to the chicken, dog, fish, fly, and turtle. Standing in front of a picket fence, the visitor saw a painting of hens and a rooster in a barnyard accompanied by the following narration: "The hens in the barnyard seem to us all very much alike. We would have great difficulty in distinguishing one from another if we did not put rings or other identification marks on their legs. But to the hen every other hen in the yard is a personality." Suddenly, the background faded and a new scene of the barnyard came into view, as seen from the perspective of the hen. The rooster was enormous; the size of the hens had also changed. The loudspeaker informed the visitor that they were witness to the social system found in the life of domestic fowl, where a clear pecking order existed: the rooster at top and the hens each occupying a dominant or subordinate place in relation to other hens within the social hierarchy. Other exhibits on display used a similar fade-out technique to illustrate the colorless world of the dog, the distorted angler seen through the nearsighted eye of the fish, and the honeycomb pattern of the fly's visual field.[8]

Formally opened to the public in 1940, the Hall of Animal Behavior sought to dramatically present the underlying processes and principles governing life in the animal kingdom. "When we tackle the great subject of animal behavior," William Douglas Burden insisted, "we are probing the very inside of life." As trustee of the museum and patron of Gladwyn Kingsley Noble's Department of Experimental Biology, Burden attempted to spearhead a major reform effort in the late 1930s to shift the focus of the museum away from its "archive function and the science of taxonomy that goes with it" to the "study of nature and man's relationship to it." "Collecting has long since reached the state of diminishing returns," Burden argued. Taxonomy was as dead as the specimens it sought to preserve. Although Burden's 1930 film *The Silent Enemy* partook in an evolutionary, taxonomic vision—the traditional perspective of the natural history museum—Burden later abandoned this perspective and set his sights on the latest developments in biological science. The end goal of biological science in the twentieth century was not classification and description, but experimental inquiry into the underlying mechanisms of life. "If the interpretation of life is not the proper major objective or goal of our Museum, then what is?" asked Burden.[9]

The interpretation of life necessitated exhibition techniques that far surpassed those of traditional museum displays. To reveal the drama of the living world and capture the visitor's attention, Burden and Noble looked to the techniques of the cinema: the synchronization of light, motion, and sound. In the central foyer of the Hall of Animal Behavior stood a large model of the three-horned chameleon that had been a featured museum exhibit at the 1939 New York World's Fair. The model, translucent in appearance and mimicking the color changes seen in a living specimen, allowed visitors to follow the movement of color through the pigment cells in a dynamic cross-section of the chameleon's skin. The museum also replaced habitat dioramas with live animals placed in naturalistic settings that combined the display techniques of the habitat group with the zoo. Convinced of the need "to make teaching so entertaining that the visitor is barely aware of being taught," Burden envisioned a future museum exhibit similar to the General Motors' Futurama 1939 World's Fair exhibit, in which individuals seated themselves in chairs and were transported along a moving platform through three-dimensional displays—incidentally, the same technique developed by Disney in the construction of Disneyland. Burden even thought that something like the same motion picture techniques of fade-in and fade-out could be achieved to "bridge the gap between different sequences" by allowing each chair to pass into "a very brief black-out between exhibits." All were experiments in what Burden described as a "new vision of learning."[10]

The Hall of Animal Behavior appeared at a controversial moment in the museum's history. Beginning in the late 1930s, Burden became embroiled in a series of major battles with administrators at the American Museum over effective exhibition techniques. Diametrically opposed to Burden's plans, Albert E. Parr, the museum's director, feared that "increased mechanization . . . [meant] increased mental regimentation and reduced individualism in the process of learning." If the museum catered to "immediate popularity" through the use of motion pictures, loudspeakers, and other forms of canned instruction, Parr believed that it might as well measure its success according to the "ticket offices of pure entertainment" rather than "brains." In Parr's opinion, mechanized education weakened the nation's "intellectual fiber" and prepared the public "for the blind acceptance of any future propaganda to which such methods might equally well be applied." In the midst of the Second World War, Parr had good reason to be suspicious of the ways in which radio and motion pictures were important instruments of propaganda in the Nazi war machine, but as Burden pointed out, the use of motion pictures in the education and recruitment of the masses had been equally put to effective use in the Allied war effort. Unsuccessful in his attempts to get the administration to embark on a program of exhibit reform, Burden became so incensed with the entrenched policies of the administration that he effectively severed his ties with the museum in the late 1940s.[11]

Burden's decade-long fight with the Museum's Board of Directors extended far beyond museum exhibition: it centered on the museum's primary objective and mission. If the museum was to "confine its major scientific pursuits to the realm of taxonomy" and its "function to that of an archive or repository for collected objects," Burden regarded its future as a scientific institution doomed. When the curator of the Department of Experimental Biology, Gladwyn Kingsley Noble, died suddenly of a streptococcus infection in 1940, Burden fought hard to keep the department intact. Through the laboratory investigation of behavior, Burden believed, the biologist was unraveling the important mechanisms that governed and controlled the living organism's activities. Animal behavior was of central importance in biological education and research because the study of behavior captured the dynamic activity of the organism while elucidating underlying biological processes and concepts. In this intimate vision of life, film as a research tool and as an effective medium for educational and promotional purposes was to play a central role.[12]

Gladwyn Kingsley Noble, like many young naturalists embarked on professional careers in the study of animal behavior, was one of those who picked up the camera in place of the gun. In an unpublished essay, "Hunting With the Camera," written while Noble was an undergraduate at Harvard, he described his adventures photographing laughing and herring gulls in an umbrella blind on Muskeget Island off the coast of Massachusetts. If only sportsmen "would exert the same effort that they do in trailing a deer, or in playing a five-pound bass," Noble reasoned, "they would experience not only the delights of acquaintanceship with the 'wild,' but they would also bring home trophies comparable with any of the prizes from the wood or the stream." Patience was the key. "He that has patience may compass anything," Noble wrote. But the experience in nature did not end for Noble upon the return to city life. In the darkroom, "one lives again for the moment out in the 'wild.'" "Great uncertainty and excitement are always present when the pictures are being developed," Noble wrote. "For instead of the bustle and the turmoil of the city sounding in his ears, once more [the photographer] feels the supreme peace and quiet of the open, the melodious many-throated cry of the seabirds as it blends with the sea's mysterious sound, the infinite murmur, solemn and profound."[13] Twenty years later, he would return to Muskeget Island, this time armed with both a camera and all the investigative techniques that early twentieth-century endocrinology and neurology could offer.

Noble had developed a fascination for studying the life histories of reptiles and birds as an undergraduate. In 1922, he received his doctorate from Columbia University for his research in herpetology and systematics. Through William K. Gregory, Noble's graduate adviser, professor of vertebrate paleontology and curator of the Department of Comparative Anatomy at the American Museum of Natural History, Noble also developed an appreciation of functional morphology and microscopical anatomy in analyzing phylogenetic relationships.[14] The close ties between Columbia University and the Museum led to Noble's appointment as Curator of the Department of Herpetology at the Museum in 1924. But Noble was not content with a career devoted entirely to systematics when the cutting edge of biology lay in the experimental disciplines of neurology, physiology, and endocrinology. Offers from Columbia University and Cornell University

Medical School in 1928 provided Noble with a bargaining platform. To stay on as curator at the Museum, Noble demanded half of the fifth and sixth floors of the Museum's African wing, then under construction, for a laboratory of experimental biology. In addition, the Department of Herpetology was to become the Department of Herpetology and Experimental Biology, with half of Noble's time given to experimental biology research.[15]

Although such a plan marked a major departure for a natural history museum that was traditionally organized around taxonomic groups rather than life processes, the Museum acquiesced to Noble's demands and in May of 1934 the city of New York completed Noble's Laboratory of Experimental Biology. It occupied the entire sixth floor and roof of the African Wing and included, among other things, an aquarium room, three greenhouses, an animal house, histology laboratory, and a physiology laboratory.[16] But because of the financial constraints caused by the Depression, the Museum could no longer maintain its previous level of support.[17] Financial backing for Noble's Department of Experimental Biology owed much to the efforts of Burden, who, along with relatives and friends, contributed funds for the operation of the department during its first five years. During the department's financial crisis, Burden constantly tried to provide Noble's department with a high public profile. He solicited friends from the *New York Times* and *Fortune* magazine to write articles about Noble's research. When, for example, Noble was able to induce egg-laying in a species of salamander through transplants of the pituitary gland, Burden urged Noble to go public. "What excitement you will evoke if the substance of your new hormone not only brings immature animals to sexual activity but increases sexual activity among the aged," Burden exclaimed. "Pull that trick and we will never have any difficulty raising funds for your research."[18] Taking Burden's comments to heart, Noble arranged to have an article published in the *New York World Telegram* on his sex hormone research.[19] Burden's public relations efforts proved effective in securing funding from the Josiah Macy, Jr. Foundation and the National Research Council Committee for Research in Problems of Sex between 1935 and 1940.[20]

Although the laboratory was originally intended to "consider many problems on the borderline between natural history and biology," Noble, facing drastic budget cuts, restricted his research to the "physiology and psychology of reproduction in the lower vertebrates."[21] From 1935 until 1940, Noble developed a program of animal behavior study that utilized the techniques of endocrinology and neural surgery to establish a detailed picture of the underlying mechanisms responsible for the evolution of social behavior in vertebrates. By analyzing the courtship behavior of fishes, reptiles, birds, and finally mammals, Noble hoped to ascertain the extent to which phylogenetic changes in neural structure were accompanied by changes in social behavior and the role of sexual selection in the evolution of courtship display. Noble understood social behavior to originate in the neurophysiological structures and processes ingrained in the individual organism as a consequence of its phylogenetic past.[22]

Because Noble's analysis depended on observations of the social behavior of organisms in their natural environment, he relied heavily on the accounts of field naturalists and amateur observers. The museum, with its network of collectors and its high public visibility, was an institution that served Noble's needs quite well. "Our unique position," Noble perceptively wrote, "is assisting local field naturalists in an analytical study of animal behavior. Some of these students are professional ornithologists, others are graduate students, and still others are business men who spend their spare time in the field. This is a service which no university laboratory is equipped to render."[23] Indeed, Noble established an impressive network of field observers that could supply him with needed information. The Huyck Preserve in Albany, New York; the Conservation Department of New York state; the Kent Island preserve off the coast of Maine; Marine Studios in Florida; and numerous individual ornithologists supplied Noble with both material and field notes necessary for developing a comparative phylogenetic study of social behavior.

Noble was not alone in his appreciation of the contributions amateurs made to the study of animal behavior under natural conditions. The investigations of bird behavior by the British biologist, Julian Huxley, for example, were greatly influenced by the accounts of English amateur ornithologists such as H. Eliot Howard and Edmund Selous. In a series of broadcasts aired on BBC radio in 1930, Huxley spoke of the craft knowledge possessed by "gamekeepers, trappers, shepherds, fishermen, amateur naturalists and country folk generally—people whose business or pleasure it is to observe Nature's happenings accurately and note them in their minds." "The watcher and the naturalist," Huxley observed, "will often make quite as important contributions as the man of science and the experimenter."[24] Similarly, Konrad Lorenz spoke of how knowledge of animal behavior could not be obtained without the amateur's love of the animal under investigation. The passion and patience for observation were traits shared by amateur and professional naturalists alike.[25]

If patience was a common trait that united the amateur and professional, the two communities also shared an interest in wildlife photography, although for different ends. As an amateur naturalist in the prime of his youth, Noble delighted in the aesthetic pleasure of photographing birds on Muskeget Island. As a professional biologist, his observational skills were wedded more to the camera's place as a scientific instrument for the recording and analysis of animal behavior than with its role as an artistic medium for capturing the beauty of nature, although the latter certainly sparked his interest in biology as a professional career.

When Noble returned in the late 1930s to the site where his early interests in wildlife photography were awakened, to the gull colonies on Muskeget, he deployed the camera as a research tool to aid in the identification and analysis of postures that functioned as signs of communication within the laughing gull's world. His silent color movie, *The Social Behavior of the Laughing Gull,* delivered as a motion picture demonstration at the 1940 meeting of the American Ornithologists' Union, is illustrative of the way film magnified an experimental and theoretical orientation in animal behavior studies toward visual communication within the animal world. It is also illustrative of the way film accentuated the dramatic moments of the private lives of animals in the wild. Although the first part of Noble's movie is meant to be a realistic portrayal of the life of the laughing gull in its natural environment, considerable dissection and analysis took place in Noble's re-creation of the laughing gull's world. The footage projected on screen was filtered through the lens of theory, experiment, and entertainment with much of the gull's life left on the editing room floor.

The Social Behavior of the Laughing Gull begins with an upward angle shot of a flock of laughing gulls landing in the dense grass on Muskeget. The dark head of the laughing gull contrasted with the white plumage of the herring gull raises the question of why the two species should develop such different nuptial colors. The audience is led on a visual tour of the habitat preferences and different-sized nesting territories of the two species. Whereas herring gulls display an ecological preference for the open dunes, the laughing gulls nest in areas of dense vegetation of beach grass and poison ivy. This preferential difference is accentuated by shots of differences in the protective coloration of the newly hatched young. The black head, reddish bill and eyelids, and dark red legs of the adult laughing gull, however, serve no protective function, but instead have a social significance. The film then provides a close-up look at the important stages in the life cycle of the laughing gull: mating, nest-building, and care of the young. In the laughing gull's world, where the outward appearance of the sexes is the same, what are the mechanisms by which individual birds are able to recognize members of the opposite sex in the midst of a colony where aggressive behavior is common? The scenes that follow help the audience recognize the repertoire of postures that have particular meaning for finding mates in the laughing gull's world, such as the pairing charge, head flagging, food begging, and the long call. If the female does not respond to the male's "pairing charge" with a submissive "head flagging" posture, followed by mutual "food begging," the male will not have achieved "sexual dominance" and the two will not pair. The audience is even witness to an alleged "rape" of an already mated female. The filter of patriarchy biologically reinforces gender roles found in human society once Noble interprets the laughing gull's world.[26]

By juxtaposing display postures used in courtship with similar types used for communication between a parent and its young, Noble visually conveys the idea that a limited repertoire of behavior patterns are found in the laughing gull, patterns that serve different functions during the bird's life cycle. The first part of the film ends by returning to a shot of the dense grass of the laughing gull's habitat with the suggestion that the contrasting colors of the laughing gull's head help accentuate the signaling cues of the head-flagging ceremony. In the published paper that followed the film's release, Noble argued that the absence of the head-flagging ceremony in the black-headed gull, a species with nuptial plumage intermediate between the dark-hooded laughing gull and the white-headed Australian silver gull, indicated that the hood was a primitive morphological structure with a social, communicative value whose function became lost in the subsequent divergent evolution of the genus from its dusky ancestral form. According to Noble, the difference in ecological preference between the laughing and herring gull for nesting sites had also led to differences in behavior.[27]

The belief that we are watching a pristine scene of the laughing gull's world is strengthened by the contrast achieved during the second part of the film. With the description of the natural environment and behavior of the laughing gull complete, nature must now be brought into the laboratory. Two men in a boat appear with a basket of young laughing gull chicks. A shot of a young gull in winter plumage against a stark blue wall indicates that we are no longer in the field. A syringe and box of cotton and the appearance of two human hands tell us we have entered the world of twentieth-century experimental biology. Noble's film has shifted from a nature film to a documentary of experimental practice. When young birds who have not reached sexual maturity are injected with testosterone propionate, they display courtship behaviors similar to those exhibited in the field. A change in head color of the treated bird is striking visual evidence that Noble has isolated the internal physiological mechanisms governing the laughing gull's behavior in the wild. This version of the laughing gull's world provides a powerful visual testimonial that nature in the field and nature in the laboratory are one and the same, a testimonial that was particularly important for Noble's efforts to convince European ethologists of the merits of his methodological approach. The Austrian ethologist Konrad Lorenz, for example, found Noble's work flawed because of its emphasis on dominance relations, which Lorenz regarded as a product of the artificial situations in which Noble studied animals.[28]

Despite Lorenz's criticisms, Noble was, before his untimely death in 1940, the most prominent professional American biologist whose research closely paralleled that of European ethologists, who were forging a new direction for the study of animal behavior. In studying the evolution of courtship display, he was interested in unraveling the visual communication signaling that takes place between mates and the underlying sensory physiology that made such communication possible. Noble made available to American biologists and psychologists a complete translation of Konrad Lorenz's seminal article of 1935, "Der Kumpan in der Umwelt des Vogels" (Companions as Factors in the Bird's Environment), an article that formed the analytical and theoretical basis of Continental ethology from the mid-1930s into the 1950s. Nevertheless, Noble took issue with many of the theoretical and methodological assumptions advanced by Lorenz and his coworker Tinbergen.[29]

Noble challenged Lorenz and Tinbergen's perception of the line between learning and instinctive behavior. Lorenz and Tinbergen argued that visual images, including colors, were inherited and functioned as sign stimuli, or visual signals that released an innate reaction among other individuals of the same species. Tinbergen, for example, explained the red patch on the lower mandible of the herring gull as a sign stimulus that directed the innate reaction of newly hatched chicks to peck the parent's bill for food. Noble believed the red patch acquired its value as a social releaser, a signal that triggered a behavioral response, through training. In *The Social Behavior of the Laughing Gull,* Noble emphasized the plasticity of behavior by showing the lack of response to and ignoring of the red spot among newly hatched and older herring gull young.[30] Influenced by the critique of Continental ethology in the early 1950s by Noble's student, Daniel Lehrman, Tinbergen later modified his views on the distinction between innate versus learned behavior and paid greater attention to the study of the ontogeny of behavior—that is, the development of behavior patterns in the individual as a result of both hereditary and environmental influences.[31]

Although Noble and Tinbergen disputed over learned versus instinctive behavior, they were united in their passion for nature photography. Tinbergen reflected on how his "special interest in the *visible* natural world" made him "turn to the camera as well as to sketching." Still photography and "cinefilming of natural animals," Tinbergen's principal hobbies, proved essential to his work "as a teacher and student of animal behavior." The development of ethology owed much to a long amateur tradition of bird-watching in which photography came to play an important role. Like Noble, European ethologists such as Lorenz and Tinbergen were themselves actively engaged in the use of film to help identify and analyze display postures that had particular meanings within the animal's world.[32]

In the early 1950s, Lorenz became associated with the establishment of the Encyclopaedia Cinematographica (EC), an international collection of 16-mm short scientific films that each document a single phenomenon or behavior of a particular species or culture. The Encyclopaedia Cinematographica is affiliated with the Institut für den Wissenschaftlichen Film in Göttingen, an institutional outgrowth of the Reichanstalt für Film und Bild in Wissenschaft und Unterricht (RWU) established by the National Socialist regime in 1934. Although cooperative agreements existed with the Ministry of Propaganda, directed by Joseph Goebbels, the RWU operated independently of Goebbels and issued only silent films that remained outside Goebbels' control. In 1945, a UNESCO commission largely absolved the RWU of its role as a propaganda arm of the Nazi Party. The commission classified only 10 percent of RWU films as "tendentious": they were more concerned about propaganda in the teaching notes provided with the films. The consolidation of scientific and instructional film production into a single institution, and the educational importance of nature protection within the National Socialist state provided fertile ground for natural history film under the Third Reich.[33]

The EC film project followed the comparative approach initiated by Lorenz in the late 1930s. Through a precise description of the instinctive behavior patterns found within closely related species and families, such as those found in the Anatidae and Cairininae families, which include mallards and Muscovy ducks, Lorenz constructed evolutionary histories of waterfowl based on the morphological similarities of their behavior.[34]

Tinbergen also utilized film for comparative analysis, but his research direction departed significantly from that of Lorenz upon his move to Oxford in 1949 to occupy a university lectureship in animal behavior in the Department of Zoology and Comparative Anatomy. Lorenz regarded instinctive behavior patterns as invariant; he believed they were little modified by selection pressures of the environment and therefore served as reliable homologies, that is similar forms observed in different species due to a common evolutionary relationship. At Oxford, Tinbergen's research moved in more of an ecological direction; he became interested in how the adaptations of closely-related species to different ecological niches led to the divergence of behavior in the course of speciation.[35]

Without film, Tinbergen noted, many of the analyses critical to his study on the adaptive divergence of behavior in gulls begun in the early 1950s would not have been possible. By the late 1950s, Tinbergen and his students had amassed a complete film record of the ethogram, the entire repertoire of displays, of nine closely-related species of gulls. Formal analysis of gull postures depended upon a "precise knowledge of motor patterns" within a species. Film was the ideal medium for acquiring this knowledge.[36] Although Tinbergen found great variation in the form that a display such as the Oblique posture assumed among individuals within a gull species, he was able to classify the posture by general appearance and by the average number of still frames each display occupied in a film sequence. The identification of gull postures marked only the beginning of Tinbergen's analysis. Unlike the obvious biological functions of feeding and copulatory behavior, the functions of many displays remained obscure. Tinbergen grouped the majority of the fourteen gull postures into two groups: agonistic displays that promoted wider spacing between gull territories, and pair formation displays that effectively reduced the distance between individual gulls.

Through film Tinbergen determined both the cause and origin of particular displays. Just as postures such as a clenched fist expressed conflicting emotional states in humans, so animals displayed signs of conflicting motivations. Tinbergen proposed that any given animal posture was actually composed of a number of movements that corresponded to different motivations. The Upright posture in gulls, for example, was composed of elements of different motor patterns that corresponded to motivations of attack and escape. Determining the derived movements from which a given display originated became an important task. Through film analysis, Tinbergen could dissect a display posture into its component parts. He found, for example, that the choking posture, which forms an important part of the meeting ceremony of many gulls, was derived from motor patterns related to the deposition of nest material or the regurgitation of food. In addition, Tinbergen and his students were able to ascertain through comparative analysis that many of the variations in gull displays among differing species were the result of adaptations to particular niches. In the case of Kittwakes, for instance, adaptation to a breeding habitat of steep cliffs, free of predators, resulted in behaviors not commonly found in closely related gull species.[37]

The work of Lorenz, Noble, and Tinbergen demonstrated that film technology afforded a more precise and exacting study of visual communication in the animal world. Yet film technology also promoted particular types of research. Because of their interest in ritualized displays, ethologists often chose organisms that had intricate courtship ceremonies, such as the black-crowned night heron or great-crested grebe, or organisms such as the jewel fish, which undergoes remarkable color changes in the process of courtship. Hence, the organisms chosen as model systems for studying behavior were also photogenic ones. By isolating the behaviors of particular animal species through careful choice of shots and editing by cutting and splicing film, ethologists erased all evidence of the long hours spent watching and waiting in the field. Both science and entertainment required that only the most spectacular and private aspects of animal life were recorded. The distillation of the natural world into a series of dramatic moments on film created an expectation of nature among lay audiences that was rarely, if ever, realized in the field.

While film became an indispensable tool in the study of animal behavior, it also played a valuable role in educating the lay public about the emerging science of ethology. Noble himself expressed an avid interest in the potential of cinema for research and educational purposes. In the 1930s, he served on an advisory committee directed by Fairfield Osborn of the New York Zoological Society that completed a comprehensive survey of natural history films and prepared production plans for a series of educational films on natural history subjects under a grant from the Rockefeller Foundation.[38]

By the late 1930s, the establishment of the Psychological Cinema Register in the United States and the Reichanstalt für Film und Bild in Wissenschaft und Unterricht in Nazi Germany made available a modest number of instructional films in the fields of animal behavior and animal psychology. Established in 1938 by Adelbert Ford at Lehigh University, the Psychological Cinema Register provided an institutional foundation for an informal film exchange network that had developed among researchers in animal behavior and psychology. Transferred to Penn State University in 1944, the Psychological Cinema Register under the direction of the primatologist C. R. Carpenter quickly became the leading distribution center in the United States for instructional films in animal behavior and psychology.[39]

Carpenter, who received his doctorate in psychology at Stanford under Calvin P. Stone in 1932, did much to establish observation field techniques for the study of primate societies under natural conditions. In the early 1930s, as a postdoctoral student of the Yale comparative psychologist Robert M. Yerkes, Carpenter conducted a field study of howler monkeys on Barro Colorado in the Panama Canal Zone. In 1937, he accompanied the Asiatic Primate Expedition to study the behavior and social relations of gibbons in Siam. One year later, in conjunction with the Columbia University affiliate School of Tropical Medicine at the University of Puerto Rico, Carpenter helped establish a colony of rhesus monkeys on Cayo Santiago. Fairfield Osborn recognized that Carpenter's study of rhesus monkeys provided a unique opportunity for the production of a film on primate behavior, and the Rockefeller Foundation advisory committee contributed the film stock to document Carpenter's expedition.[40] In his own classes on primate behavior, taught at both Penn State and the University of Georgia, Carpenter stressed the importance of motion picture films for simulating field observations of primates and providing students training in observational skills at a time when "realistic experiences in observing primates in their natural habitats [had become] progressively more impractical and expensive."[41]

Carpenter's research focused on the study of communication in primate societies. His wartime experience working on the production of military training films for jungle survival extended his research interests beyond communication among nonhuman primates to include humans as well. As director of the Instructional Film Research Program, sponsored by the U.S. military, and as head of the Instructional Television Research Program, funded through the Ford Foundation, Carpenter pioneered psychological studies in the use of mass media of communication for instructional and motivational purposes. He became a leading advocate in the development of film and television for educational instruction at both a regional and national level. In his championing of instructional television, Carpenter helped create an important outlet and market for educational films on animal behavior in the postwar years.[42]

Before the advent of educational television, however, animal behavior researchers relied upon motion picture demonstrations to proselytize their new field. Noble believed that motion pictures offered the best means for telling "the technical story of animal behavior in a way the public will understand." The value of scientific nature film lay in its ability to be simultaneously informative and entertaining. In the prewar years, at Konrad Lorenz's Altenberg home near Vienna, Tinbergen and Lorenz documented on film collaborative experiments on the egg-rolling behavior of the greylag goose and escape reactions of various species of fowl to cardboard dummies of predators flown overhead. These experiments formed part of Lorenz's "Ethology of the Greylag Goose" film, imported by Noble and shown to the public at the American Museum of Natural History in 1938. Noble remarked that the film was of "great popular interest" and "of great value to serious ornithologists who have read Dr. Lorenz's papers and who want to know more of the details of his experiments." The image of Lorenz being followed on land and in water by goslings who had imprinted on him was endearing to popular audiences, yet also conveyed information about the experimental and theoretical orientation of his research.[43] Karl von Frisch was another European researcher whose films were a featured part of the motion picture demonstrations in the American Museum of Natural History's Hall of Animal Behavior. Von Frisch, a University of Munich physiologist who shared the 1973 Nobel Prize in Medicine and Physiology with Lorenz and Tinbergen, produced a number of films beginning in the 1930s on sensory recognition in bees and fish. These were masterpieces in the illustration of experimental technique and played an important role in convincing both scientists and lay audiences throughout the world of the validity of von Frisch's theories regarding bee communication.[44]

Noble's enthusiasm for film as both a tool for research and a tool for popularization was shared by others in the developing field of animal behavior studies. Julian Huxley, whom Konrad Lorenz referred to as one of the founding fathers of ethology, asked Noble in 1937 about the automatic miniature cinema housed in the museum's Hall of Reptiles that projected footage of Komodo dragons in action next to the diorama exhibit. As Secretary of the Zoological Society of London, Huxley hoped to install a number of these machines in the Zoological Gardens on a penny-in-the-slot principle "to show scenes of animals in the wild, in contiguity to the same animals shown in captive cages."[45] In the absence of moats that would free animals from their cages and more naturalistic habitats pioneered by the famous German animal dealer Carl Hagenbeck in the early twentieth century, Huxley deemed motion pictures a useful means of re-creating a vision of animals in their free and wild state. In addition to these nature nickelodeons, he also planned a zoo cinema to exhibit natural history films, news reels, and animal cartoons such as Walt Disney's Mickey Mouse. Huxley believed that parents with tired children would find the cinema a welcome and entertaining respite. He raised five hundred pounds in seed money for the establishment of a film unit to produce natural history films for the zoo cinema and arranged for the Zoological Society of London to be the main repository for natural history films collected by the British Film Unit. However, Huxley's ambitious plans were shot down by the governing council of the society, who argued that such endeavors would demean "the prestige of the Zoo as a learned society." To the members of the Zoological Society, serious instruction and mass entertainment did not go hand-in-hand.[46]

Twenty years earlier in 1914, Huxley had written a paper, "The Courtship Habits of the Great Crested Grebe," in which he first introduced the idea of ritualized behavior into ethology. In this same paper Huxley questioned the use of the camera in natural history pursuits such as bird watching.[47] "In these days the camera almost monopolizes the time and attention of those who take an interest in the life of birds," wrote Huxley. "It has rendered splendid service, but I believe that it has almost exhausted its first field," he continued. "At the present moment both zoology and photography would profit if naturalists for a little time would drop the camera in favor of the field-glass and the notebook." Huxley was referring to the practice of taking snapshots of birds as an alternative to the nineteenth century practice of capturing prized trophies of rare birds. In the early twentieth century, however, English amateur ornithologists such as Edmond Selous and Eliot Howard became interested in studying the habits of more common birds. Huxley, whose research on the display behavior of birds owed much to this amateur tradition, believed that still photography had limited the naturalist's view because it oriented the eye to animal form, rather than motion and therefore animal habits. Only when the naturalist had discovered the "wealth of interesting things" that "lies hidden in and about the breeding places of familiar birds" through patient observation would the photographer find new subjects and new avenues of exploration. Before the First World War and the invention of lightweight motion picture equipment such as the Akeley camera, the use of motion picture technology in the field for the scientific study of behavior—the focus of Huxley's interest—was limited. By the early 1930s, both Huxley's attitude toward the camera and the technology itself had changed significantly.[48]

Huxley's first venture in natural history film, *The Private Life of the Gannet,* surpassed all expectations and won him an Academy Award for Best Short Subject (one reel) in 1937. Plans for *The Private Life of the Gannet* emerged in April of 1934 when Huxley, while staying at the home of his naturalist friend, Ronald M. Lockley, on the island of Skokholm off the coast of Wales, visited the nearby island of Grassholme, home to one of the largest breeding colonies of gannets in the world. It was a breathtaking experience for Huxley to see thousands of these graceful white birds nesting on this small island. Never had he seen "any spectacle of living nature to rival it, except perhaps for the half-million flamingoes on Lake Nakuru in Kenya." The subject was both interesting and exciting and the impression so vivid that Huxley and Lockley decided to capture it on film.[49]

Huxley was acutely aware that not all thrilling experiences of nature were readily translated onto film. The smells, tastes, and sounds of nature all played their part in rousing the emotions; a simple record of such an experience would never produce in theater audiences the same emotional response felt by naturalists in the wild. Such a response could only be achieved through masterful techniques in cinematography and editing. Huxley managed to secure the two most prominent filmmakers in interwar Britain: Alexander Korda and John Grierson.

In 1934, Korda, director of London Film Productions, was still reeling from the international success of his 1933 film *The Private Life of Henry VIII,* a film considered by historians to be one of the most important in the development of British cinema. He considered himself a patron of the arts and sciences and agreed to finance Huxley's film on the condition that the film be titled *The Private Life of the Gannet.* Korda lent his photographer Osmond Borrodaile to shoot the footage on Grassholme. In addition, Huxley asked his friend, John Grierson, the leader in the establishment of the British documentary movement, to shoot the film's finale—a spectacular slow-motion sequence of gannets plunging from great heights into the ocean waters for fish. The combination of Korda and Grierson symbolized Huxley's interest in both the entertainment and informational dimensions of nature film. Korda, a producer ensconced in the economics of commercial cinema and the Hollywood studio system, linked nature to the theater. Grierson, critical of the way cinema had been "driven by economics into artifice," sought in film an intimate view of reality that could educate and motivate the public toward a revitalized citizenship.[50]

The Private Life of the Gannet ran with Korda's feature release, *The Scarlet Pimpernel,* a comedy about the rescue efforts undertaken by English aristocrats during the French Revolution. The publicity the gannet film received prompted the owner of Grassholme, Mr. W. F. Sturt, to threaten legal action against Huxley and Lockley for their failure to acknowledge his ownership and permission in the film's title and for ignoring his claim to a portion of the film's royalties. Sturt granted Huxley and Lockley permission to visit the island for the purpose of making a film of scientific interest, not entertainment value. But the distribution of *The Private Life of the Gannet* to commercial theaters suggested to Sturt that the film was meant for entertainment and therefore apt to reap a substantial profit. Sturt had visions of grandeur not matched by the actual sales figures, which revealed the lack of markets for educational nature films during the early 1930s. After an eighteen-month run, *The Private Life of the Gannet* made a net profit of only £220. Sturt settled for £30, an amount considerably less than the settlement he initially proposed.[51]

Although linked to the theater, *The Private Life of the Gannet* drew largely upon conventions of drama developed within the documentary tradition. Huxley followed Grierson's principle that the story had to emerge from the raw material of life, rather than be superimposed from without, as in the case of Hollywood film. Documentary had to "master its material on the spot, and come in intimacy to ordering it," wrote Grierson. But documentary involved more than simply capturing nature in the raw. What distinguished documentary from the lecture film, Grierson argued, was the distinction between drama and description. The documentary film went beyond the description of the "surface values of a subject" and "explosively reveal[ed] the reality of it." Although *The Private Life of the Gannet* starts off in a descriptive format, the film's ending sequence reveals the drama in nature found on Grassholme.[52]

The Private Life of the Gannet begins with a map of the British Isles; the camera slowly closes in on the location of Grassholme and then cuts to an aerial shot of the island. As the eye is brought closer to land, the white patch on the island seen from afar transforms into a large colony of nesting gannets. The camera pans right to reveal thousands of these graceful white birds sitting on their nests, followed by close-up shots indicating how the bird's anatomy makes it well-adapted to its pelagic lifestyle (a toothed beak to hold slippery fish, membranous tissue that is drawn over the eyes when diving). The most intimate and private moments of the gannet's life are then revealed, from the mutual courtship ceremony, to the hatching and feeding of the young, to the fledgling's first flight over the ocean. Birth, adolescence, courtship, and rearing of the young would become featured dramatic events in nature found on television and motion picture screens in the 1950s. But Huxley made no attempt to utilize such intimate scenes as the ceremonial billing display to bring out the individuality of these birds or to anthropomorphize their emotions as Disney would masterfully do in the postwar years.

In *The Private Life of the Gannet,* the moral lesson is linked more strongly to a panoramic, ecological vision. Later in the film, Huxley chose a wide-angle lens to capture the real drama of the gannet's life. The music changes, now corresponding to the wonder and beauty of nature captured in the scenes that follow. Slow-motion photography and artfully composed shots that highlight the snow-white plumage and dark-winged tips against an expansive sky bring out the majestic, graceful flight of these birds. "Round and round the gannets fly in streamlined travel," recites the narrator, "the very poetry of motion. The circling flight goes on all day. The wondrous resources of nature afford no more beauteous scene than the graceful motion of this handsome bird in full flight." The aesthetic pleasure summoned in watching these birds is further emphasized by slow-motion photography of the gannets diving at all angles to retrieve fish below the water's surface. Reminiscent of Flaherty, Huxley sought to convey a romanticism for nature, made all the more real, all the more necessary, by the film's concluding remarks: "Whether this bird will maintain its present secure position is a question no one can answer. Great storms are its worst enemy, floating refuse oil from oil-burning ships cause thousands to die miserable deaths. Nature and man may one day observe a truce with all such grand forms of wildlife toward the continued prosperity of that which gives us pleasure because of its grace and beauty." In the late 1950s, as part of a concerted campaign for the international conservation of wildlife, Huxley returned to this theme that he first introduced in *The Private Life of the*

Gannet: the pleasure found in observing nature's drama, heightened by the knowledge of its fragile existence in the face of human expansion.[53]

The release of *The Private Life of the Gannet* in 1934 corresponded with the first meeting of Julian Huxley and Konrad Lorenz at the International Ornithological Congress in Oxford, where Huxley gave Lorenz a reprint of his 1914 paper on the great-crested grebe. After the late 1920s, Huxley contributed little to actual research in the emerging discipline of ethology; instead, he used his influence to support and promote others active in the field. *The Private Life of the Gannet* appeared before any real discipline of ethology had emerged. Not until the 1950s did the methodological, theoretical, and institutional dimensions of the naturalistic study of behavior coalesce into a recognizable discipline. Educating the lay public about this new science rested largely in the hands of Lorenz and Tinbergen.[54]

Unlike Huxley, and later Tinbergen, Lorenz did not extensively use film as a tool for mass education. Lorenz relied more on the printed word, in books such as *King Solomon's Ring* and *Man Meets Dog,* than on film to promote developments in this new field. In a visit to the United States in the winter of 1954–1955, however, Lorenz did capitalize on the medium of television to advance the image of his science in the public eye. Lorenz was the featured guest on an episode of *Adventure,* a CBS half-hour informational program based out of the American Museum of Natural History that aired on Sunday afternoons beginning in 1953. The show's host, Charles Collingwood, contrasts the opening sequence, which focuses on laboratory experiments on learning in squirrels, with the study of animals in their native habitat. The shift to the naturalistic study of behavior is the cue to introduce Lorenz, who is surrounded by peeping ducklings on a table in front of him. Lorenz appears at ease in front of the camera. The imprinted ducklings follow Lorenz across the table as he calls. By imprinting geese on him, Lorenz is able to "keep geese in full freedom, studying tame animals that are not captive animals." Lorenz's emphasis on the need to study animals in the freedom of their natural environment coincided with the outlook of Fairfield Osborn, president of the New York Zoological Society and director of the New York Zoological Park (Bronx Zoo), where portions of the Lorenz episode were filmed. Osborn, who likened caged zoo exhibits to symbols of totalitarianism, was at the time developing more naturalistic habitat settings for animals on display that firmly linked the zoo with research in conservation and animal behavior.[55]

In Collingwood's show, footage from Lorenz's 1937 film *The Ethology of the Greylag Goose* is spliced with interviews between Collingwood and Lorenz. As imprinted goslings follow Lorenz around his Altenberg estate, swim with him in a pond, and trail after him while he canoes on the Danube, the audience receives an impression of Lorenz as a caring mother to his adopted feathered friends. The parenting theme struck an especially responsive chord in a decade when domesticity and parenthood were idealized on television as routes to personal and social fulfillment. Like all parents, Lorenz understands the language of his infant offspring. When Lorenz mimics the goose warning call upon seeing a cardboard eagle overhead, the goslings gather under him for protection. Through television, Lorenz introduces the audience to various behavioral displays, such as the greeting ceremony, and instinctive nest movements that have meaning in the world of the greylag goose. Lorenz even adds touches of humor: The elaborate greeting ceremony of the Andean goose, in which the gander puffs out its breast to large proportions, is accompanied by equally pompous music. And in footage demonstrating the instinctive egg-rolling behavior of the greylag goose, Lorenz substitutes a wooden cube in place of an egg adjacent to the nest. The goose rolls the cube back into the nest and hopes, in Lorenz's words, to "hatch a cubistic gosling." Asked by Collingwood why he has chosen this research path, Lorenz concludes the program by saying he hopes to learn about laws of nature governing laws of animal behavior. In the end, he suggests, these laws might also hold true for human behavior and perhaps instill a sense of humility in humankind.

Parenthood and freedom are the dominant motifs of this *Adventure* program with Lorenz. To the parents of the baby-boom generation, the popularity of animal behavior subjects on the 1950s television screen corresponded to their own concerns with child-rearing and the need for wholesome family entertainment. By the 1960s, when Tinbergen actively entered the world of filmmaking for educational and entertainment purposes, the backdrop against which he crafted an image for the science of animal communication had changed.

Inspired by *The Private Life of the Gannet,* Tinbergen devoted himself to the making and producing of films on animal behavior oriented toward a general audience. Reflecting on why he chose a pursuit so far afield from his research endeavors, Tinbergen expressed his desire to "kindle interest in observing animals." Films on animal behavior had much to teach the nonprofessional, Tinbergen believed, because through the study of animals, humans would learn something about their own behavior. Furthermore, such films could help the cause of conservation in a period when the environmental movement was on the rise. Like Huxley before him, Tinbergen added a wide-angle ecological perspective to the intimate view of life furnished by the study of animal behavior. Thirty years after *The Private Life of the Gannet,* he was fortunate to find that the market for educational natural history film had significantly expanded.[56]

Signals for Survival, Tinbergen's finest film achievement, revealed the extent to which public television had created a market for the production of animal behavior films by the late 1960s. The film, aired on BBC television on a Sunday evening, December 8, 1968, proved to be a huge success. By 1971, television rights had been sold to sixteen foreign countries, and McGraw-Hill's Film-Text Division picked up the renting and after-sales rights of copies for teaching purposes. McGraw-Hill sold prints for $600 and predicted estimated sales of $120,000 over five years. Rental fees for the film were $44 per day. A book spin-off followed. Advertised in the popular press as "Sex and the single gull," the book sequel sold over 4,000 copies in the first year.[57]

Commissioned by the BBC for its color documentary television series *The World About Us, Signals for Survival* had its origins in the early 1960s when Hugh Falkus, a freelance writer and film producer, began making a number of films on animal behavior for television that featured different aspects of Tinbergen's research program at Ravenglass. On the basis of this early collaboration, Tinbergen and Falkus sketched plans for a film that relied upon Tinbergen's keen knowledge of animal behavior to document a story of the Lesser Black-backed Gull on the northeast coast of England. Tinbergen spent six weeks in the spring sitting in hides in a gullery to obtain footage for the film. The Natural History Unit of the BBC also spent many hours in the field recording the sounds of the Lesser Black-backed Gull, which were then painstakingly dubbed, syllable by syllable, to synchronize with the birds' movements. In the summer of 1967, Tinbergen wrote to Huxley, "nothing would please me more than getting an approving nod" from the man who inspired and encouraged him in the making of natural history film. Tinbergen didn't need Huxley's approval. The film won the 1969 Prix Italia award for best television documentary and the 1971 American Film Festival's Blue Ribbon Award.[58]

In *Signals for Survival,* the intimate view of life revealed by the camera and the science of animal behavior did not convey the idyllic scene of courtship, parenting, and childhood that American audiences found so appealing in the private life of animals during the 1950s. The opening shot of Tinbergen with clenched fists and a scowl indicates the universal gestures of aggression to the viewer. It is followed by shots of the nesting colonies of the Lesser Black-backed Gull, a great bird city where thieves and murderers abound. Graphic scenes of gulls eating eggs and chicks found in neighboring territories reveal that the nesting colony of the Lesser Black-backed Gull is a place where aggression is waiting to explode. What prevents this bird city from becoming a world of anarchy and chaos is a highly complex system of communication comprised of movement, posture, color, and sound illustrated by different postures that have meaning within the gull's world. The film then follows the birds from the formation of male territories, to courtship, to nest-building, to rearing of the young, and details the signaling system that makes such social interactions possible.[59]

Apart from the immediate message of the role communication plays in the world of the Lesser Black-backed Gull, one finds in *Signals for Survival* a pressing environmental concern. In the midst of the environmental movement in the 1960s, Tinbergen had become increasingly preoccupied with the lessons behavioral ecology could teach humankind. He was pessimistic about the future of *Homo sapiens* and felt the species was headed down a path of "disadaptation." Tinbergen believed that this trend was fueled by overpopulation, which was rapidly depleting available resources and would eventually lead to increased aggression and destructive warfare. During the breeding season, the Lesser Black-backed Gull lived in extremely crowded conditions similar to human urban environments. Social stability was maintained in the Lesser Black-backed Gull through clear communication, but it was a delicate balance: any misread sign could easily erupt into acts of aggression that would undermine the social structure of the group. The constant analogies throughout the film leave little doubt in the viewer's mind about the implications of aggression in the gull's world for human survival.[60]

The success of *Signals for Survival* prompted Tinbergen to launch a thirteen-part television series entitled *Behavior and Survival* in conjunction with the BBC and the German firm Windrose-Dumont-Time. "Mass education," Tinbergen suggested, "was creating a world-wide demand for these educationally stimulating and scientifically correct films." The series aired on Thursday nights on BBC television in the spring of 1973 under the title "Their World." American television rights were purchased by Time-

Life. Although Tinbergen made two of the films and supervised production of the other eleven, he was less than enthusiastic with the final outcome. He and Falkus argued over the making of a film on the oystercatcher for the series. While Tinbergen felt that Falkus had not done the story justice, Falkus complained of the poor quality of the footage Tinbergen had shot. Tinbergen also felt the scriptwriter hired by Windrose-Dumont-Time for the series often used "Walt Disney style" and failed to grasp "key points in the arguments." Furthermore, a dispute over after-sales rights of the films for nontheatrical purposes erupted between Tinbergen and the BBC. Under a royalty agreement with McGraw-Hill, Tinbergen received 25 percent of the retail price on every copy sold of *Signals for Survival.* By the 1970s, the rights for after-sales and rentals, in addition to book rights, had started to become a lucrative venture that generated additional income beyond the short-lived appearance of nature films on television. Thus, when the BBC offered Tinbergen one percent of the retail price of after-sales copies sold for his films in the *Behavior and Survival* series, he became furious. Although Tinbergen eventually acquiesced, the dispute indicates the extent to which nature on screen had become thoroughly commercialized.[61]

Near the end of Tinbergen's life, Lady Juliette Huxley, wife of Sir Julian, wrote to Tinbergen and credited him as the "real beginner of that never-ending study which now David Attenborough pursues with such triumph." Lady Huxley was rather overzealous in honoring Tinbergen as the pioneer of nature documentaries.[62] In his use of the camera as an instrument of research and education, however, Tinbergen did help bring to the public an intimate view of animal life, one that increasingly became appropriated and marketed by television and the commercial entertainment industry. Under the physical strain of carrying seventy-five pounds of camera equipment in the field and the emotional stress of dealing with television companies, Tinbergen gave up making further films at the age of sixty-four, although he continued to supervise and consult with groups such as Oxford Scientific Films.

In utilizing film to dissect and analyze animal movement to gain a precise understanding of visual communication in the animal world, ethologists zoomed in not on the mundane, but on the dramatic. The inestimable patience required in the field to observe and photograph the private life of animals was completely hidden when nature appeared on screen. In *Signals for Survival,* Tinbergen had to manufacture a scene of chicks pecking at the red spot on the parent's bill by starving them for a few days so that they would respond on cue. "One cannot give too much time for repeating statistics in brief film shots," Tinbergen reasoned. The naturalist might be patient; film audiences were not. The nature documentaries that Huxley, Lorenz, Noble, and Tinbergen made were films of "canned behavior." "Carefully selected and organised," nature on screen, Tinbergen acknowledged, was quite "different from the seemingly chaotic phenomena" that the observer would "see when confronted with the animals themselves." Formatted to meet the conventions of television, animal life became increasingly active and condensed.[63]

Through the camera lens, the science of animal behavior captured and brought to the public an intimate look at the dramatic events found in the life histories of individual species in the wild. The popularity of animal behavior as a science in the postwar years was itself due in part to what nature film and television revealed about human behavior in a period when topics regarding marriage, home, and parenting were at the forefront of public interest. Television provided an important outlet for broadcasting the moral messages found in this vision of nature into the American suburban home. In *The Private Life of the Gannet,* however, one finds an inkling of drama in nature on a more grand scale than the science of animal behavior. The aerial view of Grassholme, the gannet's island breeding ground, indicates the importance of a wide-angle perspective on all of nature. In encompassing an entire landscape, the panoramic vision embraced an ecological aesthetic, one that provided a glimpse into the complexity of the interrelations among organisms and their environments. Wedded to the science of ecology, the camera would become an important ally in the study and management of wildlife and in promoting the cause of conservation across the globe. In the hands of television and the commercial entertainment industry, it would also find a market among American audiences in the years after the Second World War.

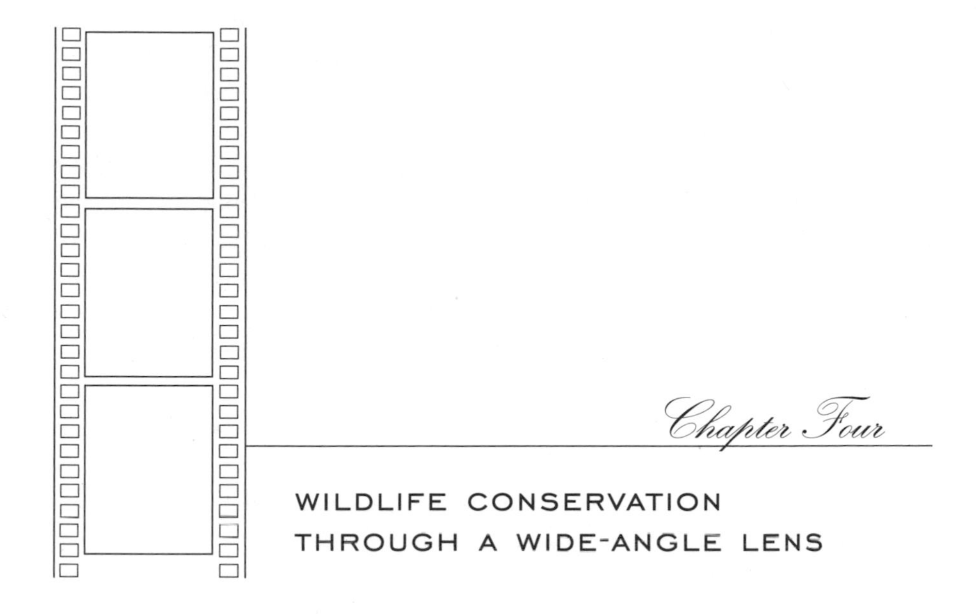

Chapter Four

WILDLIFE CONSERVATION THROUGH A WIDE-ANGLE LENS

In 1940 the New York Zoological Society, under the direction of its president, Fairfield Osborn, launched a major change in animal exhibition at the New York Zoological Park. The year marked the "beginning of the end," in Osborn's words, "of exhibiting our animal collections behind bars." While zoos in the past had displayed animals according to "man-made classifications" of orders and families, by the spring of 1941 a visitor to the Bronx Zoo could witness a scene of animal life on a realistic African veldt. Entering from a gate suggestive of an African village, the spectator encountered a savannah with zebras, warthogs, numerous species of antelope, cranes, and ground birds of various shapes and hues traversing the plains or refreshing themselves at the water hole. On a high rock outcrop, one might see lions. Like the concealed enclosures used to film animal scenes in films such as *Chang* and *The Silent Enemy,* the area was large enough to allow free movement of the animals, yet small enough to ensure every visitor an experience of African animal life unobstructed by bars or fences, but without the danger of safari. The large size also helped conceal the "synthetic habitat planting" of flat-topped elms, Texas water locusts, mountain holly, and a multitude of exotic and native plant species carefully selected and manicured to convey the look of Africa.[1]

Nearly 85,000 visitors walked through the gates of the African Plains exhibit on opening day, 1 May 1941. It was the largest single day's attendance since the opening of the New York Zoological Park in 1899. Over 3 million people came by year's end. Disturbed by the pounding guns and death cries of a battle-scarred Europe, visitors on opening day found solace in Osborn's address. "We are here to greet this sight," Osborn declared, "and millions of others will do likewise before the year is out, grateful for an hour of recreation, snatched from these troubled days. We can be refreshed for a while from the spectacle of man's cruel and needless destruction of himself. We should have no patience with those unthinking persons who rant that man, in his present cruelties, is reverting to primitive nature—to the so-called law of the jungle. No greater falsehood could be spoken. Nature," Osborn intoned, "knows no such horrors."[2]

Only a decade earlier the law of the jungle had pervaded the presentation of nature in natural history museums, zoos, and travelogue-expedition film. Filmmakers had portrayed the spectacle of death and the struggle for existence as life's central drama in part to attract box-office crowds. But in the midst of the Second World War, a more peaceful side to nature appealed to both scientists and the public. The human species provided enough violent drama. Nature offered a reprieve from a war-torn world.[3]

The African Plains exemplified this benign vision. The exhibit marked, in Osborn's words, the "opening of a new vista to the wonders of nature." It was a panoramic viewpoint that offered the public insight into the "interdependence and social relationships" of living forms and provided "a miracle-story" centered on the intricate web of life.[4] Only through a comprehensive, wide-angle view of the landscape could one begin to discern the interrelationships among organisms and their environment integral in maintaining the balance of nature. By the late 1940s nature films produced by the New York Zoological Society in the interests of conservation also reinforced this perspective.

Osborn, as one of the most outspoken leaders of conservation in the aftermath of the Second World War, saw the African Plains exhibit as a starting point in making the public aware of the need for conservation—not just of individual species, but of the natural environment and resources of the area occupied by wildlife.[5] In striving to recreate the ecological integrity of a particular habitat, the African Plains exhibit marked a departure from past efforts to develop naturalistic habitat displays. Contrary to Osborn's claims, the pioneering techniques used to create the African Plains exhibit were not new. In 1907, the animal dealer Carl Hagenbeck introduced naturalistic habitat displays using moated and barless enclosures in his famous Tierpark in Stellingen, Germany, which included an African "panorama." In mixing species of different zoogeographic regions within a single exhibit, however, neither Hagenbeck nor American zoos such as those in Denver and St. Louis, which adopted Hagenbeck techniques in the early 1920s, were attentive to capturing the authenticity of ecological relationships among animals and their natural environment.[6]

When Osborn took over the presidency of the New York Zoological Society in 1940, he embarked upon a new mission to expand the Society's role beyond the operation of the New York Zoological Park and Aquarium to include the promotion of conservation education and research on forests, soils, and water—the living resources upon which the existence of all life depends. Through natural habitat displays in zoos, wildlife exhibits in national parks, and nature films, Osborn believed, the conservation movement could educate the public and enlist their support for environmental causes on an unprecedented scale. Others feared that in bringing nature to the masses, the conservation movement would succumb to the very forces of commercialism and mass production that the preservation of wilderness was meant to counter. The prominent conservationist Aldo Leopold, for example, considered the "mores of advertising and promotion . . . to defeat any deliberate effort to prevent" wilderness areas from "growing still more scarce." In catering to the tastes of the "motorized ant," he argued, conservation organizations had pushed "the whole gamut of conservation techniques to the artificial end," diluting wilderness and "artificializing" its trophies to mere trinkets. Was the re-creation of nature in zoos, in the national parks, and on the motion picture screen turning nature into artifice or rendering a public service in the interests of environmental education, recreation, and research? The answer depended on the observer's focal point and whether humans, in the balance of nature, were inside or outside the panoramic frame.[7]

In 1945, four years after the opening of the African Plains exhibit, the New York Zoological Society embarked upon another experiment in public outreach and education in the interests of conservation: a wildlife park in Jackson Hole, Wyoming that would enable the travelling public to view the "magnificent big game animals of the West" usually witnessed only by hunters and wilderness enthusiasts.[8] The support of Laurance Rockefeller, son of John D. Rockefeller, Jr., proved instrumental in the Society's involvement in Wyoming. Laurance was a generous patron of Osborn's conservation interests. He served as chairman of the executive committee of the New York Zoological Society when Osborn was elected as president in 1940 and was also a trustee member of the Conservation Foundation established by Osborn in 1948. As president of the Jackson Hole Preserve, Inc., a company established in 1940 to oversee the management and development of his father's land holdings in the Jackson Hole region, Laurance, along with Osborn, envisioned a game park, accompanied by an information center containing material on the habits, management, and conservation of wildlife, that would be a "gathering point for naturalists and wildlife enthusiasts and an area for scientific study in wildlife conservation, propagation and management on a scale unparalleled in the nation."[9] Integral to the story of wildlife conservation in Jackson Hole was another story, a story of American heritage dear to Rockefeller's heart: the winning of the west, and the preservation of an environment crucial to "the struggle of a courageous, self-reliant, energetic people in the building of a free and prosperous nation."[10]

With the endorsement of Governor Hunt of Wyoming and Game Commissioner Bagley, Osborn and Rockefeller planned a twelve-hundred-acre game park, bordered by state highways, that would display in a four-hundred-acre section big game animals including elk, moose, mule deer, pronghorn antelope, and bison in a "semi-natural environment." Through the construction of an eight-foot-high, eight-mile-long steel fence designed "to follow the tree lines and contours of the area to make it fit naturally into the landscape," animals were to be "confined in such a manner that they themselves will hardly be aware of the restricting boundaries of their ranges."[11] Like the African Plains exhibit, the Jackson Hole Wildlife Park guaranteed the public a look at wildlife in a naturalistic setting that approximated the environmental conditions and ecological relationships found in the animals' natural habitat.

Although Harold Fabian, who represented the interests of the Rockefeller family in Jackson Hole, had warned Laurance that criticism was likely to come not only from cattlemen opposed to game conservation, but also from ecologists who would "frown upon us as interfering with the processes of nature," the governor of Wyoming had reported early public reaction to be favorable. The group was thus dismayed in November of 1945 by the negative publicity surrounding the project generated by the noted wildlife biologist, conservationist, and Jackson Hole resident Olaus J. Murie.[12]

Throughout most of his professional life, Murie was employed as a field naturalist for the U.S. Biological Survey. A graduate of Pacific University, Murie spent his early years with the Biological Survey in Alaska, where he developed a great fondness for the far north, and was later influential in establishing the Arctic National Wildlife Range. In 1927 he was transferred to Jackson Hole to conduct a study of the life history of the elk in the greater Yellowstone area. A member of the governing council of the Wilderness Society from 1937, Murie became director of the Society in 1945 and served as its president from 1950 to 1957. The Murie's home in Moose, Wyoming became something of a mecca for wilderness enthusiasts, and Murie became a powerful political force in promoting the cause of wilderness preservation in the United States.[13]

A member of the board of directors for the Jackson Hole Preserve, Inc., Murie resigned in November of 1945 when word came of Rockefeller and Osborn's impending plans. The issue, as Murie expressed it, was "showmanship vs. natural presentation."[14] Murie dashed off articles to the local and national press criticizing the project. "Proposed Rockefeller Zoo Rapped by Director of the Wilderness Society" read the headline of the *Jackson Hole Courier,* while the *National Parks Magazine* titled Murie's broadside "Fenced Wildlife for Jackson Hole." Murie argued that while "a menagerie is a wonderful service to the public in larger cities where the animal's environment is not available, it is a ludicrous intrusion in a place like Jackson Hole, where we have the real thing, and where the interested visitors can get their chief pleasure in discovering the animals for themselves."[15]

Plans for the annexation of Rockefeller lands into Grand Teton National Park meant the proposed Jackson Hole Wildlife Park would eventually be placed under the management of the National Park Service. The plans dated back to 1927, when John D. Rockefeller, Jr. with the support of President Coolidge began acquiring property in the Jackson Hole region with the understanding that it would be donated to the federal government and made a part of the national park system. Adjoining the Grand Teton National Park to the west and lying just south of Yellowstone National Park, Jackson Hole was deemed an essential part of these national preserves. The lobbying efforts of cattle ranchers and the state of Wyoming's interests in game management and Rockefeller's tax dollars, however, became major stumbling blocks when Rockefeller attempted to give the 34,000 acres, at an estimated value of $1.5 million, to the federal government. Unable to pass a bill through Congress that would make the region a part of Grand Teton National Park, President Roosevelt in 1943 created an executive order that made 221,610 acres of combined federal lands, Rockefeller properties, and private holdings a national monument. Over the next seven years, the Jackson Hole National Monument became embroiled in a bitter political controversy that effectively delayed the transfer of Rockefeller lands to the National Park Service until 1950, at which time it became a part of Grand Teton National Park.[16] Because of the planned incorporation of Rockefeller lands into a national park, Murie feared the building of a "road-side zoo in the Jackson Hole National Monument" would be a setback to current trends in national park policy toward the public presentation of wildlife in a "wholly natural" manner.[17]

Two years previous to the controversy over the Jackson Hole Wildlife Park, Murie conducted a study of bear ecology in Yellowstone critical of the Park Service's past management of wildlife for the purposes of public enjoyment. The proclivity of bears to frequent refuse sites, a behavior pattern likely established in pre-Columbian times as bears came upon buffalo jumps and Indian encampments, proved an asset in the promotion of the national parks. With the founding of Yellowstone National Park in 1872, watching bears feed at garbage dumps quickly became a popular tourist attraction. The National Park Service exploited this long-standing tradition and built an amphitheater at the Otter Creek dump near the Canyon Hotel. Every evening, upwards of five hundred people gathered to watch as many as fifty bears feed on garbage while a ranger delivered an informative lecture, protected by a chain-link fence. The practice continued until 1942, when the superintendent closed the last of the "bear shows" within the park, but this did not diminish the common sight of bears begging along the roadside for food.[18] In presenting wildlife to the public, Murie argued, the National Park Service contradicted its mission to preserve nature "unimpaired." In a letter to Park Service director Newton Drury, Murie expressed concern that the feeding of bears, particularly grizzlies, had altered their natural food habits and thus undermined the Park Service's mandate to preserve the "primitive character and value" of a region and its inhabitants. "Natural food habits, reactions to environment, distribution as affected by the associated fauna and flora, and seasonal weather conditions are all parts of the ecological story offered by an animal like the grizzly," Murie reasoned. Why then, he asked, "should we destroy these values in an animal that has become so scarce in the United States, simply for showmanship and the satisfaction of an artificially created human appetite?"[19]

Critics of wildlife presentation in the national parks characterized the urban dweller as expecting an experience of animals in the wild comparable to the city: "the same concentration of animals . . . in the paddocks of the zoological garden, the same personal safety in feeding the tamed animals, the same convenience of driving to a known place at any convenient time to see what he wants."[20] Zoos had become popular tourist attractions within the national parks during the interwar years. The zoo in the Wawona area of Yosemite kept not only cougars and other animals of the region in captivity but also included a display of Tule elk, a non-native species imported from the San Joaquin valley. Yellowstone's superintendent, Horace Albright, looked to make bison more visible to the public by bringing them closer to the main highways. In each instance, wildlife were brought into settings contrived by humans that bore little resemblance to their natural habitats. Within these public spaces, the resemblance of wildlife to trained animals and domesticated pets constituted their greatest entertainment appeal.[21]

To Murie, the Jackson Hole Wildlife Park symbolized the side-show circus elements of wildlife presentation that he and other biologists in the service of the national parks administration were determined to eliminate. Osborn aimed to "bring the animals to the people." Murie, in contrast, believed that only by bringing "people to the animals" could one "awaken the public to the value of the forest itself."[22] Murie's sentiments were in keeping with a series of policy recommendations put forward by National Park Service biologists in 1933 to avert future problems posed by the influx of people into the national parks.

Visitors needed to be re educated to appreciate the value of wildlife not according to the standards of the zoo, where tameness and the "fat certainty of an easy living" comprised the animal's charm, but according to the principles of nature, where wildness, the primitive struggle for survival, and "the marvelous interrelations of all living things" endowed the animal with a priceless, unique quality increasingly rare in a commercialized, mass society where everything became reduced to the commonplace. A shift from an "urban to a wilderness concept of the presentation of wildlife" was needed in the national parks. Displays that drew animals "out of their natural haunts" and offered a spectacular and intimate look at wildlife with the least expenditure of energy lacked respect for the ecological integrity of the animal in its natural habitat and demeaned the sanctity of wildlife. Such displays needed to be abandoned, Murie and others argued.[23]

Accommodating people to nature, rather than nature to tourists, park service biologists believed, would lead to a greater appreciation for the ecological web of life. The urge to awaken the public to the value not just of individual species, but the biotic community as a whole signified a widespread shift in ideas about natural resource management under Franklin D. Roosevelt's presidency. An ecological awareness of land use kindled by the Dust Bowl, the federal government's Civilian Conservation Corps and Soil Conservation Service, and the fledgling science of wildlife ecology all prompted National Park Service officials to consider other factors beyond immediate economic interests in the management of national park lands. Critical to this emerging perspective was a shift away from the management of individual species solely according to their commercial value as game crops or public spectacles to a broader vision best expressed in Aldo Leopold's words as the "integrity, stability, and beauty of the biotic community" as a whole.[24]

In the case of the Jackson Hole Wildlife Park, Murie believed that educational exhibits of big game mammals within a fenced enclosure offset the trend in conservation toward appreciation of natural habitat and the interdependence of life. But the fundamental issue was not an ecological approach to land management. Through the New York Zoological Society and the Conservation Foundation, Osborn championed the land as the basic value of conservation. The real debate centered on issues of conservation, democracy, and mass society in the postwar world.

As director of the Wilderness Society, Murie sought to safeguard the remnants of primitive wilderness from the onslaught of mass recreation and tourism fueled by the automobile, increased leisure time, and an expanding economy. While Murie had no doubt that Rockefeller and Osborn had a sincere appreciation for the outdoors, to them a recreation area was meant to be "all things to all men." "They think in terms of the masses," Murie wrote, "rather than of individuals" and thereby "compromise with forces that combat national park standards."[25] In presenting wildlife to the public, they had reduced nature to a common denominator and held it up to the standards of commercial success. Murie compared the Jackson Hole Wildlife

Park to a trinket, like the postcard photographs purchased in so many tourist gift shops. On countless occasions, he emphasized the importance of quality over quantity and the inherent dangers in commercialism and mass society. "It seems unfortunate if people shall become statistics, categories, a labor commodity, a bloc of votes, or any other medium for manipulation," Murie wrote.[26] The same could be said, Murie believed, of wildlife. In manipulating wildlife in the interests of promotion, Murie and others feared that the rights of the individual to solitude and escape from the "artificial contrivances" of mechanized society were being usurped and the value of encounters with animals demeaned.[27]

Suspicious of "assembly line methods" and "large scale pamphleteering" for instilling public appreciation of nature, Murie believed that a "sensitivity" toward the natural world had to arise from individual experience. Although Osborn readily admitted that individual encounters with animals in the wilderness produced the "greatest thrill," it was a luxury he thought few people could readily afford. "The Conservation movement," he insisted, could not succeed unless it was "fortified by mass education on an unparalleled scale." This was the goal of the Jackson Hole Wildlife Park and the African Plains exhibit. In Osborn's view, mass communication offered the most effective means to educate the public and enlist their support in environmental causes. And motion pictures, more than naturalistic zoo displays or wildlife parks, held the most promise of reaching a mass audience.[28]

Long-time Jackson Hole residents were well aware of the power motion pictures could wield in arousing public sentiment for wildlife conservation. The fate of the Jackson Hole elk, which Murie was assigned to study beginning in 1927, was intimately bound to the camera. In the early 1900s the largest remaining herds of elk were concentrated in the Jackson Hole–Yellowstone region (human settlement patterns had blocked the traditional autumn migration route to their wintering grounds in the Green River Basin). In Jackson Hole, insufficient forage resulted in periodic starvations of the herd. The harsh winter of 1909–1910 took a severe toll on the elk population and prompted Stephen Leek, an early settler in the region, to publicize the problem. A rancher, fishing and hunting guide, and one-term member of the Wyoming legislature, Leek had developed a fascination for wildlife photography and around 1907 purchased a Pathé motion picture camera to film elk in motion. After the winter of

1909–1910, Leek went on the Orpheum vaudeville circuit showing lantern slides and footage of the starving elk. His illustrated lecture showed "graphically the needless suffering and death among these noble animals," one reporter commented. His "pictures tell eloquently much that words cannot express concerning the real situation." Due in large part to Leek's efforts, state and federal appropriations were secured for the yearly winter feeding of the Jackson Hole elk herd and the National Elk Refuge was established in 1912.[29]

In his lecture tour, Leek combined photographic portraits of starving fawns and fighting bull elk in rut with panoramic views of the Grand Tetons and Jackson Hole valley. Through his motion pictures and still photographs, Leek, known to audiences as "father of the elk," created an image of the Jackson Hole elk as familiar, semi-domesticated pets. Many reviewers remarked upon how tame these wild animals appeared: so famished in winter, men could "walk up and pet them." This domestication of elk on screen paralleled the narrative conventions of nature writers like Ernest Thompson Seton and wildlife displays in the national parks. In each case, wildlife was made familiar. Each animal had its own personality, not unlike the family cat or dog.[30]

In this domesticated nature, some species were prized above others by virtue of their recreational and entertainment value. Leek, a contributor to sporting magazines such as *In the Open* and *Outdoor Life,* wrote disparagingly of the coyote, "a sneak and thief" of the wild, and of the "cowardly" mountain lion, whose "career" he proudly helped end with the rifle. Leek's attitude toward predators accorded with the federal government's massive eradication programs dedicated to the reduction and elimination of predators thought to threaten domesticated livestock or valued game species. The reduction of "nuisance" species in the national parks, including cougar, wolf, coyote, fox, mink, otter, and pelican to name a few, was considered necessary to protect the more popular wildlife species such as bison, deer, elk, and trout. In a story Leek wrote for the Jackson Hole newspaper, *The Grand Teton,* about the birth, growth, and maturation of White Patch into monarch of the elk herd, a story reminiscent of Felix Salten's 1926 novel *Bambi: A Forest Life,* Leek championed the elk as a noble creature deserving the respect and admiration of humankind. Through his films and short stories, Leek reinforced a conservation approach that focused on the economic value of individual species as resources for outdoor recreation and enjoyment.[31]

While the press lauded Leek's vaudeville performance as a strictly "humanitarian" effort in the conservation of America's natural heritage, by the 1920s Leek himself capitalized on the market for nature as a leisure pastime through outdoor recreation, nature study, and wildlife photography. In 1925, he established Leek's Camp and Camp Teton for Boys, a tent colony on the north shore of Jackson Lake that attracted adolescent boys from upper- and upper-middle-class families in the urban centers of the East Coast and Midwest. Through instruction in outdoor life, which included wildlife photography, the camp aimed to train the body, fortify the mind, and mold character. Ernest Thompson Seton declared that the camp would "do more to equip [boys] with robust physique and manly ideals than any other experience that might be in sight." At a season rate of six-hundred dollars per boy, Camp Teton may not have gone far in fulfilling President Calvin Coolidge's hope in 1926 to put "out-of-door pleasure within the grasp of the rank and file of our people." The camp did, however, turn nature into a profitable commercial venture and enabled Leek, with the permission of the U.S. Forest Service, to build a permanent summer camp and resort of log buildings readily accessible by automobile.[32]

The automobile and increased leisure time proved a boom to nature tourism in the 1920s, particularly in the national parks. In less than a decade after Stephen Mather took over the National Park Service upon its creation by a Congressional act in 1916, automobile ownership nearly tripled: 17.5 million automobiles were registered to Americans in 1925. Under Mather's administration, the construction of scenic highways and the building of trails, camping facilities, and water and sanitation systems became top priorities for the National Park Service. Landscape architects, the largest professional group employed by Mather's agency, carefully designed routes that provided access to the park's major attractions and furnished a diversity of scenic experiences framed through the car windshield. Auto-camping, commonly known as sagebrushing, became a favored middle-class leisure activity.[33]

While "motor camping [was] a pleasure in itself," it became "doubly delightful when combined with camera hunting," suggested Arthur Newton Pack, president of the American Nature Association. Founded in 1923 "to stimulate public interest in every phase of nature and the out of doors, and devoted to the practical conservation of the great natural resources of America," the American Nature Association through its popular monthly *Nature Magazine* offered its readers a cornucopia of products advertised to enhance their outdoor experiences; everything from horticulture and garden supplies to kennels and pet products to sporting goods, summer camps, and travel packages to the national parks. Advertisements for Bell & Howell's FILMO movie cameras appeared regularly starting in 1926. For a hundred and twenty-five dollars, middle-class Americans could share in the pleasure that J. Pierpont Morgan, W. K. Vanderbilt, and "hundreds more of like prominence" experienced taking personal movies of their travels. Recognizing that to nature lovers "each member of the animal kingdom is just as individual as any of the 'humans' [they] know," Bell & Howell offered the means to bring back a record of "communions with Nature" into the "living room at home, for a lifetime of enjoyment." Interchangeable telephoto lenses made it possible to "take close-ups of . . . shy friends from a distance." In an article that reads like promotional copy for the national parks, Pack instructed prospective sagebrush photographers "that the greatest thrills for the camera hunter lie within the borders of the National parks, where the wild animals are less afraid of man."[34]

Portraying animals as individual friends in the advertising copy for Bell & Howell's FILMO cameras readily conformed to the presentation of wildlife in the national parks, such as the bear shows at Yellowstone, and to prevailing attitudes toward management and cultivation of prized game species. Motion pictures that were produced in the interests of conservation, nature appreciation, and promotion of the national parks before the Second World War also drew upon and reinforced the tame and comic characteristics of individual animals in the wild.

To members of the American Nature Association and to audiences nationwide, the film lectures of William Finley did more to promote wildlife conservation through images of wildlife as cute, adorable human companions than did the works of sentimental nature writers or the practice of feeding bears in the national parks. Finley was enough of a recognized celebrity among nature lovers by the late 1920s that Bell & Howell featured him in their ads, shooting wildlife with his FILMO camera while a tame bear cub pawed gently at his arm for attention. A prominent figure in wildlife conservation on the West Coast, Finley helped establish Oregon's first Fish and Game Commission in 1911. He held various public offices—State Game Warden, State Biologist, and Commissioner for Fish and Game—in the preservation and management of Oregon fauna, directing a popular and successful elk reintroduction program that transplanted animals taken from the Yellowstone herd into the Wallowa Mountains of Oregon. In the early

1900s, he redirected his passion for bird collecting as a youth into bird photography, which resulted in his first book, *American Birds*. Inspired by Edward S. Curtis's motion pictures of native Americans shown at the Lewis and Clark Exposition in 1905, Finley shifted his photographic interests to wildlife movies in the early 1910s. His photographs and stories of wildlife, often written with his wife Irene, appeared in popular family and nature magazines, including *Country Life in America, Ladies Home Journal, National Geographic, Bird Lore, Colliers,* and *Field and Stream*. The wild animals the Finleys and their dog Pete befriended, such as Dinty the pet porcupine, became featured animal celebrities in the pages of *Nature Magazine*.[35]

In 1926, the American Nature Association sponsored the Finleys on a motion-picture lecture tour that played to city clubs, women's clubs, hunting and fishing clubs, Boy Scout organizations, conservation societies, and natural history museums across the United States. "As funny as a circus," Finley's 1926 motion-picture lecture included comic footage taken during an American Nature Association expedition to the Tucson mountains. Shots of Bill Finley disguised in a saguaro cactus moving across the desert in pursuit of a scared jackrabbit and of Finley being chased by a saguaro in a dream sequence made audiences laugh and captured a lightheartedness in nature that became a Finley trademark. Based on the success of the Finley lectures, the American Nature Association sponsored a number of camera hunting expeditions undertaken by William Finley and its president Arthur Newton that furnished entertaining material for the magazine's pages and footage for Finley's annual lecture tours.[36]

Finley's 1929 motion-picture lecture, "Camera Hunting on the Continental Divide," based on the Pack-Finley expedition to Glacier National Park and along the Continental Divide, combined entertaining wildlife portraits with advertisements for the American Nature Association and the national parks. Using a number of one-reel short subjects, Finley introduced audiences to wildlife on the Continental Divide, including moose, elk, bears, marmots, bighorn sheep, and animals befriended by expedition members: Franklin grouse docile enough to be petted, a mule deer named Emma, and Chippie the chipmunk. Some of the most hilarious scenes appeared in the one-reel motion picture, *Getting Our Goat,* which documented the expedition's search for the elusive mountain goat in Glacier National Park. Disguised in a white goat costume that included little horns and a goatee, Finley crawled along the rock ledges of steep precipices to get within close range of the animals. Shot-reverse shots of Finley in disguise looking at the mountain goats with close-ups of goats looking inquisitively back at the imposter brought the film to a comic end. Stills of the mountain goats were also used in advertisements for a twelve-day study tour of Glacier National Park offered by *Nature Magazine*'s tour department in cooperation with the Great Northern Railway and the National Park Service publicized during the spring season of Finley's lecture series.[37]

"Camera Hunting on the Continental Divide" not only promoted tourism in the national parks and fostered intimate associations with wildlife based on their appeal as wilderness pets, it also endorsed conservation efforts undertaken by state and federal agencies. A one-reel short subject offered spectacular footage of pronghorn antelope filmed from a pursuing automobile at forty-five miles per hour that included stop-action photography to illustrate the position of the antelope's legs in its twenty-foot stride. The antelope footage also featured scenes of antelope fawns captured by government biologists, nursed by hand, and shipped by airplane to other regions as part of a re-establishment effort undertaken in Western states. In this reel, coyotes were portrayed as the antelopes' menace; men scoured the sagebrush plateau in search of coyote dens and were shown retrieving litters of coyote pups. The contrast between adorable, bottle-fed fawns and an antelope carcass with only the head intact implicitly linked to coyotes offered compelling evidence in support of the government's predator control program. In a *Nature Magazine* article that featured still frames of the antelope film and Bill Finley's pet baby antelope Buck, Finley tempered this message by suggesting that the antelope's greatest enemy was now man. Having destroyed the coyote, man "substitutes himself as an enemy never provided for in Nature's delicate balance between life and death." Facing hunters that pursued their quarry from automobiles, "the antelope," Finley ended, "lacks a real sporting chance."[38]

In *The Forests*, a one-reel motion picture short in the Continental Divide series, humans are once again nature's enemy. In this film, unlike his others, Finley's camera records more comprehensively the delicate balance of relationships among animals within a single habitat. Through skillful editing, the animals in the forest offer a conservation lesson. Cuts between two flickers in a mist-shrouded primary forest, scenes of a logging operation, forest fires, floods, and a deeply eroded landscape, and titles such as "After cutting do they replant?" or "Forest fires. What next?" illustrate what can happen if humans do not practice sound conservation techniques. In contrast to humans, the beaver is the most industrious and skilled practitioner of soil, water, and wildlife conservation in the film. Introduced by Finley as he holds it by the tail near shore, the beaver is shown establishing important habitat for other wildlife through its dam-building activities. "'I furnish homes for waterfowl and fish,' says the Beaver," followed by shots of ducks in a pond and an angler catching a fish in the waters downstream. "'I supply water for forests and farms' says the Beaver," followed by scenes of a healthy pine forest and prosperous farmstead. Although the beaver is portrayed as a tame wilderness pet, the conservation message is less focused on the preservation of individual species for their entertainment value. Instead, the animals are filmed to convey a message about the importance of wise land-use.[39]

For its management of watersheds and creation of wildlife habitat through dam construction, the beaver became a popular icon of integrated land management in conservation films of the 1930s and 1940s. Planned land-use and conservation became a consistent and uncompromising theme of New Deal policy in Franklin Delano Roosevelt's administration. Through the coordination of federal bureaus and the creation of agencies such as the Civilian Conservation Corps and the Soil Conservation Service, Roosevelt aimed to integrate and manage natural resources, including soils, waters, forests, wildlife, and recreation, to stimulate economic and social recovery and contribute to long-range conservation. The creation of the Tennessee Valley Authority (TVA) in 1933 offered Roosevelt and many conservationists a new model for the federal coordination of land-use and resource development on a geographically unified basis. Entrusted with a broad vision of regional development along the Tennessee River, the Tennessee Valley Authority extended its reach beyond electric power production to promote flood control, navigation, agriculture, conservation, recreation, and economic and social well-being. The beaver, before its decimation by the fur-trading industry, played a pivotal role in the management of waterways in the Mississippi valley, a task taken over be the Tennessee Valley Authority, the Civilian Conservation Corps (CCC) and the Army Corps of Engineers.[40]

Pare Lorentz's 1937 documentary film *The River,* produced for the U.S. Farm Security Administration, helped shape public opinion in support of the TVA and also marked a turning point in the use of film for the promotion of conservation according to the principles of integrated land-use. *The River* followed upon the success of Lorentz's highly controversial film, *The Plow that Broke the Plains.* Made in part to enlist support for the relief programs of the Resettlement Administration (renamed the Farm Security Administration in 1937), *The Plow that Broke the Plains* conveyed a message of the interdependence of people and soil through its historical portrayal of land abuse on the Great Plains and the subsequent plight that farmers faced in the Dust Bowl years of wind, drought, and poverty that swept the region. While some criticized the film as New Deal propaganda, others celebrated the film's potential to "awaken our citizens to the necessity for immediate steps in conservation."[41] *The River* played upon a similar theme, in which economic and social prosperity relied upon the integrated and planned management of water, soil, and people. A map of the Mississippi River and its tributaries and aerial shots of the Rockies to the west and the turkey ridges of the Allegheny Mountains to the east provided the panoramic frame in the opening film segment. This wide-angle view showed the extent of the land area that drains into the great Mississippi—described as "out of joint." Cotton farming and the pillaging of natural resources, including lumber, iron, and coal in the industrialization of the nation after the Civil War had left the mountains and hills slashed and burned and the soil spent. Footage of eroded hillsides, deep gullies, and deforested mountains reinforced the meaning of the film's lyrical prose. Floods raged through the Mississippi valley, including a disastrous Ohio River flood during filming that left farms ruined and families "aimless, footloose, and impoverished." But the film ends on an optimistic tone, documenting the work of the TVA—the dams built to control the flood waters, the efforts of the CCC to put the "worn fields and hillsides back together," the application of soil conservation techniques to the tilling of farmland, and the furnishing of hydroelectric power to "model agricultural communities" established in the restored valley with the help of the Farm Security Administration.[42]

Commercially distributed by Paramount, *The River* underscored the value of the land in conserving the Earth's living resources. By focusing on the interdependence of humans and nature, *The River* offered an expansive picture of conservation, one attentive to the ecological fabric of the entire ecosystem rather than a limited valuation of isolated parts. In contrast, the films of Leek and Finley, like wildlife displays in the national parks of the interwar years, promoted conservation through the domestication of wildlife. The value of individual species in these earlier films stemmed not from the integral role these animals played in the ecological relationships of the biotic community, but from their entertainment value. In the comprehensive view of the landscape offered by such films as *The River,* the interrelationships of the Earth's living resources came into focus only when the close-up spectacle of animal life receded.

The interrelationships of soil, water, forests, wildlife, and people became a common theme in the conservation message broadcast by Fairfield Osborn to the public throughout the 1940s and early 1950s. Osborn looked upon the TVA as a model in the coordination of "land resources into a unified program" that "harmonize[d] human needs with the processes of nature." He also recognized the significance of documentary films like *The River* in enlisting public support for conservation causes. Like the African Plains exhibit and Jackson Hole Wildlife Park, Osborn looked upon the motion picture as a method, indeed the most effective one, "of re-establishing contact" with nature, since "it is viewed by the millions and combines the qualities of reality, action, sound, and dramatic effect." "Modern man has become separated from the natural world," Osborn wrote. "Consciously or subconsciously, he desires intimacy with it." Through natural history film, Osborn hoped to capture that desire for intimacy and turn it into impassioned support for environmental causes.[43]

The establishment of the Jackson Hole Wildlife Park was one conservation crusade in which Osborn utilized film as an ally. Murie had been successful, despite Osborn's pleas, in getting the Izaac Walton League and the American Society of Mammalogists to publicly denounce the plans for the Jackson Hole Wildlife Park at their annual meetings in 1946.[44] Osborn, however, could not afford such a public relations disaster. To win the approval of the biological community, Osborn sold the park as a field station for animal behaviorists and ecologists to conduct biological research similar to that of tropical research field stations like Barro Colorado Island in the Panama Canal Zone. At its 1947 meeting, the American Society of Mammalogists withdrew its objections and "commended the officials of Jackson Hole Wildlife Park for their emphasis on research."[45] In addition, to heighten interest and support for the project and to attract visitors to the region, Osborn recruited the expertise of Clarence Ray Carpenter in the making of the film *The Jackson Hole Wildlife Park.* Carpenter organized the summer research and training programs that brought upwards of twenty investigators and students each year from 1947 to 1953, at which time ownership of the Jackson Hole research station was transferred from the New York Zoological Society to the University of Wyoming. While Carpenter's work on field methods in the study of animal behavior added an important dimension to the training program of the Jackson Hole Biological Research Station, his psychological research on the use of film for instructional and motivational purposes proved equally valuable in the production of the film. Designed to improve the public relations of the park with local and state residents and to attract visitors from the east coast, *The Jackson Hole Wildlife Park* was distributed through community service groups such as Rotary and Kiwanis clubs and local Chambers of Commerce in the state of Wyoming and was shown to chapters of the Garden Club of America in New York.[46]

In keeping with trends in conservation philosophy and management, Osborn sought to instill appreciation not just for individual wildlife species, but for the land as a whole. Although Murie objected to the Jackson Hole Wildlife Park as a departure from prevailing ecological attitudes toward wildlife conservation, *The Jackson Hole Wildlife Park* film illustrates how Osborn's plans for the development, management, and promotion of the region relied upon a panoramic vision of multiple land use.

The shot that opens *The Jackson Hole Wildlife Park*, a wide-angle view of the Jackson Hole valley with the Grand Tetons in the background, gives a comprehensive perspective on the "geographic, climatic, and biological features of the region [that] provide favorable habitats for many varieties of wildlife characteristic of the West." Footage of moose, elk, mule deer, antelope, Canada geese, and sage grouse introduce audiences to the resident wildlife of the area. The film then presents a visual tour of the different vegetational communities within the park, including lodge pole pine, quaking aspen, and mixed habitat associations of sage brush, grass, and aspen that contribute to the "biotic equilibrium" of soil, plants, and animals found in the region. Beaver ponds, sedge, grass, and willow, for example, provide an ideal ecological setting for moose, while the plant community of sage brush and grass in the north end of the Park furnishes "suitable pasturage for Mule Deer . . . and the cautious and swift Antelope."[47]

The panorama of Jackson Hole offered a holistic picture of the regional ecology that was essential in understanding where more detailed investigations should begin. From here, the camera zooms in on individual studies being conducted on the ecology and behavior of the sage grouse, raptors, and moose; the nesting, breeding, and rearing behavior of Canada geese; the social organization and behavior of elk; and "beaver colonies and their effects on the ecology of the area." Humans were also considered an integral part of the ecosystem subject to scientific study. Footage of a trained psychologist interviewing a Jackson Hole resident on attitudes and interests with regard to development of the area reinforced the "philosophy of public use and education" in keeping with the Park's overall goals. However, the final outcome is only achieved once the parts are reassembled; nature must be made complete. Balancing human needs with those of nature constituted the Park's most important task. In the re-creation of nature, Jackson Hole Wildlife Park sought to integrate education, entertainment, recreation, research, and nature into one. And so the movie ends, as the camera pans across the Jackson Hole valley and more distant Grand Tetons: "This is the foundation, this is the beginning. This is the stage and you have seen a few of the actors. This is the foundation: soil, plants, animals, and people, enveloped in a mountainous western climate. This is the record of the beginning of one constructive regional development—the Jackson Hole Wildlife Park."[48]

The Jackson Hole Wildlife Park marked the beginning of a concerted effort by Osborn and the New York Zoological Society in the development of film for conservation education and promotion in the late 1940s. In 1948, in conjunction with the publication of his best-selling book, *Our Plundered Planet,* Osborn founded the Conservation Foundation, a nonprofit corporation with initial financial backing from the New York Zoological Society and a board of trustees that included affluent members of New York society, such as Laurance Rockefeller and Childs Frick. With a stated mission to "initiate and advance research and education in the entire field of conservation—soil, water, forests, vegetation, and wildlife"—the Conservation Foundation relied upon a widespread educational campaign to promote a conservation ethic among adults and children alike.[49] The foundation launched a highly ambitious and successful educational outreach mission that included a film series for schools and adult groups, a thirteen-week NBC radio program on conservation hosted by Hollywood star James Cagney, and the establishment of a Chair of Conservation at Yale University filled by ecologist Paul Sears. By the early 1950s, the foundation had produced four film series on ecology and forest, soil, and water conservation, which were distributed by Encyclopedia Britannica Films. The Living Earth Series on soil and water conservation had the largest sales of any comparable series sold by Encyclopedia Britannica to schools and adult groups, and by 1952 the gross revenues of all Conservation Foundation films exceeded $450,000.[50]

While *The Jackson Hole Wildlife Park* had a limited audience, the films produced by the Conservation Foundation reached approximately 2 million adults and children in the United States annually during the late 1940s and early 1950s. In addition to their school distribution, the films were often featured as part of Osborn's many speaking engagements to local chapters of the Garden Club of America, state conservation and forestry associations, sportsmen organizations, civic clubs, and universities. Efforts were made to extend the reach of the Conservation Foundation films beyond the predominately white upper- and middle-class audiences that composed the membership lists of conservation, garden, nature, and sporting clubs found in urban and suburban areas. In the summer of 1950, for example, The Living Earth series was shown to a small community of African-American farmers near Covington, Georgia in the yard of an old tenant house. The audience allegedly broke out in laughter when a tenant farmer shouted "Wish we could grow a crop that fast" in response to a scene showing a time-lapsed sequence of sprouting bean plants. Humorous as such scenes may have been to farmers, those working the land comprised a key audience if the message of these films, which underscored the importance of soil and water conservation in enhancing and maintaining the quality of life, was to take root.[51]

The dependence of the quality of life upon the quality of land and the need to abide by nature's laws, particularly the principle of interdependence, comprised the central theme of films produced by the Conservation Foundation from The Living Earth series made in 1948 to The Web of Life series completed in 1950. In *This Vital Earth,* the second short subject in The Living Earth series, the opening scene of topsoil blowing from a human hand makes evident that conservation begins with the land. To convey the theme of interdependence, an animated sequence illustrates the different links found in the chain of life—nature's raw materials, the sun's energy, plants, topsoil, and organization—that give rise to a coordinated and balanced community upon which the well-being of plant and animal life depend. Footage of different vegetational types, from water lilies to bay trees, is used to tell the story of plant succession from marsh to climax forest. Shots of butterflies, caterpillars, white-footed mice, hawks, and owls are woven together into a story about the web of life. "To prosper," the narrator asserts, "nature must achieve a balance. And upon the perfection of this balance depends the amount of life the land and water can support." Nature's principle of balance is illustrated again in a pond community comprised of a simple food chain that includes plankton, bream, bass, and man and is extended to agricultural systems as well.[52]

Significantly, the films produced by the New York Zoological Society and Conservation Foundation during the immediate postwar years never excluded humans, either as a destructive or constructive force, from images of nature re-created on screen. Osborn looked upon humans as important biological agents in changing the face of the earth. As such, the human species occupied a place in the natural landscape and was included within the panoramic frame that enveloped the interdependent web of life. The final scenes of *This Vital Earth* made this point clear. Returning to the animated chain of life, to footage of eroded farm land, and then to healthy cows and ranchers on good pasture, the film concludes: "Only so long as we work in harmony with nature's laws may we continue to reap her bountiful harvest. Who shall have the right to disregard these laws? Who shall have the right to misuse this vital earth?" Harmonizing humans' needs with the rest of nature marked Osborn's greatest challenge in conservation research and education.

In Jackson Hole, that challenge was exacerbated by an expanding postwar economy and increased leisure time that fueled a tourist boom. By 1950, annual visitation to Yellowstone National Park had exceeded 1.1 million. At Grand Teton National Park, attendance figures increased from 192,000 visitors in 1950 to over one million in 1955. Mass tourism placed similar pressure and demands upon wildlife management in the Jackson Hole region. The transfer of Rockefeller lands to Grand Teton National Park not only reopened the controversy of wildlife display at Jackson Hole Wildlife Park, it also created new problems in the management of the Jackson Hole elk herd, since the National Park Service's no hunting policy prevented the harvesting of elk on park lands.[53]

To facilitate the resolution of these controversial land-use issues, the Conservation Foundation and the New York Zoological Society sponsored a biological and economic appraisal of the Jackson Hole elk herd, completed in 1952 by wildlife biologist John J. Craighead. John, along with his twin brother Frank, had pursued a joint project on the ecology of raptor predation at the University of Michigan that utilized information they collected during two breeding seasons at the Jackson Hole Wildlife Park. Respected wildlife photographers, the twin brothers contributed raptor footage and the map of vegetational communities for *The Jackson Hole Wildlife Park* film. The vision of land use articulated by John Craighead in his study of the Jackson Hole elk corresponded to the ideals of integrated land-use management projected in Conservation Foundation films.[54]

In the visions of conservation shared by the Craigheads and the Conservation Foundation, humans were considered "a natural and integral part of the environment." "Policymakers of the future must realize that the people of Jackson Hole are as much a part of the total environment as the elk and the other wildlife," argued John Craighead. "The elk problem," he continued, "resolves itself into a human problem, and a satisfactory solution lies not in 'buying out' the landowner and producer but in integrating his requirements with those of the wildlife and the American public."[55] John Craighead considered an assessment of the entire watershed in the Jackson Hole region necessary to assess and to balance livestock interests, recreational needs, and wildlife values, and thereby to achieve the maximum economic and aesthetic use of the land. Looking out "through the picture windows of [their summer] cabins, across the sagebrush flats and the Snake River cottonwoods," Frank and John Craighead "saw a thousand-eyed dragon writhing along beneath the Teton ramparts—the procession of automobiles carrying visitors to this enchanted landscape." They did not retrench in horror, but instead looked upon it as a challenge to wildlife management. "It takes the wise use of land to produce and maintain wild-animal populations," they wrote. "It takes even wiser management to integrate the human element that cannot be divorced." But such an approach was necessary, they reasoned, if the region was to "endlessly fulfill its promise as a place where city-weary Americans may literally re-create body and spirit in a land of rare splendor and rich resources in living things, including man himself."[56]

The creation of naturalistic habitat zoo displays, the emphasis on natural presentation of animals in the national parks, and the production of conservation films meant to instill values for the land as a whole indicate the extent to which a concern for the authenticity of interrelationships found in the ecological web of life became an important aspect of wildlife appreciation, conservation, and management by the close of the Second World War. In books such as Sally Carrighar's *One Day at Beetle Rock* and Rachel Carson's *Under the Sea Wind,* authors similarly began to fashion a genre of natural history writing in the 1940s that encompassed the entire chain of life in its tales of animal life. In her 1947 book, *One Day at Teton Marsh,* which the naturalist William Beebe described as "authentic" as the scientific studies being conducted that same year at Jackson Hole Wildlife Park, Carrighar used the ecological relationships found in a beaver pond habitat in Jackson Hole to weave together the stories of individual animals, including the beaver, osprey, otter, cutthroat trout, mink, hare, moose, mosquito, and trumpeter swan.[57]

Earlier in the century, the promotion of nature tourism, recreation, and conservation relied upon the appeal of wildlife as tame wilderness pets. The popularity of bear shows at places like Yellowstone, the similarities of zoos in both urban and national park settings, and the domestication of wildlife on screen through the films of the American Nature Association all point to the value naturalists, conservationists, and the American public placed on individual species as spectacles and resources to be enjoyed, managed, and preserved. Cute, adorable, and tame wildlife species continued to be an important asset in the public campaign strategy of environmentalists after the Second World War, but it was a convention that existed alongside and sometimes in tension with a more holistic approach to resource management articulated by federal government agencies, wildlife biologists, and environmental organizations.

In the study and presentation of wildlife through a wide-angle lens, however, a crucial question remained: what place did humans occupy in this web of life? From Osborn's aesthetic and political vantage point, humans were a part of nature's balance, even if that balance had gone awry. This theme resounded through the research projects and educational films sponsored by the New York Zoological Society and Conservation Foundation. Although Osborn thought preservation to be an important goal, he also looked to "European concepts of how Nature and man may survive together." With human civilization permeating "virtually every living area of the earth's surface," environmentalists could no longer afford to focus on the preservation of pristine nature as their sole cause. Humans had become thoroughly enmeshed in the ecological web of life.[58]

As president of a national environmental organization devoted exclusively to the preservation of wilderness, Murie looked upon the relationship between humans and nature in much more polarized terms. Although he realized humanity was at the "threshold of a synthetic civilization," Murie believed there was still room to turn back. When Struthers Burt, a long time resident of Jackson Hole, remarked that "the entrance of man upon the scene [in Jackson Hole] immediately change[d] the balance of nature," Murie queried: "Should we not try to do the best we can to preserve *what remains* in a civilization that has let go all holds and has turned neurotic?"[59] Others shared Burt's sentiments. To Murie's criticism that the National Park Service had put an urban veneer on nature in its efforts "to make a national park be everything to all people" and thereby demeaned "the spirit of Jackson Hole" by turning it into a cheap roadside display, National Park Service director Connie Wirth responded in a slightly ironic tone. "Certainly," Wirth suggested, "the area has been so substantially modified by man that it long ago lost its *wilderness* aspects." In addition to the highways and roads that crisscrossed the region, Wirth pointed to the "large dam and artificial lake, a fact which led to much opposition by conservationists to making the area a park at all." Osborn had raised similar objections to Murie's denunciations of the Jackson Hole Wildlife Park by pointing out that the artificial feeding of the elk in the wildlife park was no different than the practice of feeding hay to the 20,000 "wild" elk every winter on the National Elk Refuge in Jackson Hole because they no longer had sufficient winter range.[60]

In the romantic vision championed by Murie and other wilderness preservationists, a clear boundary separated primeval nature and synthetic civilization. Increasingly, however, pristine nature was becoming a part of America's nostalgic past. In contrast to documentaries like *The River* and *This Vital Earth,* which attempted to efface the boundaries separating humans from the natural world, a new genre of nature films appeared in the late 1940s that capitalized on the romantic desire for pure nature unpolluted by human touch. In the masterly hands of Walt Disney, the wide-angle lens that captured the interdependence of nature would be combined with intimate views of animal life to create a fantasy of pristine wilderness on screen. In re-creating a nostalgic look at American wilderness, Disney drew upon the talents of a wide range of naturalist-photographers, two of whom, James Simon and Tom McHugh, directed and conducted research at the Jackson Hole Wildlife Park.

Figure 1. The 1910 production of *Roosevelt in Africa,* one of the first "authentic" natural history films commercially released, helped begin America's long relationship with wildlife on screen. The urban middle and upper classes were the primary patrons of natural history films, despite movie posters designed to attract a wider audience.

Figure 2. Civic leaders looked to natural history film as a respectable form of entertainment for the expanding urban working class. Advertisements like this one for Paul Rainey's *African Hunt,* the highest grossing wildlife film of the 1910s, promoted nature as a subject simultaneously entertaining, educational, morally uplifting, and profitable.

Figure 3. Nature photographers were continually torn between their desire to bring a genuine re-creation of wildlife to their audiences and the need to fabricate a more intense drama to elicit thrills and commercial success. William Douglas Burden and his wife Katherine simulated the hunt of dragon lizards in the studio, but even this artifice failed to gain them a producer. Yet, in the hands of Hollywood, Burden's East Indies expedition would later be refashioned into the blockbuster classic *King Kong*.

Figure 4. By the 1920s, developing film conventions had already begun to influence museum exhibits. The dramatic postures of these Komodo dragons in the Hall of Reptiles at the American Museum of Natural History were intended to invoke primeval monsters—a response to the success of the 1925 Hollywood adaptation of Sir Arthur Conan Doyle's novel, *The Lost World*.

Figure 5. Filmmakers like Martin and Osa Johnson in the 1920s combined science with showmanship to produce commercially successful travelogue-expedition films. The Johnsons drew upon comic and dramatic conventions of Hollywood entertainment to make nature appealing to popular audiences.

Figure 6. Wilderness pets like Bimbo, a white gibbon that appeared in *Chang,* became objects of popular affection. Nature films, like the realistic nature stories of writers like Jack London, charmed audiences with depictions of animals' individual personalities.

Figure 7. Ethnographic films such as *The Silent Enemy* (1930) offered a romantic, nostalgic vision of an America where racial purity and pristine wilderness were closely linked. But the boundaries separating authenticity from artifice were difficult to maintain. When it was discovered that Chief Long Lance *(left)*, who played a great hunter of the Ojibway, was a not a Blackfoot Indian but of mixed African American, white, and Lumbee descent, producers worried that his heritage might jeopardize the film's claim to authenticity.

Figure 8. Naturalists feared that films like the 1930 picture *Ingagi,* in which gorillas cohabitate with women, would undermine the educational value of natural history film by sacrificing truth for Hollywood sensationalism and financial gain.

Figure 9 A series of exhibits in the American Museum of Natural History's Hall of Animal Behavior, which opened to the public in 1940, used cinematic techniques to engage audiences. Here, a scene of hens in the barnyard *(above)* faded out to a "hen's-eye view" of the same scene *(below)*, illustrating the dominance hierarchy perceived by domestic fowl.

Figure 10. Motion picture technology enabled scientists to achieve a more precise understanding of visual communication in the animal world, such as this mutual billing ceremony—a series of courtship behaviors performed by gannets—captured in Julian Huxley's 1937 Academy Award–winning film, *The Private Life of the Gannet.*

Figure 11. Films made by ethologists like Konrad Lorenz, featured here in his 1937 film, *Ethology of the Greylag Goose,* were also of interest to lay audiences, thereby contributing to the growing popularity of animal subjects, particularly after World War II.

Figure 12. Through his motion picture lectures for the American Nature Association during the 1920s and early 1930s, conservationist and naturalist-photographer William Finley focused on the tame and comic characteristics of wildlife to instill in the public an appreciation for nature and to promote tourism in America's national parks.

Figure 13. As America urbanized and modernized, motor camping and amateur camera hunting for wildlife had become such a popular pastime among the middle and upper classes that Bell & Howell encouraged nature tourism through advertisements, such as this one with William Finley.

Figure 14. The African Plains exhibit, a naturalistic habitat display that opened in 1941 at the New York Zoological Park, represented a shift away from the perception of wildlife as pets, and focused on the significance of the ecological web of life. Wildlife management in the 1940s aimed to preserve natural habitat and the entire biotic community of living organisms.

Figure 15. This wide-angle shot from the 1947 film *The Jackson Hole Wildlife Park* is representative of a conservation ethic centered on integrated land management. During the late 1940s and early 1950s, such films reached over 2 million war-weary Americans who sought landscapes free of world politics.

Figure 16. In the immediate postwar years, Disney's True-Life Adventures monopolized the mass market for nature on the big screen. Many Disney photographers like Herb Crisler viewed nature through their own lived experiences, not through a scientist's lens. Drawing on naturalist-photographers with a wide range of talents, social backgrounds, and educational training, Disney sought to make nature directly accessible to all who encountered it.

Figure 17. While shooting Disney's feature film *White Wilderness* (1958), Herb and Lois Crisler spent eighteen months in the remote regions of Alaska. Many Disney photographers like the Crislers idealized wilderness as a place of frontier values. Only in the complete freedom of the wild, Lois Crisler suggested, could one escape the conformist trends of mass society and know the true individual nature of oneself and others, like the wolf she named Trigger.

Figure 18. Despite the frequent portrayal of nature in harmonious balance in True-Life Adventures, reviewers and naturalists still criticized especially violent scenes, such as this battle between a red-tailed hawk and rattlesnake in *The Living Desert* (1953).

Figure 19. Marlin Perkins and animal friends like Heinie II became celebrities through the television show *Zoo Parade,* which offered wholesome entertainment and promotional ad campaigns for the entire family.

Figure 20. Jewel Food Stores, one of the local sponsors of *Zoo Parade,* gave cards to customers that featured a *Zoo Parade* animal like Icicle *(above)* on the front and marketed a product like Canfield's root beer *(right)* on the back. Demand for these cards escalated to 150,000 per week in the early 1950s. By appealing to America's growing market for pets, Perkins helped domesticate nature on the television set for America's baby-boom generation.

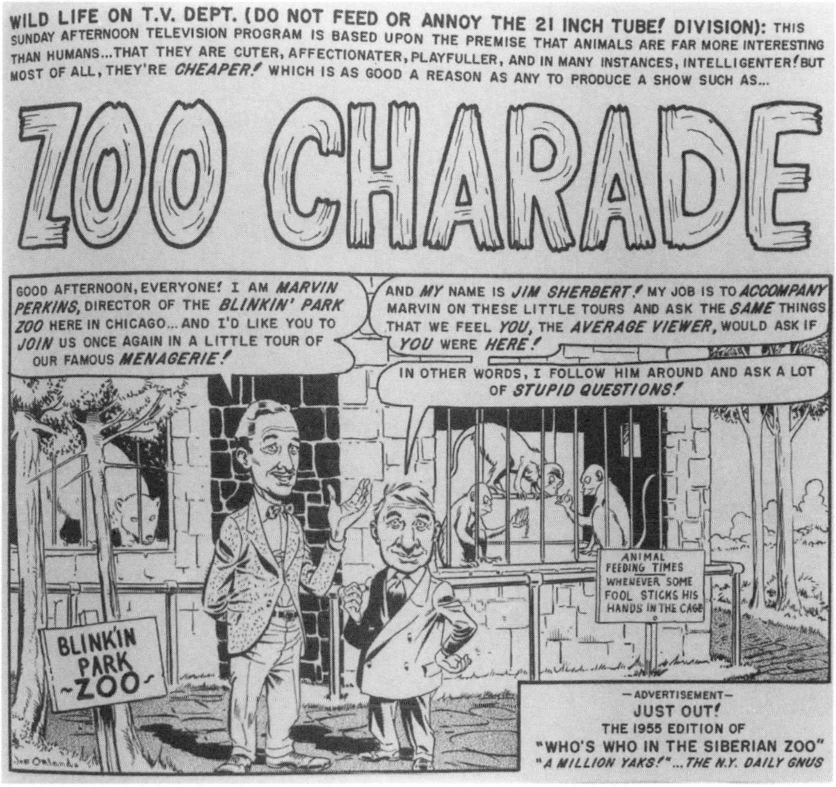

Figure 21. With its version of a lustful and violent nature, *Panic* magazine satirized *Zoo Parade*'s popularity in American culture at a time when parents longed to preserve childhood innocence and looked to a sentimental nature for moral lessons about the importance of family life.

Figure 22. When *Zoo Parade*'s ratings began to drop in the mid-1950s, the show experimented with filming on location at popular tourist attractions and incorporating more daring camera-work, two elements that proved important to the sequel of *Zoo Parade—Wild Kingdom*—which first aired in 1963.

Figure 23. Marine Studios, which opened in 1938 near St. Augustine, Florida, demonstrated the cumulative result of nature film production over the previous three decades. It was designed as a place of public entertainment, as a studio to film underwater action against an authentic background, and as a research site.

Figure 24. Marine Studios was pivotal in constructing the contemporary image of the dolphin as a loveable, gentle friend in the sea, despite its sexual aggressiveness in captivity. Science and showmanship contributed equally to the dolphin's star appeal, which attracted both tourists and funds for scientific research. Keeping the dolphin's private sexual life out of public sight, both in the tanks and in the popular media, would become an important public relations task in managing its career.

Figure 25. Flippy, the first trained dolphin, attracted such a following that in 1953 Marine Studios built a one-thousand-seat stadium just for his show. Flippy's act, which included towing a woman and a little fox terrier around the tank on a surfboard, marked an important step in the domestication of the dolphin in American culture.

Figure 26. *Life* magazine advertised that like Sandy in the motion picture *Flipper,* one could have a pet dolphin for a mere three hundred dollars. In places like Marine Studios, Flipper's appealing character had been typecast in America well before the movie production began.

Figure 27. During the 1950s, a change in Western images of Africa's landscape accompanied the continent's struggles for political freedom: the "Dark Continent" was rapidly transformed into a place of threatened ecological splendor.

Figure 28. Ecologists regarded poaching, an expanding human population, and ethnic nationalism as the greatest threats to Africa's wildlife in the late 1950s and early 1960s. Bernhard Grzimek, Director of the Frankfurt Zoo, was among a number of biologists who used documentaries to arouse public sentiment for Africa's threatened species and to help raise money for international conservation efforts.

Figure 29. Films such as *Wild Gold,* produced by the New York Zoological Society in 1961, portrayed East Africa as a land in crisis, and championed eco-tourism as its salvation. Through the advice and technical assistance of Western conservationists, these nomadic pastoral lands were turned into recreational tourist destinations.

Figure 30. In 1960, Joy Adamson's best-selling book, *Born Free,* and the movie version that followed portrayed the Adamsons and Elsa as ideal figures in Africa's peaceable kingdom. This image of African paradise reinforced the message that wildlife were a part of the world's heritage for all to share and enjoy, and fulfilled the dream of many Americans to make intimate contact with nature. Yet regions of the world that appeared pristine to Western tourists were also places of livelihood for other peoples, who did not necessarily regard nature as an innocent playground nor wildlife as a global resource that belonged to all.

Chapter Five

DISNEY'S TRUE-LIFE ADVENTURES

Four days before Christmas in 1948, Arthur Levoy, exhibitor at the Crown Theater in Pasadena, took a chance. Instead of showing the traditional double bill, Levoy accompanied the feature presentation, MGM's *The Three Musketeers,* with a short twenty-seven-minute film on the life of Arctic seals. The week before Levoy had attended a private screening of this new venture in educational entertainment at Walt Disney Studios and had agreed to give the film a two-week trial run at his theater during the holiday season. Besides pleasing Levoy's audiences, the booking made the film eligible for a 1948 Academy Award. At the awards event in the spring of 1949, Disney's *Seal Island* won an Oscar for best two-reel short subject. Hollywood's endorsement was even more notable because the film defied conventional categories. The *Motion Picture Herald,* a weekly trade magazine for motion picture exhibitors, created a unique heading for the film in its product digest section: "short feature." Within a few months, the magazine listed the next two Disney short subjects, *Beaver Valley* and *Nature's Half-Acre,* under a new category, "True-Life Adventures," the same name Walt Disney Studios gave its innovative series of nature films. By the release of the last True-Life Adventure, *The Jungle Cat,* in 1960, the series included ten short subjects and four feature-length films. With its True-Life Adventures, Walt Disney Studios succeeded in capturing and monopolizing a mass market for nature on the big screen throughout the 1950s, thereby expanding the packaging of nature as entertainment in postwar American society.[1]

"Completely authentic, unstaged and unrehearsed," *Seal Island* and the other True-Life Adventures portrayed a fantasy of pristine nature far removed from the commercial world of modern, industrialized America. Disney took great pains to emphasize that in these films nature wrote the screenplay; the eloquence, the emotion, and the drama were nature's own. The revolving globe and bold, authoritative music that opened *Seal Island* mimicked the dramatic start of newsreels such as *The March of Time* and wartime information films like *The War in Action.* But the shift to pastoral music and an animated paintbrush moving across the screen to reveal a cartoon panorama of Alaska in the "shadow of the North Pole" linked *Seal Island* to the well-known Disney fantasy classics. A "story strange as fantasy yet a story straight from the realm of fact," the "saga of the fur seals" was billed as a timeless story of their migration each year to the Pribilof islands, generation after generation, to bear their young, mate, and return again to the sea and faraway lands.[2]

In their search for pristine nature, naturalist-photographers and American conservationists found in Disney's True-Life Adventures a place of renewal to offset the oppressive conformist trends of an affluent consumer society and a means to increase public appreciation for wilderness areas they sought to preserve. To a wider public, Disney's nature—benevolent and pure—captured the emotional beauty of nature's grand design, eased the memories of the death and destruction of the previous decade, and affirmed the importance of America as one nation under God. The panoramic frame that marked the beginning of each True-Life Adventure set the ecological stage upon which the drama of life took place. In *Seal Island,* for example, an overview of the region's flora and fauna gloriously painted the moods and colors of Arctic spring.[3]

While Disney's True-Life Adventures revealed the purity of nature through a wide-angle lens, they simultaneously purified nature through anthropomorphic conventions that introduced familiar portraits of animal life. As the female seals arrive on shore in *Seal Island,* they are accompanied, for instance, by variations on "Here Comes the Bride." Disney, *Variety* joked, turned his audience into "peeping-tom naturalists." In *Seal Island,* the private lives of seals marked the subject of this voyeurism: from the arrival of the bulls in May to the coming of the females in June; from the contests over territories to the formation of harems; from the birth and raising of pups to the return to the sea at summer's end. Editing brought the dramatic elements of animals' private lives to the fore, but it was a story made to conform to the expectations and tastes found in the constructed family ideals of 1950s American culture. The studio carefully screened out footage deemed to offend "the feelings of women," a large portion of the audience whom they hoped to attract. Footage of baby seals trampled to death by male bulls on the Pribilof islands was left on the cutting room floor. Careful editing and narration also helped soften the violence of *Seal Island*'s climax—a fight scene between a young bachelor and an old bull. In purifying nature, Disney presented a sentimental version of animals in the wild that sanctified the universal "natural" family as a cornerstone of the American way of life.[4]

Seal Island and later True-Life Adventures had their origins in the animation classics and wartime information films of Walt Disney Studios. After the success of Disney's first animated feature, *Snow White and the Seven Dwarfs,* which grossed $8 million during its 1938 release, Disney embarked on three feature-length animation films—*Pinocchio, Fantasia,* and *Bambi*—each of which proved to be a financial failure. Five years in production and at an estimated cost of over $2 million, *Bambi*—a film about natural innocence, the renewal of life, and the alienation of humans from pristine nature told through the coming-of-age story of a young deer—lost $1 million during its first run in 1942.[5]

Based on the 1926 novel written by Felix Salten, the pen name of Austrian novelist Siegmund Salzmann, and translated into English in 1928, *Bambi* illustrated Disney's increasing preoccupation with realism in the art of animation. To portray in a natural manner deer and other wildlife central to the story, Disney hired the wildlife artist Rico LeBrun to teach art classes on animal anatomy and locomotion and built a small zoo in the studio to house two live fawns and numerous other species. Since captive animals did not give a completely accurate rendition of wild animals in their free state, Disney purchased footage from naturalist-photographers and sent his own cameraman to the Maine woods to photograph the change of seasons and the wildlife therein. Disney's ambition to approximate reality in *Bambi* led one animator to remark that "he might as well have gone out and taken pictures of real deer." It also earned him the criticism of one reviewer who scorned Disney's "tendency to trespass beyond the bounds of cartoon fantasy into the tight naturalism of magazine illustration." "The free and whimsical cartoon caricatures," the anonymous critic continued, "have made way for a closer resemblance to life, which the camera can show better . . . [I]n trying to achieve a real-life naturalism [Disney] is faced with the necessity of meeting those standards, and if he does, why have cartoons at all?" After *Bambi,* Disney followed his critic's advice and chose the camera over the animator's pen to realistically depict wildlife on screen.[6]

The high production costs of the animated features, combined with poor box-office returns, the closing of European markets during the Second World War, and the building of a new Burbank studio, placed the production company on financially hard times. Desperate for quick returns, Disney exploited the studio's experience making government information and propaganda films during the war to produce educational and industrial films contracted by corporations such as Firestone and General Motors. The low costs and rapid production schedules of information films promised fast returns on modest investments and offered a way out of the $4.3 million debt owed to the Bank of America in 1946.[7]

Dissatisfied by their lack of entertainment appeal, Disney quickly abandoned production of industrial documentaries despite their potential profits and embarked on a plan to make "sugar-coated" educational films. Intrigued by Alaska as America's last frontier, Disney in 1946 hired Alfred and Elma Milotte, a husband-and-wife team from Ketchikan, Alaska who ran a photographic studio and dabbled in wildlife cinematography, to photograph scenes of human and animal life in Alaska. The Disney studio began receiving thousands of feet of film from the Milottes, which included sequences of the salmon and lumber industry and shots of Anchorage and Fairbanks—standard travelogue fare. In 1947, Walt Disney followed his family physician's advice to take a summer respite and headed to Alaska, accompanied by his adopted daughter, Sharon. Upon his return, Walt instructed his production supervisor, Ben Sharpsteen, to make a film for theatrical release out of fur seal footage shot by the Milottes as they accompanied a government-supervised hunt of three-year-old male seals on the Pribilof islands.[8]

Neither Walt's brother Roy nor Howard Hughes, who in 1948 had acquired controlling interest of Disney's distributor RKO, were enthusiastic about the economic prospects of a True-Life Adventure. Hughes refused to distribute *Seal Island* through RKO, since he believed that audiences would not settle for a short subject in place of a second full-length feature. Hughes's response prompted Disney to arrange the two-week booking at the Crown Theater where over five thousand questionnaires were distributed to poll audience reaction to the film. Favorable audience response provided Disney with sufficient leverage to force distribution of *Seal Island* by RKO. Disney felt further vindicated when *Seal Island* won its Academy Award.[9]

Disney believed that packaging a True-Life Adventure with one of the studio's animated or live-action features would appeal to theaters with double-bill showings and would avoid the exhibition of Disney features with lesser-quality films produced by other studios. The combination of a True-Life Adventure with a Disney animated feature contributed substantially to the success of the nature film series. When *The Olympic Elk,* the fourth nature short-subject, was played with a re-release of *Snow White and the Seven Dwarfs* in 1952, a reviewer for *Natural History* remarked that to "find fantasy so real that it seems like fact, or fact so charming that it seems like fantasy is an unusual experience; but to have both of these combined in one motion picture program is very rare indeed."[10]

Distribution was not the only problem for the Disney studio in marketing nature on screen. Publicity also posed major obstacles. The lack of name-drawing stars or characters in the nature films made publicity stills an ineffective means for attracting the attention of potential audiences. The absence of advertising ploys through spin-off products, such as toys, books, popular tunes from the musical soundtrack, and other merchandise also made it difficult to pique public interest in advance of the film's release. To overcome these marketing obstacles, the Disney studio targeted a new and previously untapped audience for his series. The studio sent mass-market mailings to grade school and high-school teachers across the country announcing the release of a True-Life Adventure and the local theater where it would be shown. Educational pamphlets to accompany the films were also distributed to teachers for classroom use. Advanced private screenings were arranged with important school officials, and in the case of *Nature's Half-Acre,* the third short-subject in the series released in 1951 that depicted the interdependent web of life in a half-acre of land, a screening was arranged at the meeting of the National Education Association in San Francisco.[11]

Disney's marketing strategy for the True-Life Adventures proved effective. In 1953, dissatisfied with the distribution agreement with RKO,

Disney formed his own distribution company, Buena Vista, and released *The Living Desert.* The first full-length feature in the True-Life Adventure Series, *The Living Desert* captured the diversity of life and the struggle for survival in the desert region that lies east of the Cascade and Sierra Nevada ranges. Although some critics took issue with the seeming violence that dominated the film, this did not dissuade the public. *The Living Desert* grossed $4 million at the box-office. The following year, *The Vanishing Prairie,* the second full-length feature, reaped similar financial returns. *Cinderella* was the only other Disney film made during the 1950s to surpass the gross revenues of *The Living Desert* in its first run. With typical production costs in the neighborhood of $350,000 for a feature-length True-Life Adventure compared to the $2 million to $6 million outlays for an animated feature, the initial returns on Disney's animated classics did not come close to the profit margins the studio realized in mining the frontier of nature on screen.[12]

Seal Island was the first of many True-Life Adventures set in the Pacific Northwest, the region most frequented by Disney photographers to frame an image of pristine wilderness. This region represented nature's last stronghold, the one place in North America unsullied by humans. Alaska's Pribilof islands in fact had a long history of commercial exploitation. Fur seals on the islands had been harvested commercially since the early nineteenth century. The Milottes obtained the footage for *Seal Island* by accompanying U.S. Fish and Wildlife Service officials to the Pribilof islands to supervise the annual killing of three-year-old bachelor males for their valuable skins. But the interwoven histories of humans and animals on the Pribilof Islands would have undermined the fantasy of pure nature that Disney sought to portray.[13]

Only in the absence of humans could nature appear pristine. *Bambi* had already established this romantic theme. Scenes of the gunshot killing of Bambi's mother, the dogs' relentless pursuit of Bambi, and the fire ravaging the forest all conveyed the destructive presence of humans. But no humans ever appear in *Bambi,* reinforcing the notion that humans stand outside nature's pale. The True-Life Adventures sought to portray an innocent and timeless past, a time before nature had been tainted by the corrupting forces of human civilization. In a directive sent to nature photographers interested in selling footage, the Disney Studio emphasized that "there must be no evidence of civilization or man's work in the picture."[14] In *Beaver Valley,* the setting is a hidden valley along the Continental Divide "where time is still measured by the passing seasons, and only nature's law prevails." *The Olympic Elk,* released in 1951, brought the primeval wilderness of the Olympic peninsula to the public eye. In the 1954 feature *The Vanishing Prairie,* the place is a landscape and time "before civilization left its mark upon the land," before the "wagon trains rolled west toward the setting sun," before the prairie was "a paradise [of the Native American] to use as his own," a "time without record or remembrance, when nature alone held dominion over the prairie realm." In the scenes that follow, the audience is shown species threatened by the "relentless advance of civilization": the whooping crane, buffalo, pronghorn antelope, bighorn sheep, and mountain lion. Through the wonders of film, Disney brought to full life once again a disappearing wilderness.[15]

Disney's focus on the "timeless" frontier region of the Pacific Northwest, and particularly Alaska, as the setting for many True-Life Adventures coincided with public campaign efforts to preserve wilderness areas in the far north led by organizations such as the Conservation Foundation and the Wilderness Society. The completion of the Alaska Highway in 1948 threatened what many conservationists like Robert Marshall, founder of the Wilderness Society, had hoped in 1938 would become a permanent place to relive "pioneer conditions" and the "emotional value of the frontier." In 1956, under the sponsorship of the Conservation Foundation, the New York Zoological Society, and the Wilderness Society, Olaus Murie, together with his wife, Mardy, the young naturalist George Schaller, wildlife photographer Bob Krear, and University of Alaska zoologist Brina Kessel, spent three months conducting an extensive ecological survey in the Sheenjek Valley of the Brooks Range in northern Alaska. The information collected and the film narrated by Olaus and Mardy Murie, *Letter from the Brooks Range,* proved influential in the establishment of the nine-million-acre Arctic Wildlife Range in 1960.[16]

Throughout the 1950s, conservationists often pointed to the wilderness as the place where the frontier values of American individualism could still be found. In the Alaskan wilderness, where people were "farther removed from the central hive," Murie and his associates found "individuality and what [they] believed was a promising outlook on the world stemming from the strong life in the wilderness." To Murie, the freedom of wilderness and the individualism engendered through the hardships of a strenuous life served as important counters to mass society and the organizational man. His belief in the values of wilderness were echoed by his friend Supreme Court

Justice William O. Douglas. In 1954, at the age of sixty-five, Murie was one of thirty-seven people to accompany Douglas on his famous 175-mile hike along the Chesapeake and Ohio Canal to fight the federal government's plan to turn the canal into an interstate highway from Cumberland to Washington. Two years later, Douglas visited the Brooks Range expedition for a few weeks at the end of July. In a series of essays published in 1960, Douglas spoke of the therapeutic values of wilderness for those uneasy with the conformist trends found in an affluent, consumer culture. "The struggle of our time," wrote Douglas, "is to maintain an economy of plenty and yet keep man's freedom intact. Roadless areas are one pledge to freedom. With them intact, man need not become an automaton. There he can escape the machine and become once more a vital individual. . . . If our wilderness areas are preserved, every person will have a better chance to maintain his freedom by allowing his idiosyncrasies to flower under the influence of the wonders of the wilderness." The image of Mardy Murie in *Letters from the Brooks Range* washing clothes by hand in frigid Arctic lake water affirmed the values of individualism forged by the hardships of frontier life.[17]

The experiences recounted by Disney's naturalist-photographers framed wilderness as a place for the recovery of freedom and individual identity. Nowhere is this more evident than in the accounts written by and about Herb and Lois Crisler, pioneer photographers traversing Disney's nature frontiers, who discovered in nature an authenticity of experience that revealed the essential identity of animals and their own selves. Their films for Disney include *The Olympic Elk, The Vanishing Prairie,* and *White Wilderness.*

The Crislers' careers as wildlife photographers began when Herb Crisler, an avid sportsman, waged a five-hundred-dollar bet with the *Seattle Times* in 1931 that he could travel across the wilderness of the Olympic Peninsula with nothing but a pocket knife and come out, thirty days later, fat. The story was reminiscent of a similar incident seized upon by the press in 1913, when a man by the name of Joseph Knowles ventured naked into Maine's north woods and went primitive for two months. Unlike Knowles, Crisler clothed himself in duck trousers and a wool shirt and carried photographic equipment, but his adventure was met with similar fanfare. Although he won the bet, Crisler came out of the wilderness thirty pounds leaner and transformed by the experience. Abandoning the gun for good, he picked up the camera in the hopes of defending wildlife and the last remnants of wilderness. His wife, Lois Crisler, left her position as instructor at the

University of Washington, and the couple ventured into the Olympic mountains and the Colorado Rockies to photograph wildlife.[18] Upon seeing their first film, *The Living Wilderness,* Olaus Murie wrote to Disney describing it "as the most beautiful picture on an outdoor subject" he had seen. "Here is a real life, real wilderness, film," he told Disney, "produced by a rare combination of wilderness understanding and imagination, that seemed to me to have many of the qualities of *Fantasia.*" Murie encouraged Disney to pursue the theme of nature's aesthetic grandeur in future True-Life Adventures, and the Disney studio contracted the Crislers as wildlife photographers.[19]

In her 1958 book *Arctic Wild,* Lois Crisler describes the couple's experiences during eighteen months spent in the remote regions of the Brooks range, with a winter interlude in Point Barrow, Alaska, to photograph wildlife, particularly caribou and wolves, for Disney's feature film *White Wilderness.* The book is both a story of wilderness survival and an ethological study of wolves. In fact, Crisler portrays her gradual understanding of wolves as dependent upon her own discovery of freedom and independence.

Before departing for the Brooks range, Herb Crisler made arrangements with a bush pilot to capture a litter of wolf pups that the couple would raise and film during their stay in the remote wilderness. In July, two pups were transported to their summer camp, and Lois Crisler's journey into the world of another species began. At first, Lois was unable to free herself from the "filaments dragging [her] heart back toward civilization." She was unable to see wolves as anything other than ferocious animals, an understanding that reflected the self-righteous fury, anger, and nervousness of humans towards wolves. Immersed in wilderness, however, she gradually realized that "[p]eople are free and of equality in a wild sunlit wood." "No man is slave, no man is master, facing the sunlight on wild wood and wild fur and eyes. Liberty seeps like health into your heart . . . [T]hese past months, I had for the first time been exposed to genuine freedom." In the wilderness, where each creature was free to pursue its own destiny, Crisler discovered that wildness was not ferocity, but "independence—a life commitment to shouldering up one's own self." Only in the complete freedom of wilderness, Crisler argued, could one come to know one's own true nature and that of another. To truly perceive the wolf as a "wild and free being" marked "the break-through out of anthropomorphism." Lois Crisler experienced the wolf's "selfness" as a flash of creative awareness realized through understanding and love. It was an exchange of pure emotion, made possible by the abandonment of security for freedom, an act that Herb Crisler understood to be as vital a part of human nature as wild nature.[20]

If Disney's naturalist-photographers emphasized the values of individualism and freedom in coming to know wildlife on intimate terms, Disney himself emphasized the photographer's craft knowledge and the ways in which nature was equally accessible to the public. Middle-class Americans bought into the Disney version partly because of Disney's open disdain for elitist views of mass society. Disney did not condescend to the public, but instead viewed them as "deserving of lavish attention."[21] Disney's populist leanings spilled over into the making of the True-Life Adventures and help explain the enthusiastic reception of his nature films by the public and scientists alike. In filming nature, the Disney studio drew upon a wide range of talents, social backgrounds, and educational training for its photographers in the field. Lloyde Beebe, for example, a photographer for Disney's *The Vanishing Prairie* and *White Wilderness,* had no formal scientific training, but the craft knowledge he acquired growing up in the backwoods in western Washington proved extremely valuable to the Disney studio in securing footage of cougars and wolverines. Herb Crisler's understanding of nature derived from years spent in the woods as an avid sportsmen, while photographers such as Tom McHugh and Arthur A. Allen were professional biologists with doctorates in wildlife biology and ornithology, respectively.[22]

Disney emphasized that these photographers were "naturalists first." Under this category, he included "scientists, teachers, Park Rangers, and reformed hunters" who shared in a love of nature.[23] What united these individuals was not a common theoretical framework—they were sent into the field with no script in hand—but a patience and passion for watching nature. In refusing to distinguish between folk authority and scientific expertise, Disney privileged an experiential knowledge acquired not through academic training but through labor in the field.[24] "We find more and more that there's an awful lot of nature lore that gets preserved . . . even though it's false," remarked James Algar, director of many True-Life Adventures. Such errors occurred, Algar reasoned, because these experts spent more time digesting each other's articles than living with animals in the wild.[25] For Disney, nature was not only a space accessible to a select group of experts, but was open to anyone who had the yearning and diligence for understanding.

Disney boasted of the many discoveries his naturalist-photographers made—exultations that were not completely unfounded. When Julian Huxley organized a Royal Society symposium, "Ritualization of Behaviour in Animals and Man," he wrote to Disney to obtain rare footage of the vertical "racing" of the Western Grebe and the courtship display of scorpions.[26] But it was the contingent, not the repeatable and predictable, that Disney emphasized most. "We hope [the naturalist-photographer] will capture those unexpected and unpredictable happenings that cannot possibly be written into such a story ahead of time," remarked James Algar. "These are the little episodes that even naturalists may see only once in a lifetime." Such scenes, like two foxes frolicking on a snowbank, were what photographer Lois Crisler referred to as "nuggets"—rare action sequences that revealed the individuality of animals in the wild.[27]

Through the use of similar techniques adopted in his animated cartoons, Disney sought to capture the "seeming personality of an animal" in order to help the audience "sympathize with it and understand its problems better."[28] Disney told Olaus Murie how he loved to observe squirrels for hours at his desert home—"to me they have personalities just as distinct and varied as humans." Murie agreed. "All Nature," he wrote Disney, "has much in common among its various forms; certain general laws, certain general reactions, and much that can be predicted under many circumstances. But, and I hope this is not too paradoxical, there are many distinct facets that have individuality."[29]

It was this individuality in nature that naturalist-photographers and Disney sought to project on screen. Jack Couffer, a freelance wildlife photographer for many True-Life Adventures, believed that every animal possessed "some quirk of personality." The task of the photographer was "to find each animal's eccentricity and to somehow exploit it and incorporate the individualism into the story."[30] The nature films of Disney were reminiscent of efforts undertaken by artist-naturalists of the nineteenth century such as John James Audubon, who sought to bring an expressive, emotional quality to wildlife painting that elevated the "uncommon or unique event over the repeatable."[31] These were the contingencies of nature, the "nuggets," that Disney photographers sought to capture. Not surprisingly, James Algar often found inspiration for his True-Life Adventures scripts in the writings of the early twentieth-century nature writer Ernest Thompson Seton, who emphasized the "personality of the individual . . . rather than the ways of the race" in crafting animal stories.[32]

Since Disney believed animal behavior revealed the "instinctive beginnings of the deepest, most basic human emotions," eliciting those emotions became an important means for getting the audience to identify with animals on screen.[33] Musical accompaniment was a critical component of the True-Life Adventures, because the emotional motifs of theme music provided continuity, while variations on the major musical themes were synchronized to the actions on screen. This technique, borrowed from animated cartoons, added personality to the individual characters. *Beaver Valley,* a True-Life Adventure undoubtedly inspired by Sally Carrighar's 1947 book *One Day at Teton Marsh,* relied on a musical score to capture the intricately related lives of animals in and around a beaver pond on the Continental Divide.[34] The music was composed around four main themes. The "nature theme," as described by James Algar, was meant to convey a feeling of both pastoral nature and the expansiveness of the outdoors. Variations on the score were played as animals such as the chipmunk, moose, marmot, and raccoon appeared. The "beaver theme" captured the persistent industriousness of the beaver's personality through an even tempo progression, and colorations were added to reflect the beaver's activities through the changing seasons. The other main character, the coyote, required a menacing musical accompaniment. Scenes of otters at play, sliding down a snowbank while the beaver works fervently to prepare for the winter, are accompanied by music that reminds one of a circus. This was an instance where the music for a True-Life Adventure was appropriated by the public. The UCLA marching band adopted the piece as their entrance march at football games, and a popular tune, "Jing-a-ling Jing-a-ling," was made by writing lyrics to accompany the instrumental score. But the most memorable comic scene of *Beaver Valley* for many reviewers was the Frog Symphony, where the "Sextet from Lucia" accompanied the croaking of frogs in spring.[35]

While the music added a human emotional dimension to the personalities of individual animals, those involved with the making of the True-Life Adventures objected to criticism that they had anthropomorphized nature. James Algar defended the technique, since it permitted the "audience to identify with the creatures." The photographer Jack Couffer also saw anthropomorphism as a helpful means for understanding. "Since no one *knows* what an animal thinks," wrote Couffer, "what an animal does must be interpreted—put into human terms—for us to understand."[36]

If critics objected to the anthropomorphic conventions found in the True-Life Adventures, it was only because Disney claimed the pictures to be real. In 1954, the famous *New York Times* film critic Bosley Crowther, who in previous reviews praised Disney's True-Life Adventures, took issue with the films. The presentation of "beavers and otters in *Beaver Valley* as though they were scampish little animated animals right out of *Bambi* or *Snow White and the Seven Dwarfs,*" Crowther argued, was a "synthetic reconstruction of nature . . . passed off as real." Crowther may have had clear guidelines for distinguishing between real and contrived nature, but even among naturalists the distinction between artifice and authenticity was not as simple as Crowther's review led one to believe.[37]

For many of Disney's photographers, authenticity emerged from their lived experience with animals in the wild. Although many of the scenes were filmed by putting animals in controlled conditions, the photographers did not believe them to be less authentic. Staging was a legitimate film technique, provided the scene was true to events either documented or deemed plausible by naturalists in the field. The Milottes, for example, had no qualms about using a wire-cage enclosure with a glass barrier separating a coyote and a beaver to obtain footage of the two animals in the same frame while shooting scenes for *Beaver Valley*. What they objected to was the claim by the Disney Studios that the films were entirely "unstaged and unrehearsed." After the 1951 release of *Nature's Half-Acre,* the public similarly took offense at Disney's claim, since many close-up insect scenes and time-lapsed photography sequences were obviously achieved under contrived conditions.[38] In later True-Life Adventures, the caption was rephrased to read: "In the making of these films, nature is the dramatist. There are no fictitious situations or characters." This statement helped assuage further criticisms. When a scene of a Clark's Nutcracker catching a mouse in the 1953 True-Life Adventure *Bear Country* was identified as staged, for example, Robert Cushman Murphy, an ornithologist at the American Museum of Natural History, stated that there was "little, if any affront to the literal truth" in this obviously contrived scene, since mice were known to be a part of the natural diet of this bird species.[39]

While professional naturalists occasionally objected to the cute commentary or criticized scenes such as the woodcock strutting to samba music in *Nature's Half-Acre,* most striking is the extent to which naturalists and conservation organizations endorsed these nature films. Disney's True-Life Adventures won the respect and admiration of the professional biological community for their aesthetic appeal rather than for their factual content. Anthropomorphism was just one means by which Disney captured the emotional elements of nature, and it was his ability to touch the emotions of viewers that biologists appreciated most. "To employ film and camera with the comprehension of the very spirit of nature behind all its exciting activities is in itself an art akin to the arts of animation," wrote Disney.[40] Viewed from the perspective of art, biologists found much in Disney's nature films to praise. Robert Cushman Murphy of the American Museum of Natural History summarized the opinion of many biologists in his review of *Water Birds,* a True-Life Adventure released in 1952 with the live-action feature *The Story of Robin Hood,* when he wrote: "A foremost aim in our branch of education is to instill a love of nature that will redound to its appreciation and protection. There is no better way to accomplish this than by taking advantage of aesthetic opportunities. This Walt Disney has done supremely well in this film. The product is even more important because of its emotional effect upon the observer than because of what it teaches in the field of ornithology. It is primarily a work of art."[41]

The cooperation of the Cornell ornithologist Arthur A. Allen with the Disney studio in supplying sound recordings and bird footage partly explains how the Disney studio gained the support of professional biologists in the interests of conservation. As director of the Laboratory of Ornithology at Cornell University, Allen was one of the leading ornithologists in America. In association with his colleague Paul Kellogg, he had developed sound equipment for the recording of bird song in the field. Through his phonograph records and popular books, he helped bring the voices, photographic portraits, and habits of wild birds to many American homes. In the early 1950s, he became a popular lecturer on the Audubon Screen Tour circuit, showing his film *Hunting with a Microphone and Color Camera* to audiences nationwide.[42]

During the production of *Beaver Valley,* the Disney studio contacted Allen in the hopes of using parts of the recordings from the album "Voices of the Night" to dub with shots of frogs. Allen thought it unlikely that the studio would find the recordings useful, since many of the frog sounds on the records were of species that only inhabited the southwestern United States. The lack of attention Hollywood filmmakers paid to making sounds authentic irritated Allen, who grimaced at hearing "Western Meadowlarks singing in *Drums Along the Mohawk* and California Wren Tits, which are found only on the Pacific Coast, singing in *Northwest Passage.*" Allen became apoplectic when he received the species list of frogs to appear in *Beaver Valley.* "You might as well plant orange trees and cypress around your beaver pond as introduce Bird-voiced Tree Frogs and Green Tree Frogs with a picture set in the Pacific Northwest," exclaimed Allen. "Indeed of all the amphibians you list," Allen continued, "the Fowlers Toad and the Bull Frog are the only ones that get as far west as Southeastern Oregon and the latter only because it has been introduced." If the Disney studio was really interested in making the film authentic, Allen would cooperate. Otherwise, he was not interested. The studio assured Allen that every effort would be made to make the films "as factual as possible." Any deviations made were "in the interests of dramatics or humor, and this only because the films are for theatrical release." The studio agreed to delete all the frog species not native to the Pacific Northwest and urged Allen to reconsider. Furthermore, they pointed out that the True-Life Adventures were meant not only to entertain, but to "stimulate interest in the value and preservation of natural resources and wildlife." It was this statement that convinced Allen to cooperate, and he contributed extensive footage and sound recordings for later films.[43]

Conservation organizations such as the Wilderness Society and the Audubon Society fully supported Disney. In the Wilderness Society's publication, *The Living Wilderness,* Olaus Murie praised Disney's ability to convey "the simple beauty of untouched woodlands and their wild inhabitants." Another reviewer in *The Living Wilderness* hailed Disney as an ally of conservation and likened him to "a sun ripening the grain for wilderness advocates to harvest!" In 1955, the Audubon Society presented Disney with the Audubon Medal, awarded "to a man of eminence, who has rendered distinguished service to the cause of conservation." In the presentation speech, Chairman of the Board Ludlow Griscom stated that through his True-Life Adventures, Walt Disney, "aware of the universal love of nature, . . . has provided thrilling entertainment of educational quality, has demonstrated that facts can be as fascinating as fiction, truth as beguiling as myth, has opened the eyes of young and old to the beauties of the outdoor world and aroused their desire to conserve priceless natural assets forever." But the words of Richard H. Pough, head of the American Museum of Natural History's Department of Conservation, were most telling. "If the areas we as a nation have set aside to be preserved as wilderness can continue to provide settings for pictures like these," wrote Pough, "it will go far to answer the objections of those who claim that only one person in a thousand will ever be able to visit and enjoy them."[44]

Disney was just the ally conservationists needed. The wilderness experience Disney provided to the masses dispelled accusations that preservation of nature benefited only those with the money, leisure, and physical stamina to experience nature for themselves. The photographers Disney sent into the field represented a mix of educational, economic, regional, and social backgrounds: a representative microcosm of groups within American society whose public support would prove increasingly important in the political campaigns of conservation organizations. In the hands of Disney, wilderness appealed not just to biologists interested in the preservation of pristine areas for field research. Disney extended the reach of wilderness beyond the limited membership of conservation groups such as the Wilderness Society or Sierra Club to include many middle-class Americans who perhaps never would have the resources or interest to venture to the last remnants of nature free from the influences of modern life. The True-Life Adventures made wilderness available to the public without any threat of despoiling the natural areas that environmentalists prized and wished to preserve from the onslaught of mass tourism and recreation. Disney not only captured the aesthetic beauty of nature, he transformed it into a commodity with a set of values pertaining to democracy and morality that appealed to the American public. His work softened the elitist tone of environmental organizations, such as the Wilderness Society, that feared democratic mass society would threaten wilderness and reduce it, in the words of Murie, to "the commonplace."[45]

Environmental organizations may have appropriated Disney for their own ends, but behind Uncle Walt lay a major corporation that contributed substantially to the creation of the very mass society that leaders of the conservation movement assailed.[46] Murie himself expressed concern that the pressure of commercialism and mass production would undermine Disney's early achievements. "There is a danger of falling into a routine technique of preparation and presentation, that deadly monotony and commonness—like certain restaurant fare that seems to have all been cooked in the same pot," he warned Disney. "Nature on the assembly line is going to suffer in the end product without great care, and the individuality of the different pictures may be dulled by too great haste in preparation."[47] It was only a few short years until Disney fell from grace among environmentalists. In 1966, when Disney revealed his plans for the Mineral King project, a $35 million ski resort in northern California with an access route through Sequoia National Park, it became apparent to them that commercial profit had always been a driving force behind Disney's interest in the exploitation of nature, either on or off screen.[48]

While Disney's naturalist-photographers ventured to the last remnants of wilderness to capture nature on screen, viewers of Disney's True-Life

Adventures had to venture only to their local theater to partake in nature's aesthetic grandeur. When Disney entered television with his show *Disneyland* in 1954, young World War II veterans and their families could partake of the same without leaving the material and moral comforts of their suburban home. During *Disneyland*'s first season, episodes included *Nature's Half-Acre* and *Seal Island.* The release in 16-mm format of select Disney pictures made other True-Life Adventures available for home use. In 1955, Cecile Starr informed the readers of *House Beautiful* that the availability of *Seal Island* and *Beaver Valley* were the "two most important reasons why no home should be without a 16-mm sound projector." For the rental price of only ten dollars, the whole family, "from gurgling baby right up to doting grandparents," could "obtain a lifetime of pleasure" by traveling to a world that few would experience "firsthand."[49]

Disney entered television principally to obtain the financial resources to complete his dream of a theme park that would avoid the tawdry, libidinous, carny atmosphere of early twentieth-century lower-class amusement parks. Built in reaction against places like Coney Island, Disneyland represented the clean, wholesome patriotic and educational values that allowed parents and children, teachers and pupils to revisit the past and dream of the future.[50] This nostalgic vision is particularly evident in his three-part series *Davy Crockett* that aired during the first season of *Disneyland,* ABC's biggest hit television show. It was a public sensation. The series theme song, "The Ballad of Davy Crockett," escalated to number one on *Hit Parade,* where it remained for thirteen weeks. More than 10 million children rushed to buy Davy Crockett coonskin caps, which were later made of Australian rabbit and mink when the supply of raccoon skins became scarce due to the unprecedented demand. In Disney's version of America's mythic past, the wild frontier gave rise to a fiercely individualistic and democratic nation, but it was also a land tamed by the institutions of family and religion that had accompanied the Western expansion of the pioneers. Disney's Crockett, the buckskin-garbed Fess Parker, is a defender of individual autonomy, the family, and populist democracy, a man who listens first and foremost to the higher law of God, and someone whose democratic values and ideas come from his experiences in the wilderness.[51]

While *Davy Crockett* reinforced the values of a family-oriented, God-fearing people on *Disneyland,* the True-Life Adventures similarly suffused nature with a morality that universalized the institution of family and revealed the hand of God throughout the animal kingdom. The proliferation of suburbs in America during the postwar years, where growth outpaced that of urban centers by a factor of ten, provided a safe haven for domestic life in 1950s America. Although Disney claimed to "avoid drawing moral conclusions" in his "camera account of Nature's works," nature in the True-Life Adventures, reinforcing the moral values of family life, traditional gender roles, and the containment of sexuality within the home, was clearly meant to imitate suburban life.[52]

For the young American families that ventured into True-Life Adventureland, their experience of Disney's wilderness could be compared to their encounter with nature framed within the suburban setting of their ranch-style homes. The low-pitched eaves, simplicity, and emphasis on continuity with the outdoor surroundings tangentially linked the ranch-style home with the organic, prairie-style architecture of Frank Lloyd Wright. The most noted feature of the ranch house was the use of large insulated plate-glass picture windows and sliding glass doors. Just as a True-Life Adventure established the audience as a spectator of nature, so the windows and sliding glass doors of the ranch-style home, which opened onto the same level as the surrounding tame and pastoral landscape, facilitated intimacy with nature through observation rather than active participation. These glass walls, like the television screen, nevertheless marked a clear boundary between the world of humans and that of nature. If nature had become domesticated in the suburban home, it had also been domesticated by Disney for the family audiences who lived on the edge of the "crabgrass frontier."[53]

The middle landscape of the American suburbs—as much as wilderness—became an important place of environmental concern and support in the 1950s. The first serious legal challenge to the spraying of DDT, for example, took place in the suburbs of Long Island, where noted conservationist and ornithologist Robert Cushman Murphy led local citizens in an unsuccessful fight to obtain a permanent injunction against the spraying of the insecticide in their beloved neighborhood. Disney tapped into the market for suburban nature with his 1951 True-Life Adventure, *Nature's Half-Acre*. Set in a familiar plot of ground—a country field, abandoned orchard, or backyard garden—*Nature's Half-Acre* won both an Academy Award and the annual motion picture accolade for outstanding contribution by the National Association for Conservation Education and Publicity.[54]

Although a half-acre was at least three to five times the average lot size of the suburban home, the wildlife featured in *Nature's Half-Acre* are familiar reminders of spring and summer and of family to those viewing nature through the sliding-glass doors of their ranch-style homes. Within the themes of seasonal change and the balance of nature lie the moral lessons of family life that accompany the symphony of spring. Nest-building constitutes the real work of the season, and we are witness to the varieties of materials used and shapes of nests built by the inhabitants. With a safe haven established, the expectant couple is soon busy feeding the newly hatched young. Shots of "mother" and "father" of different bird species follow: grosbeaks, robins, woodpeckers, hummingbirds, waxwings, and cardinals attentively care for their young. "In nature's half-acre," we are told, "mother love is expressed in patience and devotion. Be it fair weather or foul, mother always stands by." But the gay and cheerful music shifts to a more somber, intimidating tone as a new species appears on screen: the cowbird. "This heartless creature lays her eggs in another bird's nest," the narrator states with scorn, "and then flies away, never to return." In *American Magazine,* advertised as "the magazine the whole family can share," Disney referred to the cowbird as "one of Nature's worst bums and free loaders," a species that did not conform to the family devotion and parental care he believed existed throughout the animal kingdom.[55]

If *Nature's Half-Acre* reinforced the domestic ideology of the family by revealing its central place within the morality of nature, it also bolstered religious values through its adherence to natural theological explanations of the balance of nature. Although species went extinct in Disney's cinematic nature, the origin of species remained vague. No mention of evolution ever appears in the True-Life Adventures. Instead, *Nature's Half-Acre* lends support to a theory of divine creation, although the writers did not endorse the strict creationist views of the interwar fundamentalist George McCready Price, whose adherence to a literal interpretation of Genesis, which reduced the history of life on Earth to approximately six thousand years, served as the inspiration for a creationist revival in the 1960s that continues unabated today. The creationist views alluded to in *Nature's Half-Acre* are more in keeping with the outlook of the American Scientific Affiliation (ASA), an evangelical organization of scientists established in 1941, whose leaders rejected Price's strict creationism in favor of a theory of organic development facilitated by divine intervention.[56] The ASA had close ties to the Moody Institute of Science (MIS), an educational outreach organization of the Moody Bible Institute. Headed by Irwin Moon, the MIS was devoted to the production of educational film with the purpose of demonstrating that "the wonders of science are but the visible evidence of a Divine Plan of Creation."[57] This outlook was in keeping with Disney's own view that in the filming of nature one was "in a personal way . . . initiated into a sphere where God's master plan for the existence of this planet is dramatically enacted every second of the day."[58] Such sentiments may have prompted Disney to hire Tilden W. Roberts, associate professor of zoology at the University of Southern California, as a biologist-consultant for *Nature's Half-Acre.* Roberts also worked as a consultant for the Moody Institute of Science, and the time-lapsed photography of plant development in *Nature's Half-Acre* is reminiscent of the first film the Moody Institute of Science produced in 1946, *God of Creation,* that used time-lapsed photography to reveal "evidence of a Divine plan in the universe."[59]

The overarching theme of *Nature's Half-Acre* dates back to the natural theological treatises of the eighteenth and early nineteenth centuries that found in nature evidence for God's grand design. Although violence is found in this backyard garden, revealed by shots of spiders killing their prey and birds devouring caterpillars, everything is for a purpose. Teleological explanations are writ large throughout *Nature's Half-Acre.* When Paul Kenworthy, a doctoral student in biology at UCLA and a Disney photographer, inquired about the teleological approach in *Nature's Half-Acre,* the producer Ben Sharpsteen responded that "to make the film otherwise would have provoked the fundamentalists, and that even an atheist would be amazed about the wonder in the balance of nature." The apparent violence and death are reconciled by the narrator's remarks that "nature is concerned not with the individual but with the preservation of the species. So some must die that others may live. In this way she keeps her world in balance, and makes it a place of order and beauty."[60]

Nature's Half-Acre is closely wedded to a Linnaean notion of the balance of nature where each species keeps another in check under the benevolent yet retributive eye of God. In the final segment of the film, the links between the book of nature and the book of God are made explicit, as winter releases its icy grip and "the re-awakening of life" begins. Time-lapsed shot after shot of leaf buds and flowers opening permit us to observe, in the narrator's words, "miracles the human eye alone could never see. In nature's book of wonders, this is the chapter of genesis."[61]

Disney's alliance of family and religious values in *Nature's Half-Acre* created a foundation for American democracy based on nature and the natural order. Nature thereby legitimated the family and church as the central pillars of democracy during the Cold War. From 1950 to 1960, church membership increased by an estimated 14 percent. President-elect Eisenhower's declaration in 1952 that the foundation of democracy rested in a "deeply felt religious faith" based in Judeo-Christianity indicated religion's integral role in containment culture. Indeed, political leaders like Eisenhower and prominent evangelical leaders such as Billy Graham were united in their belief that the prevailing distinction between democracy and Communism was one based on religious faith. Congressional approval of a change in the Pledge of Allegiance in 1954 to read "one nation *under God*" reaffirmed the belief expressed by Louisiana Congressman Overton Brooks that religious faith "was the one thing separating free people of the Western World from the rabid Communist."[62] Convinced that a 1941 strike at his studio had been the work of Communist agitators, Disney became an FBI informant and vice-president of the Motion Picture Alliance for the Preservation of American Ideals, a conservative organization devoted to "combat the film industry's domination by Communists, radicals, and crackpots." In this Cold War climate, Disney offered a persuasive vision of the American way of life rooted in individualism, traditional family values, and religious morality.[63]

Even Disney, however, who had a keen sense of the desires of middle-class American culture, did not always correctly assess public sensibilities. Although *The Living Desert* was a box-office success, some reviewers agreed with the movie critic Bosley Crowther that the endless scenes of mortal conflict, such as that between a red-tailed hawk and rattlesnake, or a wasp and tarantula, placed far too much emphasis on the struggle for survival in nature. In their successful adaptation to some of the most inhospitable landscapes on the North American continent, animals in their struggle for survival might reaffirm American values of competition, individualism, and industriousness, but not if that struggle entailed violence.[64]

Even though Disney was careful never to allow the death of an animal on screen with whom the audience was meant to identify, "the repetition of incidents of violence and death," remarked Crowther, "eventually tends to stun the keyed-up senses." Successful nature writers such as Sally Carrighar, Rachel Carson, and Joseph Wood Krutch had softened the violent hues of nature's palette in the postwar years through narratives that drew attention to the interdependent web of life. Americans had witnessed enough death and destruction in the Second World War. Olaus Murie criticized Disney's desert portrayal precisely because it presented "a 'feeling' of the desert" that "was one-sided." "There was much killing and harshness . . . There was not the *balance* with the beauty of the desert," Murie argued.[65]

In *The Vanishing Prairie*, released one year after *The Living Desert*, the "conflict and the rawness of nature" had been muted, but Disney nevertheless came face to face with the state censorship boards. In 1954 he was asked by the both the New York State and Maryland State Board of Motion Picture Censors to delete twenty-seven feet of film showing the birth of a buffalo. Although the New York censorship board retracted its request, the incident demonstrates to what extent nature was appropriated to support the domestic ideals of 1950s America. Disney shunned the graphic violence of lions maiming humans and allusions to primal sexuality common in 1920s travelogue-expedition film. Explicit scenes of predators killing prey were rare in the True-Life Adventures, and footage of animal copulation was nonexistent. Footage of courtship displays was considered enough to presage the act of sex. Courtship implied marriage in human terms and thus sexuality was conveniently contained. But in discreetly exposing female genitalia in the birth of a buffalo calf, Disney had stepped dangerously close to the edge of moral indiscretion. Nature had to be domesticated, but as the repeal of the New York board's decision made evident, this was a case where the government and public agreed that accommodating nature to the moral values of family life had perhaps gone too far.[66]

In recreating a vision of America's wilderness, Disney's True-Life Adventures opened a new frontier for exploitation and conquest. Mining the frontier of nature on film had existed as an extractive industry since the beginnings of cinema, but only on the scale of the individual entrepreneur, which in no way matched the financial resources, personnel, or marketing savvy of a corporation like Walt Disney Productions. The True-Life Adventures helped establish a mass market for nature as entertainment in the postwar years and cultivated an appreciation for wilderness as a source of aesthetic value beyond the limited membership of conservation organizations within the United States. In bringing nature to the masses, Disney also established film as an important propaganda tool in the enlisting of public support for environmental causes. Or so environmental groups first thought. In their rush to embrace Disney as a champion of conservation, organizations such as the Wilderness Society or Audubon Society failed to appreciate that while nature films could help rally support and mitigate the elitist overtones of wilderness preservation, they did so at a cost. Disney never side-stepped the fact that his nature films were meant as entertainment. And in framing nature as entertainment, the True-Life Adventures reinforced a tourist and recreational economy that placed a much greater demand on the very areas that conservationists were trying to protect from the influx of people and the values of consumer society.

This predicament had its origins in the dichotomy between nature and civilization that both Disney and conservationists sought to maintain as a symbol of American democracy and its exceptionalism within the history of the free world. Preservation of pristine wilderness in places like Alaska or on the motion picture screen served as a continual reminder of the free and fertile ground upon which the characters of individualism and egalitarianism supposedly took root. It was a therapeutic reminder for a culture stifled by an ideology of containment and disquieted by conformist trends within a consumer and corporate society. Disney's nature offered an escape, but it was not a release of wild abandon. Audiences found on the motion picture screen a nature tempered by an admixture of family and religious values, a sentimental nature for all ages. Although a few True-Life Adventures made their way onto the television screen and 16-mm rental films, the primary venue for watching Disney's nature remained the theater. The production times and costs typical of a True-Life Adventure could not meet the budget and scheduling demands of the television format. Outside the influence of Hollywood, individuals like Marlin Perkins would make more effective use of television in domesticating nature on screen.[67]

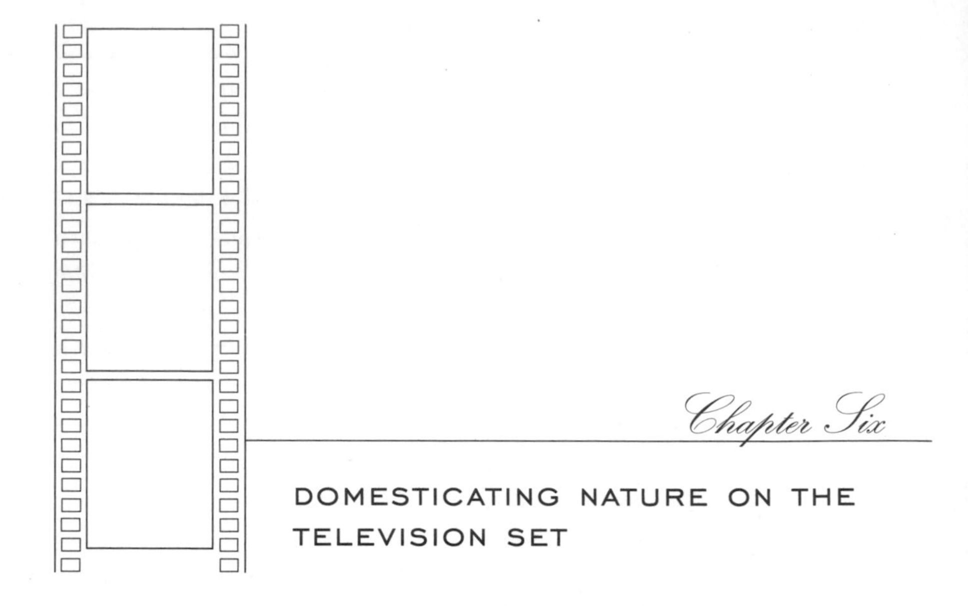

Chapter Six

DOMESTICATING NATURE ON THE TELEVISION SET

In 1945 Marlin Perkins loaded some of the more popular small animals from Chicago's Lincoln Park Zoo in his automobile and drove downtown to the studio of WBKB, the city's experimental television station. Recognizing that the "lifeblood of the zoo is publicity and promotion," Perkins looked upon the new medium of television as a way to revitalize the Lincoln Park Zoo, which he took over as director in 1944. Perkins's flair for showmanship went back to his earliest days at the St. Louis Zoo, where he got his start sweeping sidewalks for $3.75 a day. Convinced that a real understanding of animals could not be taught in the classroom, but had instead to be experienced firsthand, Perkins left the University of Missouri at Columbia as a sophomore majoring in zoology in 1926 and placed himself at the services of George Vierheller, director of the St. Louis Zoo and one of the premier zoo showmen in the United States. Within a matter of weeks, Vierheller promoted Perkins to curator of reptiles and the understudy quickly learned from the "maestro" the art of public relations in attracting crowds. Perkins orchestrated public performances every other week in which two reticulated pythons, the Maharanee of Wangpoo and Blondie, were each force-fed fourteen pounds of rabbit meat using eight men and a two-and-a-half-inch rubber pumping hose inserted into the python's throat. Advertised in advance in local newspapers, the feedings attracted thousands of visitors and were featured in the pages of *Life* magazine. Neither Perkins's theatrical verve nor the animals, however, adapted readily to the demands of television in its early years. Agitated from the car ride and by the intense heat of the studio lights, the animals were unpredictable as Perkins handled them and attempted to deliver off-the-cuff remarks on their adaptations and habits. With only three hundred televisions in the entire city of Chicago, Perkins called it quits after fifteen shows.[1]

Five years later, television's prospects had improved significantly. In 1949, completion of the first coaxial cable between Chicago and New York City linked Midwest cities with the Eastern Network. By 1950, 9 percent of American households, roughly four million homes, owned a television set. Reinald Werrenrath, a producer at the NBC affiliate in Chicago, approached Perkins in 1949 to inquire whether he would be interested in a one-half-hour daytime program broadcast live from the Lincoln Park Zoo. Perkins seized the opportunity. Just a few years earlier he had tried unsuccessfully to get WBKB to shift its broadcast from the studio to the zoo. With a mobile unit, Perkins felt he could capture the diversity of life at the zoo, and that the animals in their familiar surroundings would be more at ease performing before television cameras and microphones. The NBC station sent a mobile television camera crew and a former Marine combat correspondent, Jim Hurlbut, as an announcer to open and close the show. *Zoo Parade* premiered on the NBC network on May 28, 1950 and quickly rose to one of the top three programs in daytime telecast ratings. One year later, the show was broadcast to over forty-one cities coast to coast. By 1952, when television reached into one-third of American households, an estimated 11 million people tuned into the show on late Sunday afternoons. Surveys revealed an audience composition of equal proportions of men, women, and children. Offering wholesome entertainment for the entire family, Marlin Perkins and his animal friends became television celebrities, recognized by people all over the country. Attendance at the Lincoln Park Zoo soared to over 4 million in 1952, as crowds flocked to see their favorite animal celebrities such as Judy the elephant, Fuad the fennec, Sinbad the gorilla, Nero the lion, Sweet William the skunk, and the chimpanzee Heinie

II. Many of the animal residents at the Lincoln Park Zoo had become adored pets with whom families regularly fraternized on Sunday afternoons in the comforts of their living rooms.[2]

The popularity of *Zoo Parade* traded upon the domestication of television in American homes. In the early 1950s, magazines such as *House Beautiful* and *American Home* helped tame this new technology by welcoming television as the newest family member, oftentimes described as "the family pet." In fact, young couples added babies and televisions as newest members to their homes in almost equal numbers in the decade after the Second World War. Over 38 million babies were born in the United States between 1946 and 1955. Nearly 41 million television sets were manufactured and sold during the same period. The acculturation of television in postwar American society and the constructed ideals of family life went hand in hand.[3] Appealing to animals as a source of family entertainment and education, Perkins succeeded in gathering a large television audience by turning wildlife into domestic pets. By 1957, the shift away from live television and the rise of action-adventure series led to the cancellation of *Zoo Parade*. In the last seasons of the show, however, Perkins and his producer Don Meier began experimenting with a new format that exploited television's appeal as a window on the world. Travelling to exotic locations, Perkins would bring the life of an adventurer working with wild animals and a message of conservation to millions—America's baby-boom generation and their parents—every Sunday afternoon. *Wild Kingdom,* which premiered on NBC television in 1963, took viewers to distant places across the globe and like its forerunner, *Zoo Parade,* tamed nature for family audiences. Syndicated in 1971, the program was broadcast to over 220 stations by 1974. With an estimated audience of 34 million viewers, *Wild Kingdom* surpassed the ratings of top syndicated shows such as *Hee Haw* and *Lawrence Welk*. To the majority of Americans surveyed in 1980, watching *Wild Kingdom* ranked second in their list of activities involving interaction with animals, just behind owning a pet. The two experiences had much in common. Both depended upon the domestication of animals in the home.[4]

In a 1954 survey sponsored by the National Council of the Churches of Christ, proponents of television asserted that the magic box was "making the home once again the center of American family life, reversing the centrifugal forces that have been dispersing the family in recent decades." Although this pronouncement ignored the ways in which the exodus to American suburbs during the 1950s disrupted the traditional multigenerational households found in urban ethnic neighborhoods, it did capture the sense in which the rapid postwar growth of television was itself nourished by and reinforced the ideals of the nuclear family and domesticity. Television manufacturers were quick to capitalize on surveys in the early 1950s that indicated that families with youngsters under the age of twelve were twice as likely to own televisions as compared with families without children. A 1949 RCA advertisement promoted a scene of family togetherness, in which mother, son, father, and daughter are nestled in a semicircle around the electronic hearth in their middle-class home watching *Kukla, Fran, and Ollie*. In 1950, when *Zoo Parade* first appeared on the NBC network, RCA advertisers could have featured Marlin Perkins and his animal friends on the television console, since the show's success also depended upon the association between television and the domestic ideal of a home filled with children as a place of material comfort, security, and hope in the atomic age.[5]

Children in 1950s advertisements represented not only the bright promise of peace and rebirth in the shadows of the atomic bomb, but also the key to personal fulfillment and happiness. Educators, psychologists, and sociologists optimistically endorsed television as a new medium that would not only bring the family together, but also offer educational enrichment in the lives of children. Such optimism was tempered by critics who viewed children as helpless victims of television, and implicated it in the corruption and debasement of the moral values and tastes of American youth. In 1954, when Senator Estes Kefauver held a series of Senate Subcommittee hearings on the rise of juvenile delinquency, 70 percent of adults questioned in a nationwide Gallup survey placed part of the blame on excessive violence and lust portrayed in comic books and television and radio crime programs. In the quest for childhood innocence, parents would have to shield their children from the corrupting forces of modern society. Like pristine nature, childhood, conceived as a time of innocence, offered a place of grace from the horrific acts of destruction and degenerative influences wrought by modern civilization. Both were hallowed spaces in American society that needed to be preserved. Much as natural history films were the respectable alternative to less wholesome films in the early motion picture industry, animal shows on 1950s television offered entertaining and educational subject matter that the whole family could enjoy. In television shows like *Zoo Parade,* the construction of a sentimental nature and childhood in postwar American society were closely intertwined.[6]

Education offered an antidote to the seductive afflictions of commercial society, but a spoonful of sugar went a long way in helping the medicine go down. Studies conducted on the television viewer in the 1950s indicated that children greatly preferred incidental learning over intentional learning from television. Although the FCC allocated licenses for noncommercial educational television stations in 1952, the perception that the prime purpose of such stations was educational could prove fatal in attracting both audiences and financial support. Like other children's programs in the early 1950s such as *Ding Dong School* and *Kukla, Fran, and Ollie, Zoo Parade* succeeded on commercial television because it was, in the words of one fan, "painlessly educational."[7] Showered with awards for best children's and best educational program in the early years of its broadcast, including the 1950 George Foster Peabody Award, the 1951 Look TV Award, the 1952 TV Guide Magazine award, and the 1954 General Federation of Women's Clubs selection, *Zoo Parade* offered basic natural history information about animals through a format that entertained audiences by drawing upon conventions of situation comedies, dramas, and quiz shows. Some elementary school teachers, such as Jeanne Gaede in Joplin, Missouri, found the program an effective means of generating classroom discussion about animals in the primary grades, and looked optimistically upon the use of educational commercial television in the schools. In 1951, the National Congress of Parents and Teachers awarded *Zoo Parade* the Sylvania Television Award for the best program suitable for children. In delivering the award, Mrs. Johnny Hays, president of the National Congress, praised the show's merits in effectively combining education and entertainment. "All children love animals," she remarked to the audience, "and these television visits through the Chicago Lincoln Park Zoo are entertaining, easy of comprehension, and most constructive."[8]

Parents welcomed Marlin Perkins into their home each week because they were assured that he would bring, in the words of Mrs. Hays, the "courtesy, consideration, and such high taste that we would expect of any visitor in any American home."[9] In the 1950s, television actors, producers, and viewers regarded the medium as the most "intimate form of mass communication." Television performers were invited family guests. Burr Tillstrom, creator of *Kukla, Fran, and Ollie,* wrote admirably of the intimate contact President Eisenhower made with viewers in his 1956 televised final pre-election speech. When Eisenhower "thanked us individually for allowing him to come into our homes," Tillstrom wrote, "I felt that he was speaking to me, and that he visualized my living room, and my family, and thought of us as his friends." Although Marlin Perkins's later bid for the presidency was restricted to a nomination by Lyle the lion in the comic strip *Animal Crackers,* television critics attributed much of *Zoo Parade*'s success to Perkins's "relaxed, old shoe approach to the animal kingdom." "Whether he's handling lions or llamas," wrote a staff writer for the *St. Louis Globe Democrat,* "Marlin Perkins never fails to seem as much at home as if he were seated comfortably in his own near North Side living room."[10] Perkins was not only an invited family guest, he also extended the homeyness of his Lincoln Park living room into television land. Jim Hurlbut, Perkins's sidekick on the show, also helped add to this cozy atmosphere with his wry humor and inquisitiveness. Representing "the average guy" next door, Hurlbut asked Perkins questions about the animals and displayed an apprehensiveness toward the pets found in Perkins's family room at the zoo, including deadly poisonous snakes such as the fer-de-lance, huge anacondas, and the unpredictable antics of more famous zoo personalities like the young gorilla Sinbad and the chimpanzee Heinie II.

The regular televised appearance of animal celebrities at the Lincoln Park Zoo added to the sense of intimacy that Perkins fostered with audiences at home. Audiences identified with animal personalities on the television screen because the individual expressions and characteristics that set these animals apart made them appear much more like domestic pets. Next to cartoons, animals were the most favored television program subjects among children through the age of ten, but animals more than any other topic had the potential of reaching across a wide age spectrum. When J. Fred Muggs, a baby chimpanzee, became a regular on the *Today* show in 1952, audiences and sponsors, who previously had shunned the hybrid newscast/variety show format, responded enthusiastically. Muggs became a national celebrity. Featured as the guest of honor at an "I Am an American Day" rally held at Central Park, Muggs was sought after by advertisers, Chambers of Commerce, and even the Navy.[11]

Chicago could boast of its own animal celebrity, the gorilla Bushman, who similarly had a national following made possible by television. Purchased for three thousand dollars in 1928 from the wild-animal dealer Pa Buck, Bushman grew in both physical size and popularity during his twenty-two years at the Lincoln Park Zoo. A regular on *Zoo Parade,* the six-foot two-inch, 550-pound Bushman commanded the attention of the public both on and off the television screen. At the time of his death in

1951, he was fondly remembered on a *Zoo Parade* episode that included footage of his last birthday party in which the gorilla gave Eddie Robinson, his keeper for twenty years, one of the celery candles off his cake. Robinson spoke sadly to viewers about how there would never by another animal to take Bushman's place in his life. To Perkins's query regarding what he would most remember about the gorilla judged by the American Association of Zoological Parks and Aquarium to be the most valuable animal in any American zoo, Robinson spoke of his "good natured, lovable, affectionate" temperament. Although Bushman had a reputation for throwing celery and grapes at some of the press photographers, Robinson insisted that it was all in good fun. "He was a different character all the time," reflected Robinson. At the end of the show, viewers were introduced to Bushman's likely successor, a three-year-old gorilla named Sinbad whom Perkins had nursed back to health during a collecting expedition to the French Cameroon in 1948. As viewers watched Sinbad on a swing in his cage, Perkins described how "the coy little fellow is always looking out of the corners of his eyes at the admiring public, wanting to make sure that they see everything he is doing. He is very friendly, quite gentle, and exceedingly affectionate to his keepers." On later *Zoo Parade* shows, Sinbad would often slap Jim Hurlbut on the head or kiss him in the eye, signs of affection that endeared him to television audiences and zoo visitors alike.[12]

Just as he encouraged audiences of *Zoo Parade* to identify with the zoo animals as pets in their home, Perkins encouraged the animal keepers at the Lincoln Park Zoo "to make pets of their animals whenever possible." In his effort to "minimize the effect of iron bars and get animals and humans acquainted," Perkins "surrounded himself with men who are good at establishing the same rapport with elephants, panthers, Patagonian cavies and such that you do with *Canis familiaris,* the domestic dog," wrote an admiring critic in *The Saturday Evening Post.* These animal caretaking practices made the television show possible and closed the gap between audiences and animals. As Perkins observed, "the animals were used to having people touch them, or pick them up if they were small, and were well acquainted with their keepers. If I had tried to develop a training program to teach keepers how to handle and move animals, I couldn't have found a better medium than *Zoo Parade.*"[13] Such techniques helped ease the nervousness of animals around the television crew, cameras, and lights on the set located in the basement of the reptile house at the zoo. Perkins seemed calm even in the presence of reticulated pythons and Yacca lizards, which crawled up his arm, and remained unfazed by the dangerous beaks of macaws and Australian leadbeater cockatoos. Not all zoo personnel exhibited such a special relationship with animals on the set, however. When Lear Grimmer, assistant director of the zoo, occasionally filled in for Perkins on *Zoo Parade,* the animals were noticeably less at ease. A badger once became quite hostile in his presence. During another episode, a small banana boa tried to strike him.[14]

Unlike Grimmer, Perkins had developed a special tacit knowledge handling animals, a knowledge acquired through experience not science. Other animal keepers at the Lincoln Park Zoo also acquired this unique relationship with animals. In a show featuring the lions of the Lincoln Park Zoo, for example, the animal keeper Willie Renner was shown playing with Dillinger, Queenie, and other big cats housed in the lion cage. Renner's ability to pet and play with Dillinger, in particular, was quite remarkable, since the lion had put its former animal trainer in the hospital during the 1933 World's Fair. When Perkins asked Renner about his special talents handling animals, Renner described himself as a "hillbilly" who grew up on a farm in southern Illinois. Perkins also identified himself as a "hillbilly," whose knack with animals developed as a boy on a small farm near Carthage, Missouri. The explicit rural, anti-intellectual connotations of "hillbilly" foregrounded experiential knowledge over scientific training. Perkins, like Disney, lessened the distance between himself and his viewers and made natural history appealing on an emotional level. The wellspring of his success was the same kind of communication needed to raise any animal as a pet.[15]

The commercial advertisements on *Zoo Parade* further accentuated the associations drawn between wildlife and domestic pets by Perkins and his fans. Family members may have thought of pets, like children, as removed from the marketplace, but the economic context of zoo animals and the pet business in America was extremely important in the success of *Zoo Parade.* Celebrities like Bushman were prized commodities for a number of reasons. Not only could these performing animals attract crowds, but since the Lincoln Park Zoo owned them, the zoo animals, unlike Hollywood animal stars like Rin-Tin-Tin, had no agents. This meant that that the show was extremely cheap to produce and represented a bargain for commercial sponsors. *Zoo Parade* cost approximately one-tenth that of family situational comedies during the 1950s.[16] And animal celebrities could sell products. The total pet population had escalated to 200 million by the late 1950s, with dogs, cats, parakeets, canaries, turtles, and tropical fish among the most favored pets owned by 56 percent of American families. In trying to attract sponsors to *Zoo Parade* and its successor *Wild Kingdom,* NBC's marketing department highlighted the "pre-selected mass market" of pet owners that constituted a "natural audience" for the show. Sponsored locally in Chicago by Jewel Food Stores and by Ken-L-Ration products on the national NBC network until 1955, *Zoo Parade* allowed advertisers to tap into the market of animal lovers in America. Jewel Food Stores launched a highly successful promotional campaign tied into the show by including with each purchase cards that featured a photograph of the animal of the week at the Lincoln Park Zoo with selected natural history information on the back and an advertisement for a particular product. For example, a card featuring Sinbad perched on the swing in his cage described on the reverse how he thrived on a "diet of pablum, milk, fresh fruits and vegetables" and fancied playthings much the same as any child. Parents now had a new ploy for convincing their little ones to eat "Aunt Mid's original transparent packed fresh vegetables," whose logo was also prominently displayed on the card. In less than fifteen months, the demand for these animal cards ran to 150,000 a week. Advertisements for Ken-L-Ration products, notably dog food and biscuits, ranged from commercial spots done live by Hurlbut to animated dog cartoons. A typical animated cartoon commercial that aired on a show focusing on "zoo mothers and their families" featured Rover Jones, "a nervous and run down" dog who "growls at his wife and children and has lost interest in his work." When Rover's master commands him to get the newspaper, the dog obliges grumblingly. While reading the paper, the owner happens upon an ad for Ken-L-Ration dog food to improve the disposition of your pet. The spot ends with a happy Rover and his perfect family, proof that "life can be beautiful for dogs that eat Ken-L-Ration."[17]

The thematic focus on animal personalities, the petkeeping practices of zoo caretakers, and the promotional advertisements for pet products all helped create a more intimate relationship between the audience and animals of *Zoo Parade.* The names of animal celebrities and the Ken-L-Ration ad starring Rover and his family also point to the intentional blurring of the distinction between animals and humans on the show. Although traditional Christianity reinforced this distinction by granting humans dominion over other creatures and by excluding animals from the realm of divine grace, in Perkins's "one big happy family," animals were the brethren of humans under the eyes of the Lord. One of the most popular *Zoo Parade* programs broadcast each Easter Sunday preached to children that animals and humans were loved equally by God. The Easter show began, as every traditional Easter Sunday in America begins, with a parade in which the animals donned their latest spring fashions. Heinie II appeared in a midnight blue tailcoat with a top hat. A spider monkey from South America, dressed in "a very handsome Spanish motif," was introduced by Hurlbut, who exclaimed over the similarities between animals and people. This elevation of animals into the human realm is developed further in the Easter story that follows. Using footage of different animals throughout the zoo, Perkins narrates in different voices a morality play in which each animal has a part. The dramatic tension revolves around the observations of a squirrel who watched an Easter service accompanied by music sung by a chorus of a thousand angels. The animals all wonder whether they will ever hear such music, composed in reverence for the Lord, who loved "man so dearly" that "he gave up his life for" him. Midway through the story, Raja the tiger asks the rabbit whether he thinks the Lord will find enough love for the animals so that they too might hear the choir of angels. The climax of the story builds as the animals wait in silence. "Suddenly they knew that something was happening. There was a sighing in the air, a quivering of the Earth, and a stillness in the trees. And as the entire zoo waited expectantly, an angel appeared, hovered a moment in the trees, and said simply, 'You too are loved. Listen.'" As the angel disappears into the trees, the show ends with the animals listening in jubilation to a rising chorus of music. By virtue of their innocence, the animals had been welcomed into the kingdom of God. In this sentimental version of the natural world, the innocence of animals and children forged a special bond that parents could look upon nostalgically in watching *Zoo Parade* with their growing brood.[18]

In sheltering young, be they animal or human, from nuclear annihilation, the threat of communism, and the more insidious side of commercial culture, Americans in the 1950s upheld the nuclear family as a safe haven. Animal behavior stories, especially those that focused on themes such as courtship, nest-building, parenting, and development of the young, universalized the family as a natural unit. The ideal of universality conformed precisely to the marketing needs of national television advertisers, who sought to project an image of the white, middle-class American family in which ethnic and class differences were homogenized. Parenting strategies observed in animals also reinforced Dr. Benjamin Spock's emphasis on the need for women to follow their "natural" maternal inclinations in his best-selling *Baby and Child Care,* the bible of child-rearing manuals among parents of the baby-boom generation. A *Zoo Parade* episode that aired on Mother's Day in 1951 illustrates the ways in which animal shows of the 1950s naturalized the institution of family and the full-time responsibilities of the female in caring for the young. Perkins praised a Canada goose for the motherly attentiveness she displayed toward her goslings and introduced audiences to the motherly devotion female opossums, skunks, and porcupines bestowed upon their young.[19]

The ideals of domesticity in the 1950s extended beyond women in the home to include the active involvement of fathers in the rearing of the well-adjusted child. At the end of the Mother's Day program on *Zoo Parade,* the animal orphans at the zoo, which included Sinbad and Heinie II, each present their male animal keepers with a bouquet of flowers. These animal keepers were both surrogate mothers and devoted fathers, neither of which seemed to challenge their masculine identity. Male identity in popular family sitcoms such as *Ozzie and Harriet, Father Knows Best, Leave It to Beaver,* and *The Donna Reed Show* centered around the responsibilities of fathers like Ward Cleaver and Ozzie Nelson in raising their children. Although ideals of fatherhood in the 1950s did not challenge the traditional gender-based division of labor in the home, the careers of these television fathers were incidental to the tasks of parenting, the real source of meaning and fulfillment in their lives. In the effort to naturalize the nuclear family as a universal way of life, however, finding male role models of loving fathers proved a more difficult task than finding examples of motherly devotion throughout the animal kingdom.[20] Perkins was quick to point out instances where the male played an active part in the development of the young. In a tour down "main street at the zoo," for example, Perkins singled out an adult pair of coyotes and their recently born pups. "Here's a case where we can leave father right in with the young ones," remarked Perkins. "We don't have to separate them like we do with the bears. He stays right in there and is a valuable asset because he helps to protect the family life too." In contrast to terrestrial animal life, the ocean world offered more prominent examples of male involvement in rearing young. The lack of an aquarium at the Lincoln Park Zoo prevented Perkins from spotlighting the fatherly achievements of fish, but viewers on late Sunday afternoons needed only to switch their dial to the CBS television show *Adventure* to find a number of examples of noteworthy male parents under the sea.[21]

Broadcast in cooperation with the American Museum of Natural History, *Adventure,* a one-hour science program, premiered at five o'clock on

Sunday afternoon in May of 1953. A variety show that took advantage of the museum's science staff and its vast collection of specimens and exhibit displays, *Adventure* ranged in subject matter from the origin of the solar system and the beginning of life on Earth to shadow plays and ceremonial dances in Bali to the mating behavior of bighorn sheep.[22] In an early episode hosted by Mike Wallace, Eugenie Clark, research associate in the Department of Animal Behavior at the American Museum of Natural History and author of the Book-of-the Month Club selection *Lady with a Spear,* introduced the audience of *Adventure* to the mouth-breeding behavior of the African tilapia. The African tilapia is one example from the ocean world in which the male tends to the fertilized eggs, hatches them, and protects the living young from predators by sheltering them within its mouth. In close-up footage of a male sea horse incubating eggs in an abdominal pouch and then giving birth to young, Clark revealed another example of male parental care in nature. When Wallace expressed shock that the father seahorse made no effort to protect its "sweet babies" from predators once they had been born, Clark allayed his fears by pointing out enough young were born to guarantee the survival of the species. The sea was not without violence, but the balance of nature assured that in the end, some offspring of the species would endure—a calming message to the human species, worried about the prospects of its survival.[23]

Animals that eschewed their parental responsibilities or did not conform to the "traditional" family model of breadwinning males and homemaking females became symbols of one of America's greatest fears: Soviet communism. On *Adventure,* the defiance of traditional gender roles by the bee contributed in part to its alleged appropriation by the Soviet state. In a dramatic episode that aired in January of 1954, Charles Collingwood alerted viewers to a film that CBS had secured from Russian scientists that "touched upon the greatest political enigma of our time." Because CBS had "no intention of showing a Russian propaganda film on American television," Collingwood narrated the film, which contains spectacular close-up and detailed pictures "undebatable on either side of the iron curtain" of life inside a beehive. The film begins in an "innocuous sylvan delight" where bees are busy pollinating woodland flowers. But the innocence of nature is gradually shattered as the audience is led into the dark interior of the hive. Collingwood described how this bee city has only one mother queen. All the important work is done by females. The male, shirking its responsibilities as family provider, is "incapable of collecting pollen, making wax, or stinging enemies." Its "sole duty," Collingwood divulged, "is to fertilize the queen." Why had the Russians spent five years and endured great physical hardship to make this film? asked Collingwood. The reason, he answered, is that when "Russian scientists think of bees, they think of themselves." Referring to a sequence that portrays the hive's takeover by a new queen upon the death of the old matriarch, Collingwood suggested that "no Russian could think of this without thinking of the death of Stalin, the succession of Malenkov, and the execution of Beria." The life of the bee appealed to Russian scientists, Collingwood argued, because it provided "a natural basis for the regimentation of the Soviet state." In its self-sacrifice for the "homeland of the hive," the individual bee subordinated itself to the good of the group. The prevalence of females in the workforce, the absence of males in supporting the social structure, and the suppression of individual dissent were an affront to the very principles and values of American democracy in the 1950s. The life of the bee offered a model of totalitarianism in the animal kingdom. To Russians, Collingwood informed his viewers, the bee demonstrated that the "cold inhumanity of the Soviet system is not the product of the ambition of men in the Kremlin but is rather the function of a profound and universal natural law." The fatal flaw in this argument, Collingwood reasoned, was that "what is true of a bee is not true of man." Although other *Adventure* programs encouraged audiences to draw universal moral principles from nature, particularly examples that reaffirmed the values of family life, when nature subverted the principles of American democracy and traditional gender roles, animal stories suddenly became less relevant as fables for humanity.[24]

Collingwood was dead serious in his resolve to contain Soviet communism from infiltrating American nature, perhaps too serious for television audiences, who seemed to prefer shows like *Zoo Parade* that instead took delight in the friendly, playful, and innocent side of animals. Backyard bomb shelters and civil defense drills in schools were enough to remind parents and children that they were living in the shadow of mushroom clouds and the Red menace. Nature need not be a reminder too. In flushing out communist critters, *Zoo Parade* took a lighthearted approach in keeping with the image of wild animals as docile family companions. Peter Parakeet, special zoo correspondent for Chicago's *TV Forecast,* learned that when "Perkins put Ivan the European brown bear on the program without once checking his professional background," the animals rebelled. "Many of us thought," the Zoo citizens replied in an interview, "that, in view of our difficulties with Russia, putting Ivan in the spotlight was pretty poor judgment. But it was finally settled when Ivan said he was a White and *not* a Red Russian bear." Humorous nature could go a long way in easing the anxieties and fears of Americans in the postwar years.[25]

Humor could also mockingly expose the conventions by which television programs like *Zoo Parade* succeeded in attracting audiences, particularly parents preoccupied with protecting the innocence of their children. During the 1954 Senate Subcommittee Meetings to Investigate Juvenile Delinquency, comic books were targeted as one of the seamiest aspects of commercial culture that contributed to the moral corruption of America's youth. When the comics industry self-imposed strict censorship in response to the Senate hearings, William Gaines, publisher of E.C. Comics, was forced to divest his company of New Trend titles, including *Tales from the Crypt,* and turned his staff to the production of a ten-cent satiric comic that became *Mad Magazine. Mad*'s success prompted Gaines to publish a companion comic, *Panic,* that similarly delighted in ridiculing anything in consumer culture marketed as innocent, among other things. Although its success was short-lived, *Panic* parodied *Zoo Parade* in one of its last issues in 1955.[26]

"Zoo Charade," hosted by Marvin Perkins and Jim Sherbert at the Blinkin Park Zoo, directly lampooned the themes of domesticity and innocent nature that made the show appealing to "traditional" families in the American suburban home. *Panic*'s "Zoo Charade" was a Sunday afternoon television program "based upon the premise that animals are far more interesting than humans . . . that they are cuter, affectionater, playfuller, and in many instances, intelligenter! But most of all, they're CHEAPER!" In "Zoo Charade," Sherbert, like Hurlbut in *Zoo Parade,* accompanies Perkins through the zoo and "ask[s] the same things the average viewer would ask, in other words . . . a lot of stupid questions." The main running gag through the comic has Perkins reaching into various cages to introduce the audience to animals, such as the little "Irish Shrew" or "Italian Broad-Tail Minx." Each time Perkins reaches in, out comes a seductive woman with large breasts who replies, as in the case of "the Eager Siberian Beaver: 'Marveen, my beeg strawng curator! You bring ouwt de bist in Tanya.'" When Sherbert asks Perkins for the "vital statistics" of these "charming creatures," like "nomenclature and habitat! In other words, what's her name and address, Marv, Boy?" Perkins responds: "Later, idiot! This is a kid's program." Backstage, the television crew is visible to comic fans, who point not only cameras but guns at a dog in an effort to force him to eat "Kam-L-Ration," a product from the commercial sponsor. In "Zoo Charade," Perkins aroused a wild sexuality in women he kept as pets at the zoo. It was a telling joke on the ways in which *Zoo Parade* hid the sexual escapades of animals in the wild. Gun-toting humans and alligators that threaten to chomp off Perkins's arm in the comic also displayed a violent side to the animal kingdom kept backstage in the real television program. Parents may not have found the comic in good taste, but that was the point. In playing off a lustful and violent nature, the *Panic* writers offered a satirical look at Marlin Perkins's popularity in American culture at a time when parents looked to a sentimental nature for moral lessons about the importance of family life.[27]

Through *Zoo Parade* and spin-offs, including a nationally syndicated newspaper column, a Dell comic book series, and a product line of Marlin Perkins's *Zoo Parade* Tooby Toy Animals, Perkins and his animal friends became national celebrities. Other animal stars soon rivaled the place of Bushman, Heinie II, and Sinbad in the hearts of children, however. In the 1954 fall season, ABC and CBS introduced *The Adventures of Rin-Tin-Tin* and *Lassie* on early Sunday evening.[28] *Lassie,* which surpassed rival *Rin-Tin-Tin* in ratings and fan mail, attracted 25 million viewers, received consecutive Emmy awards in 1954 and 1955 for the best children's program, and sold oceans of Campbell's soup. Although neither *Lassie* nor *The Adventures of Rin-Tin-Tin* competed with *Zoo Parade*'s late Sunday afternoon time slot, this new television format of sentimental adventure told through the exploits of a boy and his dog plucked at the heart-strings of children and nostalgic parents and outclassed the emotional appeal of even the most adorable zoo animals. Sinbad or Sweet William might draw laughs, but they couldn't easily evoke an "honest-to-goodness" cry, which Robert Maxwell, producer of *Lassie,* gave as the reason why audiences found the program so memorable.[29]

In focusing on particular charismatic animal species that had pet-like appeal, Perkins subtly preached a conservation message by helping audiences identify with animals in the natural world. In the early years of *Zoo Parade,* Perkins often debunked myths and stereotypes about various animals such as coyotes, wolves, and crows deemed to be either harmful or a nuisance to humans.[30] In contrast, *Lassie*'s occasional perpetuation of prejudice against the value of certain wild animals led the American Nature Association to mildly rebuke the show for having "muffed" an excellent opportunity to teach children the lessons of wildlife conservation. In a 1956 *Lassie* episode, trouble came to the farm when young Jeff tamed a red-tailed hawk. After Gramps discovered the hawk inside the barn and his chickens killed, Jeff accepted the wisdom of his grandfather's adage that the "only good hawk is a dead hawk" and poisoned the bird himself. Thinking at first that the script writers would have Gramps discover "that a weasel or large rat had killed the fowl, thus absolving the hawk, the most beneficial of his kind," the editor of *Nature Magazine,* when proved wrong, felt "disappointment" and "disgust" in the nature lessons learned. As television shows like *Lassie* offered entertainment through more familiar animal companions, *Zoo Parade* began foregrounding other approaches in educating television audiences about the value of wildlife, approaches that didn't rely solely upon wild animals' similarities to domestic pets.[31]

Other changes in television programming also prompted a shift in *Zoo Parade*'s format. By the mid-1950s, *Zoo Parade*'s audience numbers fell from its peak of 11 million viewers in 1952. In the fall of 1956, when television reached into 72 percent of American homes, the audience for *Zoo Parade* had fallen to an estimated 7 million fans. Ken-L-Ration dropped sponsorship of the program in 1955. The show went through three different sponsors—Quaker Oats, American Chicle Company (makers of Beeman and Dentyne chewing gum), and Mutual of Omaha insurance—before its cancellation by NBC in 1957. The sponsorship difficulties *Zoo Parade* faced were in keeping with a larger trend, as advertisers slowly abandoned children's programs in favor of family shows when the networks started to return the early evening slot (4:30 p.m. to 7:00 p.m.) to affiliates. *Zoo Parade,* which aired on Sunday afternoons from 4:30 to 5:00 p.m. during the 1952–55 seasons and from 3:30 to 4:00 p.m. in its final two years, straddled the afternoon/early evening market and, therefore, was vulnerable to this trend of sponsor abandonment. In addition, the number of live television shows started to plummet after 1955. Children's programming followed this pattern. Action-adventure series aimed at children increased from 39 percent in 1952–53 to 79 percent in 1958–59. These factors all played a significant part in the gradual transformation of Perkins's Doctor Dolittle image in the early years of *Zoo Parade* to that of daring, worldwide explorer in search of adventure, his memorable role on *Wild Kingdom.*[32]

Struggling to retain audiences and sponsors, *Zoo Parade* departed from its "pretty well-worn" formula in 1955 and began to switch from a live to recorded program. Although the move was significant for *Wild Kingdom,* precedent for televising wildlife in their natural habitat had been established in earlier *Zoo Parade* shows that focused on animals of particular geographic regions. Initially, these episodes simply picked a continent or region as their focus and then discussed the distributional range, taxonomic relationships, and adaptations of species found within a specific area, choosing representative animals from the Lincoln Park Zoo as examples.[33] Within a few years, the *Zoo Parade* shows began to use maps and incorporate stock footage of wild animals in their natural habitats with a voice-over by Perkins. Disney's True-Life Adventures offered a source of footage shot in the wild that Perkins included on his television show. On a 1953 *Zoo Parade* episode that featured animals of the hot regions and their adaptations to arid climates, for example, Perkins introduced a scene from *The Living Desert* that illustrated the food habits of the burrowing ground squirrel and also put in a promotional pitch for the Disney film. With the 1954 premiere of *Disneyland* on ABC, however, Disney Studios retained the exclusive right to their wildlife footage. If *Zoo Parade* were to increase the number of shows focused on animals in their natural habitat, it would have to travel directly to "unspoiled" nature.[34]

Before television became a fixture in every American home, manufacturers promoted its ability to take viewers to the far corners of the globe. An early 1944 Dumont Television advertisement captured the sense in which television promised to make everyone "an armchair Columbus." "You'll sail with television through vanishing horizons into exciting new worlds," declared the promotional copy. Parents, educators, and psychologists also believed in the educational benefits of television for children as a window on the world. By the age of ten and eleven, reported one study of the 1950s, the child is "ready to explore and conquer the world." In the mid-1950s, the first cohort of America's baby-boom generation had reached this adventuresome age. On television, children and their parents could go places. Television, like the family vacation, was an adventure. In taking family audiences on their first *Zoo Parade* adventure, it is not surprising then that Perkins and Hurlbut traveled to one of the most popular nature tourist attractions in the 1950s: Jackson Hole, Wyoming. Already familiar to many American adults and their children, Jackson Hole offered an exciting look at wildlife in its natural surroundings not too far from home. Perkins gradually extended his reach beyond the domestic setting of the Lincoln Park Zoo, traveling first through familiar natural landscapes in North America before taming nature across the continents.[35]

"Jackson Hole Country," sponsored by Mutual of Omaha with travel arrangements made possible by United Airlines, aired on *Zoo Parade* in two parts. The use of a single location for multiple shows helped keep costs down and also made for an effective hook to attract audiences to the following week's show. In the fall of 1955, when *Zoo Parade* left North America in search of more exotic wildlife in East and South Africa, seven episodes were planned. The photographic safari included shows on the migration of wildlife across the Serengeti, the capturing of wild animals for zoos, and visits to the Port Elizabeth snake farm and Krueger National Park. Like the economical tourist, *Zoo Parade* planned its travels for family audiences to ensure the biggest bang for the buck. The promotional ads for United and Scandinavian Airlines and the focus on popular tourist attractions such as Grand Teton National Park and Krueger National Park also made the wildlife in these natural surroundings seem all the more familiar to "arm-chair explorers" who traveled with Perkins.[36]

In the last season, while *Zoo Parade* tried to regain audiences and sponsors, the show's producer, Don Meier, and his staff proposed a number of concept ideas for a new series that included titles such as "Wide, Wide Animal World" and "World Safari." Amidst the variety of suggestions, which included features such as guest pets, odd personalities in the animal kingdom, and filmed or remote pick-ups from wildlife reporters that would bring viewers the latest in news headlines from the animal world, a consistent theme emerged that drew upon experiences shooting *Zoo Parade* on location. The show would "embark on a global venture" to "wherever a date with *authentic* adventure in the wild" could be made. "Authentic" was a "key word." With no contrived plots, "these real adventures" would blend a "live, direct-to-camera approach with dramatic narration, scored with original music" to give an "even greater impact of reality and thrill of immediacy for the audience."[37] By launching each show from "Animal Central," the "familiar" studio setting of *Zoo Parade*, with props such as a map of the world or a particular pet animal, Perkins would conduct the program. In this way, the series would retain the main appeal of *Zoo Parade*, featuring Perkins and his animal friends, while adding a new look in keeping with the shift in television programming to filmed outdoor action-adventure series.[38]

NBC began promotion of *Wild Kingdom* six months after *Zoo Parade*'s cancellation in the fall of 1957. To potential advertisers, the network highlighted the program's "broad audience appeal, the track record of *Zoo*

Parade and the increasing importance of family pets in the contemporary American scene." But four years elapsed before a sponsor appeared. In 1962, at a banquet of the Zoological Society of Omaha where he was the guest of honor, Perkins met Mutual of Omaha's board chairman, V. J. Skutt, and discussed the new series concept. A previous sponsor of *Zoo Parade,* Mutual of Omaha expressed interest in the new program, provided the commercial spots were integrated into the show and Perkins delivered them. NBC signed a contract with Mutual of Omaha and Don Meier Productions for thirteen episodes to air on late Sunday afternoon beginning in the 1963 spring season. *Mutual of Omaha's Wild Kingdom,* which in its first ten years won four Emmys and over a dozen honors for best children's program and outstanding contribution in the service of wildlife conservation, had been born. In the first production season, "Animal Central" moved from the Lincoln Park Zoo to the St. Louis Zoological Garden, where Perkins assumed the responsibilities of his former director George Vierheller. In returning to St. Louis, Perkins also returned to a certain flair for showmanship in *Wild Kingdom* that highlighted the thrill and danger in working with wild animals he cultivated as a young man. As the first wave of America's baby-boomers came into adolescence during the early 1960s, adventure became an important element in *Wild Kingdom*'s success.[39]

Don Meier emphasized a "strong sense of immediacy" and a feeling of "personal involvement" as the two most important factors that kept *Wild Kingdom* vibrant and interesting to audiences during its eight years on NBC and as a series on syndicated television during the 1970s. Unlike *Zoo Parade*'s live television format, *Wild Kingdom's* reliance on filmed segments made it difficult to create an impression of communicating with the viewer. Many critics praised the unique, participatory qualities of live television, but these were lost in the change to telefilm when Hollywood entered television production beginning with Disney's arrival on ABC. In the live production era, bloopers were a memorable part of the shared experience and contributed to the special bond forged between viewer and program. The unpredictability of animals on *Zoo Parade* added to the spontaneity and immediacy of the show. The most vivid incident recalled in the collective memory of *Zoo Parade* audiences happened on April Fool's Day in 1951. During a Sunday afternoon rehearsal, Perkins was bit by a timber rattlesnake while directing the camera crew where to position their lenses in photographing a venom extraction scheduled in the program. Although the cameramen recorded the rattler sinking its fangs into Perkins's left finger, the mishap did not air on television, despite the insistence of many fans, who wrongly said they saw the accident happen. What audiences did see that day was Judy the elephant throw Leer Grimmer clear across the floor when he stepped into her cage to open the show for Perkins, who had been rushed to the hospital twenty minutes earlier.[40]

To retain the sense of immediacy conveyed on *Zoo Parade, Wild Kingdom* resorted to a number of techniques. The switch from "Animal Central" to filmed segments, a technique first tried in the Jackson Hole episode on *Zoo Parade,* played off the convention of news broadcasting that used an anchor in the studio as the continuity in the switch to filmed newsreels and correspondents. Misadventures with animals, particularly those that highlighted the thrill of danger, also became important elements of the show. In *Wild Kingdom*'s first season, for example, an episode that featured the fur seals of the Pribilof islands included footage of a half-ton irate bull seal charging Perkins, who fled straight toward the TV cameras. A later program that matched the April Fool's Day episode of *Zoo Parade* in the mythic memory of Marlin Perkins's fans included a death-defying struggle between Perkins and a twenty-two-foot anaconda. On location on the Dadanow Ranch in British Guiana, Perkins stepped into what he thought was a shallow cattle pond to capture the huge snake that posed a risk to newborn calves. Though lassoed by Stan Brock, the ranch manager and co-host of *Wild Kingdom* from 1968 to 1972, the anaconda pulled Perkins into deep water and began wrapping its muscular body around his chest. Eventually, they pulled the snake to shallow water and succeeded in wrestling it into a burlap bag, but Perkins was noticeably shaken. These near mishaps included in the program led fans and critics to praise the "hard factual realism" in *Wild Kingdom.* As a television critic from the *San Francisco Chronicle* noted in 1966: "One of *Wild Kingdom*'s admirable features is its honesty about its subject. This is nature as it is. . . Perkins and Fowler are active participants in each episode, not merely narrators of wildlife films. They are THERE."[41]

Wild Kingdom also added to the sense of immediacy by creating a sense of personal involvement with the wild animals. In *Zoo Parade,* Jim Hurlbut's role as the average guy and his interactions with Perkins and his animal friends allowed viewers to believe that they were on the set with the animals themselves. But Hurlbut's character, which added to the humorous, sentimental elements of *Zoo Parade,* didn't fit with the action-adventure theme of *Wild Kingdom.* In his place, Don Meier cast Jim Fowler, an athletic, six-foot six-inch former professional baseball player with a deep bass voice.

In his early thirties, Fowler as aide and assistant to Perkins appeared to be an all-American boy, an image that appealed to viewers of all ages and education. A falconer, whose knowledge of birds was acquired as a youth growing up on a large family farm in Georgia, Fowler and his birds of prey became the subject of the pilot for *Wild Kingdom.* Known for having captured one of the world's most powerful predatory birds, the rare South American harpy eagle, Fowler enabled viewers to vicariously experience "the sense of adventure and excitement" of working with wild animals.[42]

Involvement shows, as they were described by production staff, created and maintained this sense of adventure. These shows made up roughly two-thirds of the episodes on *Wild Kingdom* and highlighted the exploits of Perkins and Fowler as they aided wildlife biologists, managers, and conservationists. The majority of these interactions centered on the capture of species "imposing in size, spectacular in action," or occupying "scenic or unusual" habitat and on the use of fast transport vehicles.[43] An episode titled "Operation Rescue," for example, took viewers to Venezuela, where a recently built hydroelectric dam threatened to flood hundreds of square miles of wilderness and trap thousands of wild animals. Perkins and Fowler aid Pedro Trebbau, director of the Caracas Zoo, in saving endangered animals stranded on islands using boats, airplanes, and helicopters as the water levels rise.

When Fowler left the show in its seventh season, he was replaced by the "barefoot cowboy" Stan Brock, an English-born rancher in British Guiana with the physique of a body-builder and the physical strength to match. Audiences did not respond favorably to the brute manhandling of animals displayed by this modern-day Buffalo Jones. While acknowledging that the program had done an immense service in raising the public's consciousness about the value of and need for wildlife conservation, by the early 1970s some biologists publicly condemned the show for its excessive showmanship and "staged danger." In the effort to entertain audiences, Perkins, in the eyes of some, had strayed too far from the educational mission of the show.[44]

Whether wrestling giant anacondas in Guiana, observing a family of Bengal tigers from a tree-top perch in India, or riding helicopters across the Canadian wilderness in search of moose, Perkins educated viewers about the value and importance of wildlife conservation. The educational qualities of *Wild Kingdom* appealed to the sense of family responsibility that Mutual of Omaha wished to promote as part of its corporate image. Mutual of Omaha's board chairman V. J. Skutt believed the program offered families a wholesome, educational, and entertaining experience and thus demonstrated to parents that the company was concerned, like them, with the moral upbringing and education of America's children. As one parent and faithful fan of *Wild Kingdom* wrote: "In a time when there are so few programs which will help and not hinder them, yours stands out as one not only good, but good for them." When the first cohort of America's baby-boom generation to grow up with Marlin Perkins became environmental activists in the late 1960s, the conservation theme also boosted Mutual of Omaha's image. In later years, Perkins was quick to promote the parallels between conservation and insurance. "We teach the need for conservation," Perkins told Mutual's policyholders. "Mutual teaches the need for adequate health insurance. And there is a parallel between the two. Disregard for our natural resources can do great harm to our environmental security. Lack of concern for the financial consequences of inevitable illness, injury and death can ruin our financial security." In its endorsement of *Wild Kingdom,* Mutual of Omaha assured viewers through Perkins and his show that the company protected future generations from the risks faced in a troubled world.[45]

Nature, once an exclusive playground patronized by the elite and upper middle class, had been appropriated by corporate America and mass culture. Two economic relationships with animals sustained nature appreciation as a commercial commodity on the television set: pet keeping and tourism. Perkins successfully capitalized on these two markets, both of which influenced the ways in which environmentalists used the media to enlist public support for environmental causes in the postwar years. The domestication of wildlife on *Zoo Parade,* achieved through the myriad associations established between wild animals and domestic pets, captivated viewers by cultivating an intimate, emotional relationship with individual animal personalities. These charismatic animal species might be used to sell not only commercial products like Ken-L-Ration, but also the value of wildlife. A bobcat, identified by Hurlbut as "a marauder like the coyote," is transformed after a commercial break into a cute, adorable pet when Hurlbut feeds a bottle to a young orphaned bobcat, cuddled in his arms. On this *Zoo Parade* episode, filmed in Jackson Hole, Perkins educated viewers about the value of this feline species in keeping the populations of rodents, birds, snakes, and frogs in check. In the 1920s, William Finley had used wilderness pets effectively on film to sell Bell & Howell cameras and to teach that generation of American children the value of wildlife conservation. Perkins did the same for the first generation of American children to be raised on the television set, although he expanded the conservation frame beyond individual species to include the web of life.[46]

By transforming wildlife into domestic pets, animal rights groups and environmental organizations could appeal to pet owners in America, a market targeted by *Zoo Parade* and *Wild Kingdom,* for support. In the 1960s, as the environmental movement gained political ground, footage and logos of charismatic species such as harp seals, dolphins, and panda bears became common in the political campaign strategies of animal rights groups and environmental organizations such as the International Fund for Animals, Greenpeace, and the World Wildlife Fund. President of the Elsa Wild Animal Appeal Foundation and co-founder of the Wild Canid Survival Center, Carol Perkins, Marlin's second wife, used the domestication of individual wildlife species to garner funds and support for environmental activist causes that ranged from the passage of the Endangered Species Act, to the abolishment of traffic in animal furs, to the establishment of a captive breeding and educational center for the red and Mexican wolf. In 1978, Richard Starnes of *Outdoor Life* vehemently attacked a list of television nature shows, including *Wild Kingdom,* for presenting the "outdoor world as a benign fantasy land" in which animals "behave surprisingly like lovable little furry people." "When the generation that has been nurtured on such pap . . . comes of age it is plain to see that the antihunting and antigun lobby will gain thousands of recruits," Starnes complained. Starnes didn't need to wait anxiously for America's next generation to grow up. Their parents had been raised on the sentimental and playful images of animals he so despised and had already gained significant political clout.[47]

Performing and captured animals were not the only displays of tame nature on *Zoo Parade* and *Wild Kingdom.* In taking audiences on location, these television programs brought distant lands and their inhabitants into the comfortable and familiar space of American homes. Distant wilderness rapidly became a prime tourists' playground. Perkins globalized the work that Disney began. Tourism would have equally important repercussions in shaping environmental politics in the postwar years.

Leisure time, either in nature or lived vicariously through film, offered escape and respite from the mechanical age. Western conservationists after the Second World War increasingly upheld tourism as the solution to save the world's wildlife heritage from extinction, particularly as more and more

Americans took flight for international vacations in the 1950s and 1960s.[48] In visiting the world's national parks, Perkins reinforced this aesthetic vision, grounded in an early twentieth-century dichotomy between the city and country, between nature as a place of play and nature as a place of work. The former produced a service economy sustained by entertainment, the latter produced an economy of goods.[49] But the two economies were incompatible in the history of twentieth-century American environmentalism. To work on the land was to defile the pleasurable experience that might be had from it as a recreational resource. From South Africa's Krueger National Park, Perkins informed his audience that such big game reserves "completely separated from the farming interests and the urban interests" of local residents, held the key to "preserv[ing] wildlife for future generations."[50]

Perkins praised the knowledge farmers acquired of animals and the land through experience, but when he extended his reach to the globe, nature came under the purview of Western scientists.[51] Under the influence of international environmental organizations, wildlife ecologists and conservationists played a larger role than other laborers in reshaping the landscape. Through their knowledge, land once deemed unproductive would be transformed into tourist attractions. Scientists would help rescue and sustain the world's vanishing wildlife for the admiration and pleasure of future generations. They had been, all along, important allies in the production of nature as entertainment.

To preserve nature and to assure a pleasurable tourist experience, concealing the artifice of production was an essential part of natural history filmmaking and wildlife management. Signs of artificiality were thought to destroy the illusion of this recreated nature as God's place of grace. It is for this reason that one-third of *Wild Kingdom* programs were reserved for "ecology shows." Such episodes told the story of wildlife in a particular habitat without any visible evidence of people. Like Disney's wilderness fantasies, these *Wild Kingdom* programs sought to sustain the illusion of "pristine" wilderness for a "hard core of the audience interested only in the observation of wildlife without the interference of man." With tour guides like Carol Perkins, middle-class Americans traveled on photographic safaris to Africa, India, Nepal, and the South Pacific in search of "pristine" wilderness and experiences similar to those watching wildlife on the television set. In dealing with "the conservation of wildlife and its habitat," *Wild Kingdom* purported to carry "no vestige of political, ideological, or governmental conflict or controversy." But the very image of an innocent nature set apart from humans, reinforced through an aesthetic vision championed by Western conservationists and television programs like *Wild Kingdom,* carried an immense amount of political baggage. Regions of the world that appeared "pristine" to the eyes of Westerners were also places of livelihood for other peoples, who did not necessarily regard nature as an innocent playground nor wildlife as a global resource that belonged to all.[52]

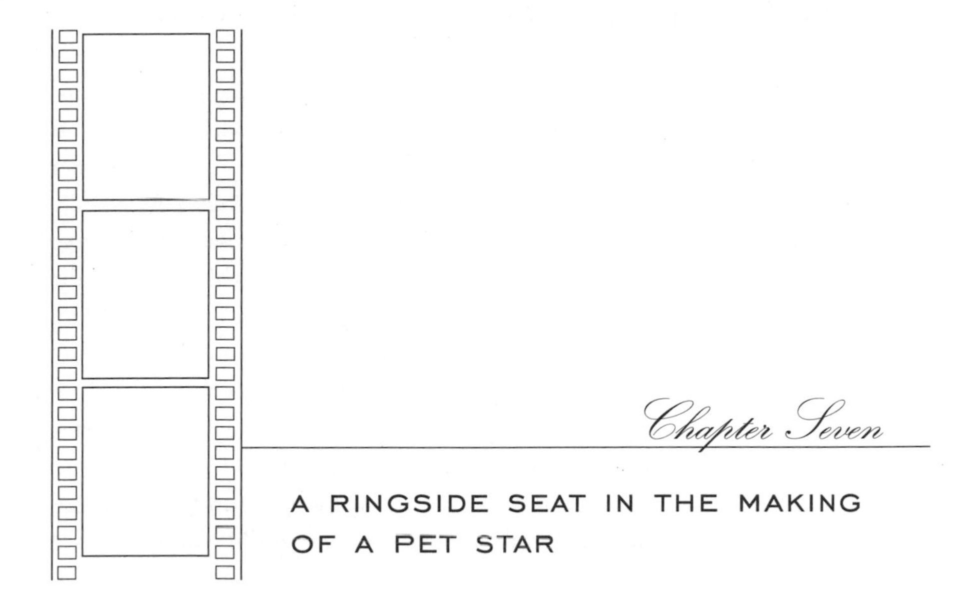

Chapter Seven

A RINGSIDE SEAT IN THE MAKING OF A PET STAR

In April of 1993, the tourist magazine *Condé Nast Traveler* ran a feature on Australia's top ten beaches, one of which, Monkey Mia, is home to a pod of dolphins that have been cavorting with tourists and beach residents for some thirty years. A two-page glossy photograph of a blonde, suntanned swimmer taking a close-up snapshot of a smiling bottle-nosed dolphin is accompanied by a jarring caption: "The mating habits of male dolphins amount to gang rape, although this is seldom publicized." It was not the first exposé of dolphin sexual behavior in Shark Bay, Australia. The story broke one year earlier, when Natalie Angier, a science writer for the *New York Times,* detailed the ongoing research on dolphin sexual behavior conducted by scientists at Shark Bay. Describing dolphin courtship as "brutal, cunning, and complex," Angier highlighted the aggressive behavior by which male dolphins "collude with their peers as a way of stealing fertile females from competing dolphin bands." "Dolphins are turning out to be exceedingly clever, but not in the loving, utopian-socialist manner that sentimental Flipperophiles might have hoped," she remarked.[1]

This Machiavellian picture of dolphins is as much the product of science, cultural values, and nature as the more familiar image of dolphins as loveable, gentle friends in the sea. Used to market everything from environmental legislation to skin-care products to tuna, the celebrity status of marine mammals is not likely to wane, despite stories revealing a seamier side to the dolphin's private life. For example, in 1994 the Keiko/Free Willy Foundation, in coordination with the Earth Island Institute, launched a huge public campaign for the rescue, rehabilitation, and eventual release of the dolphin's close cousin, Keiko, the orca star of the Warner Brothers' film *Free Willy*. In his time of need, Keiko attracted $8 million in corporate and public financial contributions, including a $2 million seed grant by Warner Bros., a $3,000 contribution from an elementary school in Kodiak, Alaska, and the donation from United Parcel Service of a C-130 Hercules plane to transport the killer whale from Mexico City to the Oregon Coast Aquarium. Donors were concerned that the adorable killer whale whom they had come to love on film was languishing in captivity in an out-of-the-way theme park in Mexico and was in desperate need of more lavish facilities and care. The ability of marine mammals, and dolphins in particular, to arouse the interests of corporations, public constituencies, the federal government, and science attests to the power that such charismatic species have within American culture. But the dolphin's affectionate image was made, not bestowed by nature, as a result of the intertwined and sometimes conflicting interests of science, the military, environmental organizations, and the commercial film and entertainment industries—an image that has had a profound impact on the shape of scientific research, environmental policy, and international relations in the postwar years.

Known as the "pig fish" or "herring hog" in the early twentieth century by American fisherman to denote its alleged voracious appetite for commercial fish, the dolphin first garnered big-time public attention in February of 1940. On the north coast of Florida, eighteen miles south of St. Augustine, a captive birth of an Atlantic bottle-nosed dolphin was witnessed by biologists, cameramen, and hundreds of tourists through the portholes of Marine Studios and suddenly became a national media event. *Life* reproduced the highlights of the first dolphin birth recorded on film in a series of stunning photographs. Surprised by the tail-first appearance of the fetus, an uncharacteristic delivery position for mammals, the staff at Marine Studios struggled unsuccessfully to artificially resuscitate the infant dolphin after the mother failed to bring the baby to the surface for its first breath of air. A flurry of letters from interested readers arrived at the publisher, asking *Life* to report on the condition of "Mother Porpoise" after her stillborn birth and offering opinions on whether the tail-first delivery was normal for dolphins or was symptomatic of abnormal births. One reader, obviously unaware that the mother died nine days after the birth, even sent a photograph taken of the mother porpoise shortly after the birth to prove that she was no longer grieving.[2]

Marine Studios was not the first aquarium to house dolphins in captivity, but it was the place where contemporary media images of the dolphin were first constructed and where detailed and sustained scientific investigations of their behavior were first performed. Designed to serve several purposes, as a place for public entertainment, as a studio for filming underwater action against an authentic background, and as a research site for the study of undersea life, Marine Studios had free admission on opening day and welcomed over twenty-five thousand visitors on 23 June 1938. The original impetus for holding sea life in captivity to make its filming feasible came from Merian Cooper's 1927 film *Chang,* in which a scene of an elephant stampede was shot within a stockade large enough not to impede the animals' natural movements and behaviors, yet small enough to allow camera operators to take action shots with relative ease. William Douglas Burden and Ilia Tolstoy had adopted the same technique for filming the caribou migration scene in *The Silent Enemy,* and they decided that a similar approach could be used to obtain underwater action shots of marine life. With financial backing from his cousin, Cornelius Vanderbilt Whitney, and the use of his personal investments, Burden, in association with Tolstoy, and guided by the scientific expertise of Gladwyn Kingsley Noble, began construction in May 1937 of what would be the largest oceanarium in the world and the first to house numerous species within a single tank.[3]

Burden recognized the power of film to capture the audience's attention, and he designed Marine Studios with the psychology of motion picture perception in mind. Emotions, Burden argued, formed the mainspring of interest in any subject; to capture them was the educator's first task. "If the exhibits in Marine Studios' tanks are of sufficient intrinsic dramatic value, if they are sufficiently graceful and vivid and striking to arouse the audience's admiration or wonder or curiosity, then it will be relatively easy," Burden wrote, "to engage the spectator's attention along more serious lines." To achieve this, Burden decided to design the oceanarium so that the "usual distractions that are so ever-present in the exhibition hall of a museum or aquarium" were absent. One needed to create the conditions of the motion picture theater. "To sit comfortably in the dark and allow one's attention to be fixed on the lighted screen requires no conscious effort," Burden reasoned. "The response is automatic. Similarly if the enclosed galleries or corridors which run at different elevations around the entire perimeter of the tanks are so arranged that each observer can sit comfortably in relative darkness in front of his own porthole—the person screened on either side by a projection or curtain that isolates him from the neighboring portholes and spectators, his attention will be more easily fixed on the moving exhibits beyond his own glass ports." The effect Burden envisioned would be as if the visitor were "in a motion picture theatre, looking out into a brilliant world of the undersea."[4]

The task faced by Marine Studios was the same task faced by the makers of natural history film: to create the illusion of reality and to immerse the viewer in the activities being viewed, which in the case of Marine Studios was life off the coast of Florida, seventy-five feet below the surface. Although scheduled feedings of animals by divers were staged throughout the day, Burden hoped that the aesthetic beauty and wonder of undersea life were sufficient in themselves to capture the visitor's imagination and attention.

However, Burden soon realized that what gripped the public's fascination was not the subtle and "extraordinarily complex relationships of life" found within the ocean, but the dramatic spectacle found in the activities of particular large species such as sharks, moray eels, manta rays, and most of all, the Atlantic bottle-nosed dolphin. Once captured and prodded to perform, the dolphin quickly proved to be a public and scientific sensation.[5] Hollywood studios such as MGM and Paramount were soon attracted to Marine Studios not just for filming underwater action scenes, but also to investigate the dolphin's potential star appeal. In the fall of 1938, MGM signed a contract with Marine Studios for moving-picture rights to film a one-reel Technicolor Pete Smith short subject entitled *Marine Circus*. Billed as "A Short with a Porpoise," the picture played in theaters across the country in the spring of 1939, offering visits to the undersea with Hollywood guiding the tour. The picture opens with a scene of the North Florida coast and then dissolves to an inside view of Marine Studios and what appears to be a giant swimming pool at a posh oceanside resort in Florida.

As "pretty girls" at poolside dive into the water, the scene cuts to a split screen in which the audience sees the young women swimming with their heads above water and "friendly porpoises swimming about with them" below the water. The next two sequences show the undersea life at Marine Studios in all its Technicolor splendor, then cut to dinner time. The ringing of the dinner bell signals to the aquatic inhabitants the arrival of a male diver who undertakes the perilous task of feeding by hand large and dangerous fish such as the jewfish, manta ray, and shark. *Marine Circus* ends with Marine Studios' main attraction, the dolphin, performing spectacular acrobatic jumps to retrieve its supper topside.[6]

Despite Burden's faith in the fascination of the authentic undersea, he viewed the Hollywood short-subject film, which accompanied feature motion pictures, as the most valuable medium for publicity for Marine Studios. Spotlights in magazines such as *Travel* and the *Saturday Evening Post,* and featured tours in travel guides such as the 1939 WPA guide to Florida helped attract visitors, but movie shorts reached a far wider audience. A Grantland Rice Sportlight, for example, made at Marine Studios in 1940, was distributed by Paramount to over 6,500 theaters throughout the country.[7] Balancing the oceanarium's scientific research mission with its public entertainment side, however, proved to be a difficult task. When Burden became aware of Pete Smith's plan for diving girls in the *Marine Circus* short, he urged Smith to reconsider. "Marine Studios is a serious scientific enterprise," he reminded Smith, and "diving girls are hardly in keeping with a serious scientific enterprise." The inclusion of such scenes, Burden feared, would give the impression that "Marine Studios [was] really nothing more than a road-side attraction," an impression that he was desperate to avoid. Although Burden eventually acquiesced, he took great pains to distance Marine Studios from the carnivalesque atmosphere that surrounded nineteenth-century animal exhibits such as the beluga whale kept, if only briefly, at P. T. Barnum's Aquariel Gardens and trained to pull a woman in a cart around the edge of the tank. Burden knew only too well from the financial loss of his 1930 film *The Silent Enemy* that nature, without sufficient drama and spectacle, would fail to attract crowds. But he was loathe to completely abandon authenticity for the sake of spectacle.[8]

Films like *Marine Circus* and articles in the popular press not only captured the public imagination and thereby spurred tourists to visit Marine Studios, they also captured the attention of scientists. Science would play an important public relations role in Marine Studios' claims to authenticity and would thereby help balance the delicate relationship so common in natural history film and of such concern to Burden in his hopes for success between science and spectacle, education and entertainment. As the primary institutional center for research on marine mammals throughout the 1940s and 1950s, Marine Studios attracted through its publicity efforts many researchers from such places as the American Museum of Natural History, the University of Chicago, Johns Hopkins University, Harvard University, and the University of Florida. Per F. Scholander, for instance, was a Norwegian biologist who conducted pioneering research in association with Laurence Irving at Swarthmore College on respiratory physiology in diving mammals. He came to the United States in the late 1930s largely because of a newsreel he had seen about Marine Studios in a motion picture theater in Oslo. Similarly, E. M. K. Geiling, a professor in the Departments of Pharmacology and Physiology at the University of Chicago, first became intrigued with the research possibilities of the oceanarium through an article he read in the *New Yorker*.[9] Geiling's biochemical research on the composition of porpoise milk subsequently validated Marine Studios as a research facility when it made headline news in the *St. Augustine Record* as laying "the ground work for the most extensive study of aquatic mammals ever attempted."[10]

In marketing itself as an institution that combined scientific research, education, and entertainment, Marine Studios needed the endorsement of the professional scientific community. Acknowledgment of Marine Studios in scientific publications was of paramount importance in this regard. When Geiling and his research team failed to credit Marine Studios in the titles of their published papers, he explained to Milton Bacon, Marine Studios' public relations director, that it lacked an established research laboratory with a board of scientific directors. Getting editors of professional journals to acknowledge it thus proved difficult. It was clear that a research lab and a consulting board of scientists were essential.[11] In appealing to Burden and Marine Studios' Board of Directors for the funding of a major research facility, Marine Studios curator Arthur McBride argued that a research lab would add to the institution's public prestige and "pay dividends at the box-office." He compared Marine Studios to General Electric, where "the organization's research laboratory" and its "reputation for know-how in the electrical industry was largely responsible for selling G.E. products." Marine Studios could profit from a similar relationship, since it was after all a commercial enterprise geared toward the marketing of undersea life.[12]

By attracting research scientists, Marine Studios not only profited from the publicity, but also gained scientific expertise essential to the operations and maintenance of the oceanarium. Marine Studios relied heavily on technical contributions from scientists in its artful mimesis of nature in the undersea world. Geiling, in exchange for the use of Marine Studios' facilities, offered medical advice in the treatment of skin lesions and diseases that appeared in captive dolphins. The exhibition of large specimens such as sharks at Marine Studios only became feasible after a series of investigative studies pursued in collaboration with Noble and his Laboratory of Experimental Biology at the American Museum of Natural History. These experiments yielded an anaesthetic, Evipal, that when administered through a hypodermic syringe attached to a harpoon head would allow sharks to be successfully captured and transported. From Noble's lab, Burden recruited research assistants Arthur McBride and Arthur Schmidt to oversee the scientific management and operations of the oceanarium exhibits at Marine Studios. Both utilized their networks within the scientific research community to solve difficulties that threatened the quality of exhibits and, hence, admission sales.

The intertwined interests of science and spectacle were manifold. In its early years of operation, Marine Studios staff constantly struggled to keep the tanks free of epibdella, a trematode (or parasitic flatworm) introduced from fish caught in the Florida Keys. The parasite eats out the eyes of pelagic fish, who are otherwise not commonly exposed to the diseases carried by coral reef fish. McBride estimated that Marine Studios lost $16,000 each year to epibdella epidemics, not to mention the labor costs involved in having divers remove the large number of infected fish each day after the tourists had gone. While Charles M. Breder of the New York Aquarium helped diagnose the problem, treatment proved elusive until McBride and Schmidt, through their wartime associations with research scientists at the Woods Hole Oceanographic Institute, discovered that copper citrate made an effective shark repellent, solved problems of water clarity created by algae blooms, and killed trematodes as well. These scientific success stories, which translated into increased visitor revenues, helped in securing support from the Board of the Directors for the establishment of the Marineland Research Laboratory, completed in 1953.[13]

While the results of scientific inquiry fostered the success of an entertaining and educational spectacle, its methods threatened to hamper it. Public relations required the building of a research laboratory to shield the public from the labor integral to the research, operation, and maintenance of the exhibit displays. When Marine Studios first opened, it was ill-equipped to handle the demands for space, equipment, and specimens placed upon it by outside scientific investigators. Samples for the initial biochemical studies conducted by Geiling on the urine, blood, and milk of dolphins were extracted from animals in the flume that connected the two main tanks. McBride managed to obtain a sample of fresh porpoise milk for Geiling by suspending the dolphin in a sling above the flume and massaging the animal's mammary glands. Such experimental work could not remain plainly visible to the public. Temporary facilities were built for the holding and dissection of dolphins until the permanent research laboratory was complete. Any hint that dolphins were being injured resulted in unfavorable public reaction and left Marine Studios open to objections by the SPCA. To recreate a realistic experience of undersea life, Marine Studios screened the visitor from the manual labor integral to the production process, just as the makers of natural history film created an illusion of reality in their filming techniques that hid the long hours spent by wildlife photographers in the field, the staging of scenes in animal compounds, and the editing and synchronization of images and sound in the production studio.[14]

Expanding U.S. naval operations and naval research during and after the Second World War augmented the study of oceanographic science, and in turn transformed Marine Studios and its star attraction. The invention of the self-contained underwater breathing apparatus by Jacques Cousteau and its further development with the establishment of the French Navy's Undersea Research Group in 1945 opened the Earth's last frontier to direct exploration, habitation, and observation. Consequently, experimental work on marine mammals took on new significance and their visibility new appeal. In 1946, Geiling pointed out that dolphins were ideal experimental organisms for comparative research on the "problems of underwater existence and activity of man." Laurence Irving remarked in 1951 that he only "dimly foresaw the applicability" of his collaborative research with Per Scholander on respiratory physiology of dolphins, begun in the late 1930s, "to problems limiting man's usefulness in the air and under the sea." Research at Marine Studios had suddenly taken on great national importance.[15]

If the ocean as vast battleground sparked military and scientific interest in oceanographic research, it also fostered in the public a sense of wonder and fascination. Rachel Carson, whose 1951 book *The Sea Around Us* escalated to the *New York Times* best-seller list in less than three months and sold well over 200,000 copies in the first year, attributed the "awakening of active interest in the exploration of the sea" to the Second World War. Her first book, *Under the Sea Wind,* published in 1941, failed to attract a wide readership. Physicists may have unlocked the secrets of the atom, but the sea, as the last great frontier, still contained mysteries and intrigues of nature. In the mushroom-cloud explosion that ends Walt Disney's 1954 movie version of the Jules Verne classic, *20,000 Leagues Under the Sea,* the Nautilus sinks forever into the ocean's shadowy depths. In the popular imagination, the secrets of the sea known only to Captain Nemo and his crew were hidden once again in the silent undersea world. The desire to explore this uncharted terrain had captured the public imagination.[16]

Hollywood was not alone in profiting from popular interest in the ocean world. Marine Studios experienced a boom in attendance and publicity during the 1950s. In 1953, attendance figures approximated 700,000 visitors at an admission price of two dollars per person. Marine Studios expanded its operations in 1954 to the West Coast, turning its name and trade secrets over to Oceanarium, Inc. for an 18 percent stock interest. Located in Palos Verdes, California, Marineland of the Pacific quickly came to rival its Florida counterpart, averaging over 5,000 people per day during its first year of operation.[17]

Ticket sales were not the only source of revenue. Concession sales were also important. The Marineland Gift Shop in 1946, for instance, could not keep up with visitor demand for Kodak print and 16- and 8-mm color motion picture film. R. J. Eastman, director of public relations, estimated that film sales, which amounted to two thousand dollars for the months of March and April, could easily have been doubled had supplies been available. The Marine Village Court, a twenty-two-unit hotel on the oceanfront adjacent to Marine Studios and the Dolphin Restaurant and Penguin Bar not only attracted tourist dollars, they also attracted film stars such as Robert Benchley and literary figures like Ernest Hemingway and Marjorie Rawlings, honorary fire chief of Marineland in 1946.[18]

The increased public demand and interest in the undersea world for recreation and entertainment in the postwar years placed considerable strain on the negotiations between science and spectacle at Marine Studios. The success of Marine Studios' star attraction, Flippy, the educated dolphin, threatened to tip the scale toward entertainment. Caught in a nearby estuary in 1950, the two-year-old male dolphin underwent an extensive training program headed by Adolph Frohn, a fourth-generation wild animal trainer who performed with the legendary *Bring 'Em Back Alive* Hollywood star Frank Buck and with Ringling Brothers' Barnum & Bailey Circus before joining Marine Studios in 1949. Under Frohn's direction, Flippy became Marine Studios' premier performer. In his act, Flippy entertained thousands through a series of tricks that began with the yank of a lanyard and raising of a pennant near his tank and included honking a horn, ringing his own dinner bell, and jumping to break through a paper-covered hoop three feet above the water. Barrel rolls brought cheers from the crowd, but none of Flippy's tricks bedazzled spectators as much as the show's climax, when the dolphin star towed a woman and a little fox terrier around the tank on a surfboard. Flippy attracted such a following that a stadium with a seating capacity of a thousand was built in 1953 just for his show. Flippy's reputation spread across the Atlantic: before his death in 1955 he was a regularly featured celebrity in the *Illustrated London News*.[19]

The appearance of Flippy with a fox terrier attested to the domestication of the dolphin in American culture. The attribution of individual personalities to animals through names such as Flippy, or Algae and Splash, Flippy's understudies, or Grumpy, the loggerhead turtle, also reinforced Hollywood screen types and played off the associations of these undersea creatures with tame pets. The naming of individual animals had a long tradition within natural history film. Martin and Osa Johnson adopted the convention from the genre of the realistic animal story crafted by writers like Ernest Thompson Seton and Jack London. Disney found it a powerful way to elicit emotional identification with animals and capture their personalities on screen. Marlin Perkins appealed to pet-keeping audiences by featuring Chicago's favorite Lincoln Park Zoo pets—each with its own cutesy name—on his show *Zoo Parade*. Although McBride may have discouraged the "personalizing" of animals in the exhibits and believed that it "detract[ed] from the fundamentally scientific purpose of the Studios," a brochure from the early 1940s reveals that even generic names such as Mrs. Porpoise and Baby, or Professor Schoolmaster and Dr. Doctor Fish conjured up images of human occupational stereotypes and personalities in the undersea world. The names of dolphin stars attracted crowds and provided trademarks for the stars' oceanariums. Bubbles, the pilot whale, became as much a featured celebrity at Marineland of the Pacific, for example, as Flippy was for Marine Studios.[20]

If Flippy's performance appeared little different from other animal shows featured in three-ring circuses, the popular press and Marine Studios put a different spin on Flippy's accomplishments. *Popular Mechanics,* for example, emphasized that "Flippy is not just a trick animal like a trained seal. He is not even on exhibit at Marineland. He is one of a host of marine experiments being conducted at the world's most remarkable deep-sea zoo, where scientists peer through portholes to watch the private lives of marine denizens. Flippy is proving that porpoises have a brain development so advanced that it may compare with that of the dog and the chimpanzee." *Natural History* similarly declared that "Marine Studios blazes the trail in the study of porpoise psychology and produces a star performer."[21]

Frohn was no experimental psychologist. Vernacular knowledge proved valuable in the development of certain stunts, such as the surfboard in tow. The idea actually came from a group of fisherman in St. Petersburg who had trapped a dolphin and fitted it with a harness attached to a row boat, in which they sat for a free ride around the lagoon. Many of the training techniques Frohn developed were based not on the principles of behavioral psychology, but on a craft tradition handed down and modified from one generation of animal trainers to the next. But Frohn's ability to train Flippy provided visual testimony to scientific claims made by McBride and psychologist Donald Hebb in 1948 that the "behavior of the porpoise is in some respects comparable with the chimpanzee's, in others perhaps with the dog's." Even in professional scientific journals, investigators such as Barbara Lawrence and William Schevill from the Woods Hole Oceanographic Institute commented that Frohn's successful education of "Flippy" confirmed that "porpoises are of a high order of intelligence." As the individual personalities of dolphins and their remarkable intelligence became more apparent, researchers like Schevill expressed misgivings about certain aspects of marine mammal research. In 1955, Schevill was prompted to free a dolphin supplied by the Studios after it became ill due to a salinity deficiency in its tank, even though an autopsy of the animal might have provided important pathological information. "We were reluctant to slice up a beast we had gotten so well acquainted with," admitted Schevill.[22]

Flippy's career as an experimental subject of psychology raised his status, and that of Marine Studios, from a side-show attraction to an authentic educational, yet entertaining, experience. The importance of science in fashioning Marine Studios' public image was evident after Frohn left in 1955 to take a position as dolphin trainer for the newly opened Miami Seaquarium, the first of Marine Studios' major competitors and the future home of Flipper. Marine Studios recruited Keller Breland to devise a new set of training techniques that utilized science rather than the craft knowledge of circus animal trainers for its dolphin show. Breland, a graduate of the University of Minnesota, had assisted B. F. Skinner during the Second World War in Project Pigeon, a government-funded research program to use pigeons as homing devices in missile-guidance systems. Inspired by the success of operant conditioning in Project Pigeon, a technique that brought observed patterns of behavior in an animal under control through reinforcements such as food, Breland and his wife Marian formed in 1947 Animal Behavior Enterprises, a commercial venture to develop applied animal psychology to train animals for film, television, and other staged shows. In a matter of two short months, Breland's assistant, André Cowan, a Marine Studios maintenance department worker, had Algae performing a stunning set of feats that far surpassed Flippy's talents. Algae "waved his tail to the audience, leaped high in the air to catch a football, tossed a basketball through a hoop 5½ feet above the water, and made a high jump to take the rubber tip off a pole suspended about 16 feet above the water." It had taken Frohn six months simply to get a dolphin to ring a bell. The success of Breland's scientific training program reinforced the image of Marine Studios as an authentic scientific enterprise unrelated to the traveling circus or roadside zoo.[23]

Breland's use of operant conditioning was one among many examples of scientific research helping to fashion the dolphin's role as simultaneous experimental subject and popular celebrity. To facilitate experimental research, McBride trained dolphins in the late 1930s to tolerate handling by trainers and to accept a harness. His early success suggested that dolphins could be educated to perform in staged animal shows. Slapping mullets on the water, a technique developed to entice dolphins to eat dead fish, provided one of the first important clues into dolphin hearing and communication and was also used in the first trained performances in which dolphins leapt out of the water for food.[24]

However, scientific research did not always lend support to the dolphin's career in the public limelight. Dubbed "quiz kid of the ocean" in a 1953 article in *Science News,* the dolphin, in its meteoric rise to fame, found its popular image shaped as much by the constructed ideals of family life found on television shows and in films of the 1950s as by science. As early as 1946, CBS made plans to broadcast Marine Studios films to an estimated 3,500 receiving sets. Television specials such as *Marineland Carnival,* a CBS show that featured Lloyd Bridges, Rosemary Clooney, and Bill Dana, and included a baseball game skit in which Clooney and Dana test their batting skills against a dolphin pitcher, helped to introduce the animal into American culture. The 1955 film sequel to *Creature from the Black Lagoon, Revenge of the Creature,* in which both Clint Eastwood and Flippy make their first Hollywood screen appearances, further familiarized mass audiences with the charismatic dolphin. To offer an entertaining and educational experience that the whole family could enjoy, however, Marine Studios had to perform some skillful editing and savvy public relations in the presentation of its star performer. While television shows such as *Adventure* and *Zoo Parade* offered visions of nature that conformed to the cult of domesticity found in the 1950s suburban home—with a focus on parenthood, traditional gender roles, and the nuclear family as routes to personal fulfillment—certain aspects of dolphin behavior were not so easily domesticated. Just as the research laboratory at Marine Studios was removed from public sight, so too were certain aspects of the dolphin's sexual life, kept purposefully behind the scenes.[25]

As early as 1914, Charles Townsend noted that dolphins kept briefly in captivity at the New York Aquarium attempted to mate frequently, although he said nothing more about their sexual behavior. In the first year of Marine Studios' operation, McBride observed rampant sex play, particularly among males in the tank. In their scientific study on dolphin behavior published in 1948, McBride and Hebb were explicit about the kinds of sexual behavior observed. "Among males," they reported, "there is a good deal of masturbation, on the floor of the tank and against other males. One male had the habit of holding his erect penis in the jet of the water intake for prolonged intervals. The males also show a good deal of sex play with sharks and turtles, with the appearance of attempted copulation. With the turtle as sex object, the penis is inserted into the soft tissues at the rear of the shell."[26] Flippy was himself known among Marine Studios personnel for his ceaseless "masturbatory practices." Often, the sexual aggressiveness of dominant males resulted in the death of other dolphins and specimens in the tank. Herman, the first pilot whale ever to be kept in captivity, for example, was severely injured in 1949 after two bull dolphins and a female continually bit and rammed him into the sides of the tank during the mating season. He died later after a second attack resulted in a broken jaw. After the first live birth of a dolphin in 1947, McBride had to administer large doses of paraldehyde to the bull male in order to curb its frequent attacks on the newborn.[27] In 1954, Forrest Wood, who became curator of Marine Studios after McBride's untimely death in 1949 at the age of thirty-four, remarked that in his experience, "male *Tursiops* [Atlantic bottle-nose dolphins] will engage in sexual activity with practically anything, male or female, animate or inanimate." Wood found it "hard to understand why a male *Tursiops* may actually show preference for a different species when his own kind is available. Or why they have such a propensity for homosexual behavior when females are on hand." Whether such sexual behavior was a product of keeping dolphins in confinement or was in fact representative of their behavior in the wild remained an open question. Arthur Schmidt suggested that the preponderance of same-sex activity observed in captive dolphins was no different from the sexual behavior they had observed among other species kept in artificially confined conditions in Noble's Laboratory of Experimental Biology. In either case, in ranking dolphin intelligence, McBride and Hebb suggested that the dolphin's preference for same-sex partners and elaborate sex play placed it closer to primates "(particularly human) than to that of cat or dog."[28] Scientific evidence aside, oceanariums knew that successfully managing the dolphin's career depended on keeping the dolphin's private sexual life out of public sight, both in the tanks and in the popular media.

Unbeknownst to the public, the variety of sexual behaviors displayed by dolphins contributed to their charismatic appeal. Through a technique of cross-drive conditioning, Breland adapted sexual behavioral patterns observed in male and female dolphins into stage tricks, such as the tossing of objects with the tail. Although articles in the popular press occasionally mentioned the sexual belligerence of male dolphins, such as a 1950 *Saturday Evening Post* piece on the "Practical Joker of the High Seas," these exposés became exceedingly rare as the demand for family entertainment soared.[29] More common was the experience of Frank Essapian. An assistant at the Marineland Research Laboratory, Essapian was asked by the editor of *Natural History* magazine to excise a paragraph on homosexual behavior in his article for the magazine on dolphin behavior in 1953. Although Essapian did mention the attacks on a sand-bar shark and Herman the pilot whale, the dolphin's aggressiveness was depicted as defense against adversaries whose presence they "either resented or considered dangerous." The main focus of Essapian's article, like the 1940 feature in *Life,* centered on mother-infant relations. In Disney's 1959 True-Life Adventure, *Mysteries of the Deep,* the dolphin footage filmed at Marine Studios similarly emphasized the "miracle of living reproduction" and highlighted the caring mother assisting her infant to the surface for its first breath of air. Any reference to the male's part in reproduction and the rearing of young was kept behind the scenes. Not surprisingly, footage of female dolphins masturbating with a beach ball by inserting it into their vaginas, shot at Marine Studios for animal psychologist Frank Beach, never made its way to the theater. The image of the dolphin was being cropped for popular consumption.[30]

Scientific research, tourism, and the entertainment industry had turned the dolphin into a highly valued commercial commodity in the 1950s. Typecast as a friendly, playful, highly intelligent mammal of the sea, the dolphin, with its charismatic built-in smile, played its part well. It was a role that capitalized on military, public, and scientific interest in the sea and conformed to the criteria of family fare and moral values widely evident in nature shows on television and the motion picture screen. How different a part it was from the place dolphins occupied in American culture at the turn of the century. When Townsend captured dolphins off the coast of Cape Hatteras for the New York Aquarium in 1914, their only commercial value was for jaw oil used in the lubrication of watches and clocks. In fact, the market for dolphin byproducts was so limited that only a few dolphin fisheries existed in the whole of North America. Even in the late 1930s, the staff of Marine Studios looked upon dolphins as a cheap and easily accessible specimen for scientific research and display. But the training of dolphins in the postwar era to assume the role of playmate and close relative to humans significantly increased their value as both experimental organisms and tourist attractions.[31]

In 1955, Wood expressed concern to Burden that too little was known about the causes and cures of dolphin diseases. "While it [was] no great problem to get new porpoises," he remarked, "with a limited number of trained porpoises," an "occasional illness and death" could seriously affect "an important part of our show." The presence of porpoises had become critical to the Studios' well-being and their treatment and presentation warranted special concern. The commercial value of dolphins had already become so significant that by 1947, when Ilia Tolstoy, a partner in the founding of Marine Studios, proposed to sell information and data pertinent to the design, construction, and operation of oceanaria to a Miami-based company, Burden suggested a figure of $100,000, with the stipulation that Miami Seaquarium not exhibit dolphins for an undisclosed period of years. Miami Seaquarium opted instead to lure a number of Marine Studios' scientific staff and its animal trainer rather than agree to such a condition.

Without dolphins, an oceanaria was financially doomed. And, as Burden had discovered, without dolphins trained to conform to popular tastes and values, the same was true. [32]

Scientists, too, profited from the vogue for dolphins in mid-century American culture. Dolphins not only attracted tourist dollars, they also attracted the interest and funds of the Office of Naval Research (ONR). ONR's interest in dolphin research originated from difficulties sonar operators experienced in detecting enemy ships from false targets during the Second World War. In the spring of 1942 at Fort Monroe, Virginia, listening equipment installed to protect the entrance to Chesapeake Bay picked up loud hammering noises that completely drowned out sounds emitted by ships in the area. Investigations revealed the sound source to be a school of croakers (fish) that had returned to the Chesapeake to spawn. The potential interference of marine animal sounds with sonar gear, acoustic mines, and underwater listening equipment became a topic of military concern, and the Navy began a concerted effort to compile a library of marine animal sounds as a reference for sonar operators. In 1943, the first commercially available recording of underwater animal sounds was issued by Folkways, and included noises made by fish, crustaceans, and marine mammals, many of which were incorrectly identified. Contrary to *The Silent World,* Jacques Cousteau's book and 1956 film with Louis Malle under the same name, the ocean was a noisy place when compared to the hum of a busy office.[33]

Marine Studios became an active center for research in the field of marine bio-acoustics after the Second World War. Aristotle had noted the shrill cries and moans of dolphins in his *De Animalium Historia,* but it was at Marine Studios where the mysteries of dolphin communication were first seriously explored. From the first day dolphins were placed in captivity, McBride noted "that variations in the whistling sounds" made by the animals were likely indicators of "different emotional states." Upon the suggestion of William Schevill, a pioneer in the field of porpoise sound recording, Marine Studios started broadcasting "the underwater chatter" of dolphins over the public address system. Schevill, in conjunction with ONR-funded research he was pursuing, had brought with him a naval hydrophone that revealed the captive porpoises to be "noisily articulate." A record containing a selection of these porpoise conversations was cut in 1952 and made available for purchase. The broadcast of these recordings over the PA system, however, had to be discontinued after McBride's successor, Forrest Wood, noticed that the porpoises in the tank reacted to the PA broadcasts. Wood had initiated experiments on the meanings of porpoise phonation and feared that the porpoises would "become accustomed to the playback and fail to respond in any perceptible manner," thus ruining his experimental design. In this instance, the demands of science and popular consumption did not converge.[34]

While whistling noises seemed to have an emotional significance—an infant separated from a mother would emit such cries until reunited—the function of what biologists variously described as a "creaking-door," "rusty-hinge," or barking sound remained elusive. McBride and Hebb had posed the question of whether vision or hearing was the primary sense by which porpoises navigated through water. Based on anecdotes of fisherman that porpoises would avoid ordinary seine nets but were readily caught in nets of larger mesh, McBride suspected in 1947 that porpoises had "some highly specialized mechanism enabling [them] to learn a great deal about [their] environment through sound." He obtained from the Navy what he loosely described as "a supersonic sending and receiving apparatus" and began trials on the behavioral responses of dolphins to ultrasonic frequencies. At the time of McBride's death, he had yet to analyze his data for publication, but visiting research scientists were aware of his speculations.[35] Schevill and Winthrop Kellogg, a psychologist from the Oceanographic Institute at Florida State University, independently established that the "rusty-hinge" sound emitted by dolphins was actually a series of rapid clicks or pings with a wide-band frequency spectrum that was used to navigate by echolocation. Decisive experimental evidence for the navigational use of sonar by dolphins came in 1959 when Ken Norris, curator of the Marineland of the Pacific, trained Zippy, a captive dolphin acquired from Marine Studios, to wear a blindfold of soft rubber suction cups. Even blindfolded, Zippy was able to locate objects from as far as thirty feet away. Schevill, Kellogg, and Norris were indebted to Marine Studios for the supply of captive porpoises and for the use of their research facilities, although each pursued their sonar research in conjunction with the Office of Naval Research. Their work, as Kellogg was quick to point out, bore directly on matters of "national defense, as a means of improving man-made sonar." Suddenly, the dolphin had become an ally not only of scientists and the entertainment industry, but of the military as well.[36]

Sonar was only one among the many national defense interests proposed for dolphins by John Lilly, Chief of the Section on Cortical Integration at the National Institutes of Health. In May 1958, at a dinner address before the annual meeting of the American Psychiatric Association, Lilly suggested that "before our man in space program becomes too successful, it may be wise to spend some time, talent, and money on research with the dolphins." "Not only are they a large-brained species living their lives in a situation with attenuated effects of gravity," Lilly reasoned, "but they may be a group with whom we can learn basic techniques of communicating with really alien intelligent life forms." Following so closely upon the heels of public hysteria generated by the launching of Sputnik in October of 1957, Lilly's remarks about dolphins serving as surrogates for extraterrestrial life made a huge media splash. The dolphin, through the image projected at Marine Studios and on Hollywood screens, had become man's invaluable friend.[37]

By the late 1950s, the ocean and outer space had each taken on the mantle of science's final frontier. Each remained a mysterious place, yet unconquered by science. Marine Studios diver Bill Zeiler remarked that his descent into the circular tank had altered his perceptions of both space and time. "Feeding time lasts about twenty minutes," he wrote, "but to me it seems almost a week. I am totally oblivious to the outside world. I feel like a person who has suddenly been transported to another planet." The opening scene of Cousteau's 1956 film, *The Silent World,* played upon similar images. Accompanied by an eerie, mysterious sound track similar to the musical scores of 1950s science fiction films, the opening credits and footage of divers 165 feet under the ocean, along with the amplified sounds of their breathing apparatus, reinforces both visually and audibly the narrator's suggestion that "these divers wearing the compressed air aqualung are true spacemen swimming free as fish." The sounds of a regulator delivering compressed air, gravity-defying movements of underwater divers, and poor picture quality combined to give an otherworldly feel to the first live undersea broadcast in television history. Hosted by Charles Collingwood and Ken Norris from Marineland of the Pacific, the February 19, 1956 episode of the CBS series *Adventure* was filmed twenty-two miles off the coast of California in the rich kelp beds of the Pacific Ocean. In drawing parallels between dolphins and extraterrestrial life, Lilly made explicit the associations between the ocean and outer space commonplace in the Eisenhower years of the Cold War. Both were alien environments—the next frontiers to be explored, conquered, and made hospitable for human habitation.[38]

The source for Lilly's belief in the possibility of establishing interspecies communication with dolphins was an expedition of eight esteemed neurophysiologists from the Institute of Living, Johns Hopkins University, Mt. Sinai Hospital, the National Institutes of Health, and the University of Wisconsin-Madison that Lilly assembled in 1955 to spend two weeks at the Marineland Research Laboratory, mapping through electrical stimulation the sensory and motor regions of the dolphin's brain. Because the dolphin's brain-to-body-size ratio was closer to humans than that of other primates, they appeared ideal experimental subjects for investigating motivational centers of the brain and did not then, at least, pose the same ethical dilemmas as arose in electrical stimulation research on human subjects. Wood, curator at the Laboratory, warned the investigators that the experiments they proposed would be fraught with difficulties. First, the skull of the dolphin was incredibly difficult to penetrate. Neuroanatomical studies of the brains of dead specimens at the lab had previously used a combination of saw, screwdriver, and pliers to get the brain out intact. Such a technique, Wood suggested, "might leave something to be desired when you want to expose the cortex of a living animal." Wood also expressed concern about the animal's reaction to anaesthesia, since his experience suggested that "when the respiratory control mechanism is sufficiently affected the blowhole valve (or valves) automatically clamp and the animal suffocates." His concern was warranted. Unaware that dolphins have a voluntary respiration system and would thus suffocate when unconscious, the group quickly killed all five porpoises Marine Studios placed at their disposal.[39]

Touched by the respect that Wood, André Cowan, and other Marine Studios staff had for dolphins, and the disapproval his invasive and destructive experimental procedures elicited, Lilly became interested in exploring dolphin intelligence and communication based upon his acquaintance with the dolphin vocalization research and training program at Marine Studios.[40] In 1957, after developing a technique on monkeys for brain electrical stimulation without anesthesia, Lilly returned to the Marineland Research Laboratory to try once again to map the motivational centers of the dolphin brain. A sleeve-guide made of hypodermic-needle tubing was hammered into the dolphin's skull until it just penetrated the cranial cavity, and then a drive control unit pushed an electrode, inserted through the sleeve guide, at one-millimeter increments further into the brain. After a day of experimenting in this manner, Lilly hit upon a region of the cortex that when stimulated caused the animal to vocalize in a lively manner. Lilly then set up a reward system whereby the dolphin could activate the positive motivational center of the brain by pushing a button with its beak. Compared to Lilly's experience with monkeys, the dolphin learned to operate the reward system in relatively few tries. After a few hours, however, the dolphin began pushing the switch in a frenzied manner, went into a grand mal seizure, and died. Although saddened by the loss of yet another experimental subject, Lilly began to suspect that dolphins possessed a complex language based upon their large brain size, rapid learning abilities, and vocalization skills. In 1959, with support from the Office of Naval Research, the National Science Foundation, the National Aeronautics and Space Administration, and private philanthropists and with a supply of captive porpoises from Marine Studios, Lilly established the Communications Research Institute on St. Thomas in the Virgin Islands in an effort to crack the code of dolphin language and establish interspecies communication.[41]

Lilly's 1961 book *Man and Dolphin,* written in a speculative, polemic style to provoke scientific and public support for the study of interspecies communication and his own institute, received widespread acclaim and ridicule. Margaret and William Tavolga, biologists in the Department of Animal Behavior at the American Museum of Natural History who pursued research on dolphin behavior and the biological significance of sound production in fishes at the Marineland Research Laboratory, denounced the book as an example of how not to do scientific research. They questioned the facile links Lilly made between brain size, intelligence, vocalization, and language, and decried Lilly's experimental handling of the animals. While Lilly insisted on the need for "bilateral kindness" in order to establish a dialogue between humans and an alien species, the Tavolgas found it difficult to understand how dolphins could "feel 'kindly' toward Dr. Lilly while they are confined in coffin-like plexiglass tanks and steel tubes are being hammered into their heads." Forrest Wood also expressed dismay with Lilly's book and was upset by the "free use" Lilly made of Wood's name.[42]

Despite these objections, biologists working on dolphins profited by Lilly's work. Wood himself noted that *Man and Dolphin* "stimulated interest in porpoises" and "possibly induc[ed] people to stop by Marine Studios when they otherwise would go past." But *Man and Dolphin* did more than increase ticket sales. Lilly offered a whole range of military applications for trained dolphins that included everything from "hunting and retrieving nose cones, satellites, and missiles," to "scouting and patrol duty" for the Navy, to the delivery and attachment of atomic nuclear warheads on submarines and surface ships.[43] The establishment of the Marine BioScience Division of the Naval Missile Center at Point Mugu, California in 1963—where Wood headed research on marine mammal physiology, hydrodynamics, sonar, and training for underwater military operations—occurred in large part because of naval interest in marine mammal research inspired by Lilly's provocative claims. And while Lilly's speculations seemed like science fiction to biologists such as William and Margaret Tavolga, the launch of Sealab II in 1965, a U.S. Navy–sponsored underwater colony two hundred feet beneath the waters of La Jolla, California, turned fantasy into reality. Headed by Scott Carpenter, the second Mercury astronaut to orbit the Earth, Sealab II employed Tuffy, a dolphin trained at the Navy's Marine Bioscience Division at Point Mugu, to shuttle packages back and forth between the underwater station and the surface and to aid in the rescue of lost divers. While some biologists criticized Lilly for sacrificing scientific integrity by pandering to the public's fascination with extraterrestrial life, they neglected to consider how the dolphin as popular spectacle contributed not only to Lilly's fame, but to opening the military coffers for scientific research. In nature as entertainment, science and spectacle are never far apart.[44]

In 1963, when MGM released *Flipper* the movie, the dolphin already had an established career in Hollywood, the military, and scientific research. Mitzie, the dolphin costar of Chuck Connors and Luke Halpin, may have been the first of several female dolphins to play the loveable, intelligent friend of the boy Sandy in the movie and television sequel that followed, but *Flipper* was not the dolphin's first stage appearance. Scientific research, the entertainment industry, and military interests each played a backstage role in the domestication of the dolphin within American culture. By the early 1960s, dolphins even became part of the pet trade. In 1963, *Life* magazine alerted its readers that they too, like Sandy in the motion picture *Flipper,* could have a pet dolphin for a mere three hundred dollars.[45]

The making of the dolphin into a glamour species within American culture was, like the making of natural history film, the result of much behind-the-scenes labor in which scientific research and vernacular knowledge, education and entertainment, and authenticity and artifice were edited and integrated into the final scenes that appeared before the public. The ending credits of *Flipper* provide a partial list of the historical actors, institutions, and cultural forces at play. Adolph Frohn, Miami Seaquarium, Marineland of the Pacific, and John Lilly are each singled out, but such names only touch the surface. Ivan Tors, the producer of the original motion picture and television series that ran from 1964 to 1968, and executive producer of such television shows as *Sea Hunt* and *Daktari,* was also a major contributor to Lilly's Communication Research Institute. Douglas Burden had long planned a feature-length film around a boy and dolphin and had even asked for Lilly's help in finding a suitable filming location in the Virgin Islands in 1959. Apparently, production plans were forestalled and Marine Studios lost out to its major competitor, Miami Seaquarium, where much of the underwater footage and training of dolphins for the motion picture and television series took place. Nevertheless, Flipper's character as a playful, communicative, highly intelligent creature of the sea had been typecast within American culture well before production at Miami Seaquarium began.[46]

The conventions that emerged and shaped scientific and public understanding of dolphins in 1950s American culture continue into the present day. Animal rights and environmental interest groups seize upon the dolphin's human-like intelligence to appeal to the public for financial support. Ecotourists flock to beaches to swim with dolphins. And the television network NBC appropriated visions of the ocean as a militarized space frontier complete with a talking dolphin in the 1993–1995 NBC series *Seaquest,* which initially included a weekly endorsement from Robert Ballard of the Woods Hole Oceanographic Institute. However, changes in the cast and screenwriting, which gave the show a "sexier" look and departed from its science-fact premise, turned the show into what outraged fans dubbed "Bayquest 90210" and "Voyage to the Bottom of the Barrel."[47]

The conventions that shape our affections for dolphins have become so thoroughly entrenched and have played such an influential role in shaping marine mammal protection that efforts to shift environmental policy away from a focus on endangered species to ecosystem-processes face difficulty in attracting public support. In August of 1997, the Clinton administration and Congress sought to downgrade the dolphin's glamour-species status in favor of a more ecosystem-centered approach to fisheries management. Fearing an international battle sparked by a complaint filed by Mexico under the General Agreement on Tariffs and Trade against U.S. environmental laws that prohibit the sale of tuna caught by purse-seine nets, which kill thousands of dolphins a year, the administration worked out an agreement that would permit the sale on the U.S. market of tuna taken by Latin American fisherman through purse-seine nets, provided that dolphin kills are limited to five thousand animals per country per year. The agreement, favored by some environmental organizations, was labeled by others the "Dolphin Death Bill." Citing concerns about the harm posed to other marine fauna caused by dolphin-safe fishing techniques, the U.S. government has favored the bill as a step toward ecosystem management. But as a reporter for *U.S. News & World Report* noted, American consumers "may care less about preserving ecosystems than about the plight of creatures as intelligent and appealing as dolphins."[48] If the dolphin's place in American culture is to be remade and new alliances in its protection are to be forged, we all must look behind the scenes from which the dolphin emerged as pet star and acknowledge our part in this seaworld creation.

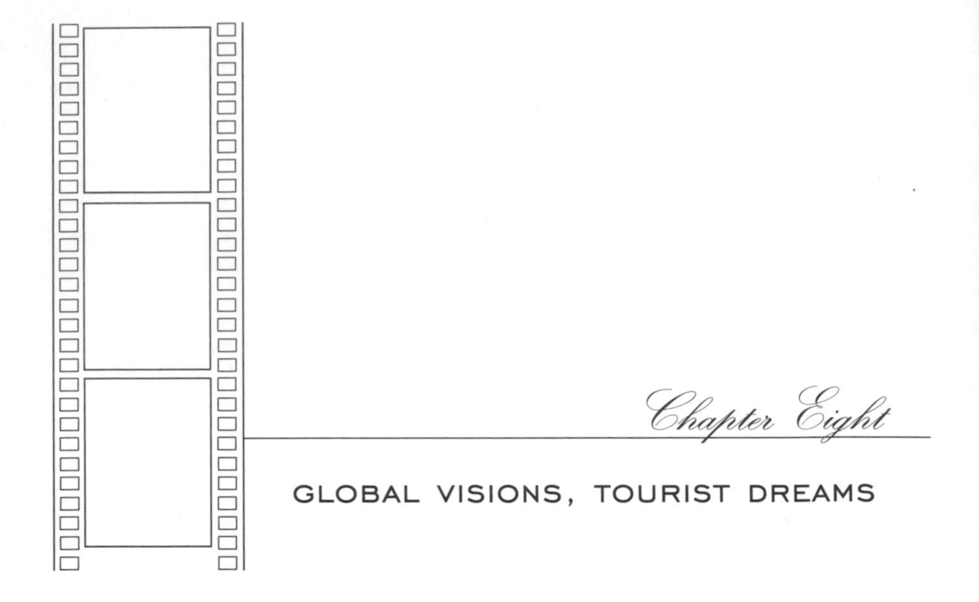

Chapter Eight

GLOBAL VISIONS, TOURIST DREAMS

At the 1992 Rio Earth Summit, BBC television adopted a popular icon to signify the global environmental agenda of this United Nations conference: "The Whole Earth" photograph taken from an Apollo spacecraft in 1972. Surrounded by total darkness, planet Earth, with its delicate white clouds swirling over radiant blue oceans and familiar land masses, is a celebrated symbol of the finite and fragile home that humans as global citizens share. But recognition of the global nature of environmental problems did not emerge with the Apollo spacecraft shot or the environmental bumper sticker, "think globally, act locally." In 1949, a scene of Earth revolving in space, presaging the Apollo image, opened the Conservation Foundation film *Yours Is the Land.* Created by a special effects studio, the footage conveyed a message of the earth as a shared home, finite in its resources and belonging to a new world order. In the immediate postwar period, conservation became part of an internationalist vision of the United States and its allies meant to insure peace and security through economic growth and prosperity across the globe.[1]

In an era of unprecedented affluence, *Yours Is the Land* reminded parents and children that they were living in a world of limited resources. At mid-century, American conservationists argued that the limits of the Earth necessitated the abandonment of isolationist sentiments at home and nationalist zeal abroad if the quality of life for Americans was to be sustained and that of developing nations was to be improved. In the face of fears that overpopulation coupled with dwindling resources would create conditions of want and need and would open the door to communism, conservation became a powerful means to champion the cause of international democracy in the name of economic interdependence. Bolstered by the research of Western conservationists, financial assistance from international environmental organizations, and tourist dollars, seemingly unproductive lands in developing nations would be transformed into gainful recreational economies. Nature as a tourist playground would stimulate economic growth and prosperity in the developing world while protecting wildlife as a global resource. Wildlife films would act as the messenger, fulfilling the dreams of Americans to experience a type of nature not found in their back yards.

The United States first articulated a limited global vision regarding management of the Earth's natural resources during the 1930s. Roosevelt's Good Neighbor Policy, with its avowed commitment to nonintervention and trade reciprocity, advocated an economic internationalism, albeit one confined to the Western hemisphere.[2] Despite the hemispheric solidarity against the Axis powers, many obstacles prevailed in achieving inter-American cooperation, not the least of which were negative cultural stereotypes. At the suggestion of Nelson Rockefeller, for whom hemispheric security hinged on economic prosperity of Central and South America, Roosevelt created in 1940 the Office for Coordination of Commercial and Cultural Relations between the American Republics (renamed the Office of the Coordinator of Inter-American Affairs, or OCIAA, in 1941) to promote inter-American exchange through cultural and information campaigns that mobilized the press, radio, and film industries.

Walt Disney was one of the first Hollywood producers enlisted by the Motion Picture Division of the OCIAA "to carry a message of democracy and friendship below the Rio Grande." While Disney's OCIAA films were intended to promote cultural reciprocity, the packaged spectacle of Latin American geography, music, and dance in fact encouraged a model of dependency in which Latin America was portrayed as a product to be consumed by North American tourists.[3] In the OCIAA film *The Three Caballeros,* for instance, Donald Duck receives three gifts from his friends south of the border. The first gift contains a movie projector and a film, *Aves Raras* (Strange Birds). Pablo the penguin was one strange bird who leaves his native Antarctica home to embark on a voyage of discovery in search of a warmer climate. In the film, animated postcard images of the geographical and cultural resources along the coast of Chile and Peru accompany Pablo's migration route to the Galapagos islands. The second and third gifts introduce the Brazilian parrot Joe Carioca and Panchito the Mexican rooster, who host Donald Duck's tour of Latin American culture. Inside the Disney film, Latin America *was* entertainment.

Animated birds were not the only feathered fauna to entertain audiences in the hopes of furthering the ideal of inter-American cooperation; real birds were enlisted as well. At the Convention on Nature Protection and Wildlife Preservation in the Western Hemisphere, opened for signature to member states of the Pan American Union in 1940, birds were just one natural resource under discussion. Ornithologist Robert Cushman Murphy took the convention as a sign of "the realization that modern problems cannot all be solved inside national boundaries."[4] In 1941, Fairfield Osborn, president of the New York Zoological Society (NYZS), wrote to Kenneth McGowan, director of production for the OCIAA's Motion Pictures Division, in the hopes of carrying the international message of birds to North American audiences through film. "The migration flyways of birds," Osborn urged, "may well be used to strikingly express the mutualities which nature has provided between the two continents." Osborn believed nature to be a universal category, the bond that unites "people of any blood or race." In the midst of Roosevelt's push for economic internationalism, birds became useful for diffusing nationalist sentiments and promoting the common ownership of natural resources.[5]

Before production of the bird migration film began, Osborn consulted with John Grierson, acclaimed British documentary filmmaker and commissioner of Canada's National Film Board. Grierson, who saw a "need and opportunity" for natural history films with well-developed themes, provided support for half of the film's production cost, while the OCIAA and the NYZS's Conservation Fund covered the remaining expenses.[6] Two versions on the migration theme were made: *High Over the Borders,* a twenty-minute National Film Board production, and *Birds on the Wing,* a ten-minute Columbia Pictures short subject.

High Over the Borders begins with an autumn farm scene in Wisconsin in which a young boy, Richie, is disheartened because his swallows have left their nest in the barn. His mother assures him that the birds will be back next spring. The scene then cuts to a shot of two young boys talking in their native language in Argentina. With shots of swallows flying overhead, we hear the following narration: "Ricardo says they are his birds, while Richie says they are his. Both of them are right, even though the swallows are the same. For the wild birds belong to all of us, to anyone who sees their graceful flight, who hears their call, to everyone who is aware of their existence." The audience then learns the follies of believing in exclusive ownership of birds through a visual tour of the flyway of the Canada goose, the hummingbird, swallow, and other migratory birds. Osborn, who disliked the idea of "humming-birds being turned against democracy," appropriated footage from a German film on hummingbirds financed by the Nazi Division of Ballistics and Aeronautics and recycled it to promote inter-American cooperation by documenting the hummingbird's trek from Florida across the Gulf of Mexico and over ancient Mayan forests in the Yucatan. Intercuts between migratory birds in flight and aerial photography of the regions over which they fly accentuates the shared cultural and natural resources of particular regions within the Western Hemisphere. The spectacular aerial photography also foreshadows how important surveillance of natural resources would become to maintain global security after the Second World War.[7]

The flyways of migratory birds in *High Over the Borders* instill a sense of unity among all nations of the Western hemisphere; birds are a common resource shared and enjoyed by all. This theme is further strengthened at the end of the film by the joyful faces of people of different racial and ethnic groups of the Americas as they follow the flight of birds overhead. The accompanying narration speaks of how the birds "mock the man-made lines by which nations separate themselves. For the birds are free, they are at home in the hemisphere. To them belong all the lands over which their wings carry them and they belong to all the people who live in those lands." Yet even in the film, reciprocity is never quite equal. For although all nations could freely appreciate the beauty of these global treasures, only the laboratory science of North America would eventually unlock the mystery of bird migration. In the second part of *High Over the Borders,* viewers learn that all information on bird migration collected by people and governmental agencies throughout the Americas is sent to the Fish and Wildlife Service in Washington, D.C., which serves as the central clearinghouse for data analysis. The segment concludes with an authoritative scene of scientists in the laboratory handling a Canada goose, with the assurance that "one day we may find the answer in the laboratory" to "the mystery that provides us with a thrilling story." As the scene fades to the finale with Ricardo and his friend riding on their burro, Ricardo tells his friend in Spanish that the birds have gone back to North America: "They all go there, and some day I'll go there to make a nest." The juxtaposition of the adult scientists and the two young friends reminds us of the paternalistic slant given to relations between the United States and Latin America during and after the Second World War. Prevailing opinion upheld Western science as the key to protecting wildlife and to preserving these world treasures for the enjoyment of future generations.

High Over the Borders was part of the National Film Board's wartime propaganda series titled *Canada Carries On*. As such, it intersects with a number of the series' themes, including the geopolitics and strategic importance of the world's resources in wartime production and the preparation for a postwar international world that would depend in part upon the planned use and conservation of natural resources, including wildlife.[8] In the emerging postwar world, conservation had taken on a distinctly different meaning than that embodied in the isolationist sentiments of a burgeoning industrial nation during the Progressive era. Conservation was to be part of a new internationalist vision, one that dispelled economic nationalism on the part of developing countries by emphasizing the common ownership of the world's resources. Global peace and security in the future were to be assured through the prosperity of all, a prosperity that would come through economic development and modernization based upon science and technology.

High Over the Borders and *Birds on the Wing* marked the beginning of Osborn's efforts to utilize documentary nature film to make citizens in the United States and foreign countries aware of the international dimension of environmental problems. "Too few people realize the worldwide character of the conservation problem in its broadest sense," Osborn wrote in preparation for his speech at the 1949 United Nations' Scientific Conference on the Conservation and Utilization of Resources (UNSCCUR). "Whether we wish to look outside, beyond the borders of our country, or prefer to turn our gaze inward, we are vitally concerned for our own well-being with improved conservation practices." With the hopes of promoting international conservation to assure the well-being of the United States and to secure nature as a global resource for all to share, the Conservation Foundation established an advisory council composed of noted conservationists, ecologists, and geographers such as Harold Coolidge, Charles Elton, G. Evelyn Hutchinson, Carl Sauer, Paul Sears, and William Vogt, to oversee a number of major research projects in its beginning years. Working in association with the Food and Agricultural Organization (FAO) of the United Nations, for example, the foundation initiated a survey of the "incidence of soil erosion in North, Central, and South America." The foundation's joint project with the FAO and Osborn's appointment to the Planning Committee of the Economic and Social Council of the United Nations in connection with the 1949 UNSCCUR meeting indicate the extent to which conservation issues occupied a central place in discussions regarding international development throughout the 1950s.[9]

Osborn believed that the best technical assistance to Third World nations would be conservation and land-use planning. Without Western aid, Osborn argued, native peoples would take the land for their own. But the United States, which had far outstripped its national resource base to sustain its economy, was too dependent on global resources to withstand intensified nationalism in the developing world. Speaking of Truman's Point Four program, aimed at providing technological skills, knowledge, and equipment to impoverished countries, Osborn insisted that such policies would only be successful if the United States "washes itself clean of paternalism and of the careless assumption that the ways of the Western world . . . are desirable for all people." It was not the Western ways of democracy that Osborn called into question, however, but faith in "industrialization as an ultimate panacea" and in "seeking ever-higher standards of living."[10] Sound agricultural practice, he felt, rather than industrial development, was the way to promote democratic values in the Third World.

Exporting conservation education and research became an important element of technical assistance offered to developing nations by the Conservation Foundation in the postwar years. In 1951, the Conservation Foundation received a $12,000 grant from the Rockefeller Foundation to make Spanish and Portuguese sound tracks of *Yours Is the Land* and *The Forest Produces,* compilation films based on The Living Earth and The Living Forest series aimed at secondary-school students and adults. Concerned about the "prospects of growing populations and diminishing resources" and the impact that population pressure on scarce resources would have on international relations, the Rockefeller Foundation looked upon the Conservation Foundation as a "convenient and effective agency" for conducting certain research projects related to the Rockefeller Foundation's interests in agriculture and human ecology, particularly those in Latin America.[11] *Yours Is the Land* proposed that poor land-use practices were a "major deterrent to economic development" and reaffirmed the need in developing countries for technical assistance programs in agriculture, like those funded by the Rockefeller Foundation.[12]

The international reach of postwar American environmentalism is accentuated by the opening shot of *Yours Is the Land.* The image of a revolving Earth in space and the mention of it as a planet of an average-sized star called the sun reminds viewers of its finitude and evanescence. The reminder that it is also our home, accompanied by a scene of an American farm and wheat field, imparts a caring attitude for the land and expands that affection beyond the North American landscape to include the globe. All is not well, however: We have not taken good care of the earth, we are told, and footage of soil erosion adds to the message. The film then takes us through a visual tour that follows the formation of topsoil and the ecological succession from pond to forest. We are witness to the interdependence of soil, water, animal, and plant life in the ever-shifting balance of nature. By instilling an appreciation for the land as a whole, *Yours Is the Land* promoted wildlife conservation through a wide-angle lens.

The arcadian scenes of nature in the film are disrupted by the arrival of humans. Stills taken of the dioramas of human ancestors from the American Museum of Natural History's Hall of Man are followed by an animated sequence illustrating the expansion of the human population around the globe and footage of deforestation, fire, erosion, and flooding created by the decimation of the land left in the wake of an expanding human population. An animated sequence comparing the relationship between population size and arable land in 1630 and 1949 emphasizes the point that "we have come at last to land's end." Only an integrated, long-range plan of land management can prevent future disaster and restore the delicate balance between humans and nature within the harmonious web of life.

Science offered technical knowledge to conserve the world's natural resources; children offered hope for a better tomorrow. The final sequence of *Yours Is the Land* begins with two children in front of a cornfield, and then cuts to idyllic scenes of wheat fields and mountains that accompany a soundtrack based on the song "America the Beautiful." But this optimistic vision is abruptly disrupted by the narrator's comment, "Yes, it's a good land, America. At least it was." Shots of burned forests, floods, trees being felled, dust bowls, and a sound track of complete silence follow. The audience's confidence is gradually restored as the narrator assures them that "it doesn't have to be like this, for today we have the technical skills, and science stands by to help." The film returns to the scene of the two children on the farm. With "America the Beautiful" playing in the background, the film ends in a spectacle of patriotism, family values, and individual responsibility. A Conservation Pledge appears on screen. It was a pledge meant to instill in children an appreciation for the environment, in much the same way the Pledge of Allegiance said by schoolchildren each morning was meant to reaffirm their faith in American democracy.[13]

Yours Is the Land, in combination with The Living Earth series reached over 2 million adults and children annually in the United States through distribution by Encyclopedia Britannica Films. Distribution of *Yours Is the Land* and *The Forest Produces* in Latin America was planned through package sales to schools facilitated by Truman's Point Four program. To help launch distribution, a premiere of the films with Spanish narration was arranged in May of 1952 at the Pan American Union. The "lovely setting" for the premiere, located on a "large balcony overlooking the inner garden" with a fountain, tropical plants, and a sliding roof that opened to the sky, seemed perfect "for any kind of film on nature." Over three hundred educators and diplomats from Latin American countries attended the Spanish premiere. With an enthusiastic endorsement from the noted Mexican ecologist Enrique Beltran, *Yours Is the Land* was subsequently shipped to fourteen distributors throughout Central and South America.[14]

Exporting nature appreciation and conservation, Osborn hoped, would contribute to "the internal welfare of many of the better developed nations" through "the continued improvement of resource use practices in the less developed nations." The well-being of the United States depended critically on the sound management of the natural resources of the globe. In the vision of integrated land management the Conservation Foundation promoted for Latin America, agriculture took precedence over other types of land use. As the Conservation Foundation expanded its reach beyond the Western Hemisphere to the African continent, wildlife came to be seen as the most valuable global economic and cultural resource of the land, a resource that needed to be preserved and sustained.[15]

During the 1950s, a change in Western images of Africa's landscape accompanied the continent's struggle for political freedom; the "Dark Continent" was rapidly transformed into a place of threatened ecological splendor. Once a land where the great white hunter affirmed both his manliness and the power of empire by subduing savage beasts, the new Africa became a place where a more intimate contact with an innocent, serene nature might be had. "This sense of peace and love and closeness to nature," reported *Reader's Digest* author Katharine Drake upon her visit to Kenya's popular tourist attraction, Treetops, was "not easy to come by in an atomic age."[16] For those who could afford to travel, Africa and its wildlife offered respite from the anxieties faced by a generation living in the shadows of the atomic bomb.

Armand Denis's film, *Below the Sahara,* released by RKO in the summer of 1953, brought to the American public a Technicolor vision of innocent nature in Africa. At the time of the film's release, few thrills were left of Africa's big-game country that American audiences had yet to see. *Below the Sahara* is a transitional film, suspended in time between the genre of travelogue-expedition films, such as Denis's 1938 *Dark Rapture,* which, in one reviewer's words, captured the "very heart of darkness, [the] immense, unknowable, and savage" Africa, and the nature documentaries and television shows of the late 1950s, such as Denis's *On Safari,* which portrayed Africa as a pristine wilderness endangered by the inevitable arrival of African independence.[17]

Born in Belgium, Denis emigrated to the United States after World War I, when a fellowship from the Belgian Relief Commission provided an opportunity to travel abroad. He abandoned his career as a research chemist and embarked upon a life of travel and adventure. With profits from a volume control device he invented for radio, he traveled to Bali, where he filmed his first successful box-office release, *Goona Goona*. After directing *Wild Cargo,* the second film featuring *Bring 'Em Back Alive* star Frank Buck, Denis turned his sights to Africa. Accompanied by his first wife Lelia Roosevelt, first cousin of Theodore Roosevelt, Denis launched in 1935 a two-year film expedition through the heart of the Belgian Congo. His friend Julian Huxley described *Dark Rapture* as "the finest African nature film ever made," an opinion with which the *New York Times* agreed.[18] Following the war, and after the release of his second African travelogue, *Savage Splendor,* MGM hired Denis as technical adviser for its blockbuster production *King Solomon's Mines*. However, by the time the film neared completion, Denis had a change of heart. He asked that his name be removed from the credits, because he felt the MGM feature presented "altogether the wrong view of wild life." "The scenes glorifying the hunter and the killing of animals," wrote Denis, "contributed to exactly that view of Africa that I disliked." When Denis returned to Africa after the war, he was dismayed by what he found. On the plains of East Africa where he had previously filmed an abundance of wildlife, he found little but "emptiness and silence." Equally troubling, in Denis's eyes, was the influence of Western "progress" on the indigenous tribes of the Congo. The Watusi king featured in *Dark Rapture* had now adorned himself in a grey pinstripe suit and the latest model Chevrolet. This progress, Denis believed, threatened Africa's wildlife, as habitat was cleared for agriculture and poachers plied their trade in the lucrative game, ivory, and rhinoceros horn markets. Denis became determined to produce films "relevant to the plight of animals" in Africa.[19]

Below the Sahara is a departure from Denis's previous films in part because of its almost exclusive focus on wildlife. *Dark Rapture* had fixed the audience's gaze primarily on tribal customs and dress found within the Belgian Congo. Except for a sequence on the capture of wild elephants for domestication, wildlife footage did not occupy center stage. In *Below the Sahara,* some footage reminiscent of the racist caricatures in *Dark Rapture* can be found. For instance, when the Maasai and other local tribes gather and dance, supposedly in Denis's honor, the narrator describes them as "emotional, fun-loving, [and in] need [of] little excuse for a jumbo dancing party." But these scenes are few. Instead, the camera's focus is on wildlife: from the inscription on Denis's Land Rover that reads "Protect Wildlife" to the spectacular footage of thousands of flamingos in flight. Wildlife provide the entertainment and endless comic fare. In a gannet rookery along the Atlantic coast of South Africa, for example, a male and female engaged in the head-flagging display that forms part of their courtship ceremony are, according to the narrator, "either [in] an argument or a budding romance. It could be both."[20] Denis had clearly learned a lesson from Disney in marketing nature as film entertainment.

As the success of both Disney and Perkins made evident, making nature entertaining on film depended in part upon the removal of any signs of human presence. In Africa, native peoples, once an entertaining spectacle of nature for American audiences, became in the 1950s wildlife's greatest threat. In *Below the Sahara,* for instance, aerial shots of jumping impala, topi, kudu, secretary birds, water buck, and wildebeest on screen remind audiences that a "wealth of wildlife . . . still survives in immense Africa." For a few fleeting frames, "the pages of time have been turned back toward a more primitive age when animals roamed the tropical earth in the countless thousands before man the enemy, man the ravager and destroyer had been born." But the ravager depicted in Denis's film is the native African. The film's climax builds as Denis and his second wife, Michaela, head off to witness a gorilla hunt. Denis emphasized that their "mission [was] not to participate, perhaps not even to record the brutal battle on film, but first and foremost to save some baby gorillas, immensely valuable to science, from being killed by the warriors." Not surprisingly, Denis saves a baby gorilla from "the ruthless vengeance of the hunters." Michaela holds the baby gorilla in her arms; the native Africans carry dead gorillas back to the village. The juxtaposition of shots triggers a reminiscence of the two Africas experienced on the journey: the one, a colorful and glamorous ecological paradise to be saved through the economic aid and technical expertise of the West; the other, a savage and violent land. The latter was an image that reinforced Western stereotypes of darkest Africa and racial fears of what the future might hold. By the end of the decade, the media consistently portrayed freedom for Africans as a threat to the freedom of wildlife.

Alan Moorehead's 1957 book *No Room in the Ark,* which was serialized in the *New Yorker,* presented to American readers a message similar to *Below the Sahara.* Moorehead, a war correspondent for the London *Daily Mail* in North Africa, returned to the continent in 1956 to travel in sub-Saharan Africa with the hope of seeing "jungle in the raw and the Africa of tribal drums." Upon his arrival, Moorehead's hopes were quickly dashed when a resident informed him that he wasn't likely to find "the noble naked huntsman with his spears and his war dances." Dismayed to find Johannesburg little different than Chicago or Memphis, Moorehead abandoned his hopes of seeing "primitive tribal life" and set out on a three-month tour in search of wildlife that began in Kruger National Park and ended in Kenya and Tanganyika. For his *New Yorker* readers, he painted a landscape in transition. In a paternalistic tone, he described Africans as "lost between the old world of the ritual dance and the new one of water-skiing tourists on the lake." He conveyed an impression of Africans in an adolescent stage, no longer children in the state of nature, yet not quite sure of themselves in the "sophisticated" adult world of whites. Moorehead's account may have weakened cultural stereotypes of Africans as part of a primitive landscape, but he simultaneously reinforced a view that stripped them of their land. The removal of humans, which included both "primitive" tribes and the great white hunter, transformed Africa from a savage and violent nature to an ecological paradise. "Once the human element is removed—or at least the prospect of danger from that source," Moorehead reflected, "an entirely new world emerges, in which the emphasis shifts from the marksmanship of the hunter to the skein of influences that govern the animal community." Yet Africans were distinctly portrayed as a threat to that ecological paradise. Moorehead's exposé of the poaching trade and sketch of Maasai pastoralism as a grave danger to the ecological integrity of Serengeti National Park could leave little doubt in the reader's mind. Amidst ads for British European Airways, expensive color-slide cameras, and convertible luggage-to-seat products, Moorehead looked upon Western tourism as one saving grace for Africa's wildlife. The influx of tourists and their dollars to witness this paradise would convince Africans to look upon wildlife as a vital economic resource.[21] Disney, in promoting Roosevelt's Good Neighbor Policy, had packaged Latin America as a tourist paradise for North American consumption. In the same way, the beginnings of ecotourism in Africa—sparked by such films as *Below the Sahara*—reinforced the United States' internationalist vision of economic interdependence with the Third World.

By 1960, the image of Africa as an earthly paradise, an ecological Garden of Eden, became a prominent feature in nature films, television shows, and articles in the popular press. In eating the fruits of Western civilization and freedom, Africans were themselves cast out of this garden. Indigenous tribes allegedly lacked the enlightened appreciation of nature found in developed nations like the United States. Their voice in the management of wildlife was silenced, just as their presence in shaping the natural landscape of Africa was effaced from film. Scientific testimony fostered this representation within popular culture, as ecologists increasingly frequented Africa to conduct ecological reconnaissance missions and to offer advice on game management and land-use planning.[22]

Through the Conservation Foundation and the New York Zoological Society, Osborn helped sponsor numerous ecological surveys of the African continent influential in the formulation of technical assistance programs within international agencies such as UNESCO and the FAO. The relationship between land use, overpopulation, and political stability concerned individuals prominent in an emerging network of international environmental organizations that included the Conservation Foundation, the International Union for the Conservation of Nature (IUCN), and the World

Wildlife Fund. While no formal conservation body existed within the United Nations, Julian Huxley, a prominent British biologist and first director-general of UNESCO, managed to bring nature protection within UNESCO's purview by suggesting that "the enjoyment of nature was part of culture, and that preservation of rare and interesting animals was a scientific duty."[23] UNESCO did not immediately enter the arena of nature protection and conservation, but Huxley utilized his position within the agency to help establish in 1948 an international hybrid governmental and nongovernmental organization, the International Union for the Protection of Nature (IUPN), later renamed the International Union for the Conservation of Nature. Throughout the 1950s and early 1960s, such agencies as the IUCN and the Conservation Foundation emphasized the need for ecological surveys in the design of U.N. technical assistance programs for developing countries.

These ecological reconnaissance missions were needed, Osborn believed, because "with unspoiled tribal life . . . rapidly diminishing, the viewing of wildlife and wild scenery [had] become the dominant attraction" and thus constituted an important economic asset for developing African nations.[24] In 1956, American ecologist Lee Talbot inventoried threatened mammals in Africa, south Asia, and the Near and Middle East. Talbot's inventory provided the initial data base for IUCN's Red Data books, the primary reference work on the status of threatened and endangered species. From the fall of 1956 through 1957, George Treichel, a University of California graduate with training in geography and ecology, along with his wife Jeanne, surveyed the status of wildlife areas in sub-Saharan Africa to produce a naturalist's guide to African national parks and wildlife reserves. The following year, Frank Fraser Darling, the Conservation Foundation's vice-president, at the request of the Kenya Wild Life Society and the Royal National Parks of Kenya, conducted an "ecological reconnaissance" of the Mara plains in Kenya, an area rich in wildlife threatened by pastoralism and African poaching.[25] By 1959, Osborn believed the wildlife situation had reached such crisis proportions that he instituted the African Wild Life Fund to support scientific, educational, and training projects that pertained to wildlife conservation in eastern and central Africa.[26]

These scientific accounts depicted an African landscape at the brink of ecological destruction. Treichel estimated that over half of the wildlife habitat in sub-Saharan Africa was decimated and that wildlife populations had been reduced by seventy-five percent.[27] Biologists singled out a number of factors contributing to this ecological crisis. A rapidly expanding population placed pressure on marginal lands unable to sustain pastoralism or agriculture. Tsetse fly eradication programs instituted by colonial governments opened previously uninhabitable and prime wildlife areas to livestock grazing and agricultural expansion. Lucrative markets for rhinoceros horn, elephant ivory, and meat led to increased native poaching both inside and outside national parks and game reserves. "Nationalism," Darling wrote, "in a world that must overcome the jejune irrationality of nationalism," comprised the greatest threat to African wildlife.[28] Native political leaders needed to be educated, biologists concurred, on the significance of wildlife as a unique economic and cultural resource of international value. In paternalistic language commonly found throughout the scientific and popular literature, Treichel expressed hope that as "the intellectual interests" of native leaders "expand, they also will regard" the national parks and reserves of Africa "as priceless living museums of natural history," the destruction of which would result in great loss to the international community.[29] Some authors questioned the ability of Africans to address the problem, and urged international control of Africa's dwindling wildlife population. When decolonization of the Belgian Congo turned violent in the summer of 1960, Osborn submitted a request to the Secretary General of the United Nations to send a task force of U.N. troops to protect the national parks during the conflict.[30]

Conservationists upheld tourism and technical assistance in wildlife ecology and range management as cornerstones of the efforts to preserve Africa's wildlife. In a 1960 ecological survey of Central and East Africa sponsored by UNESCO, Julian Huxley argued for the preservation of African wildlife "as an object of study and a spectacle to be enjoyed."[31] The former legitimated the presence of ecologists in African countries; the latter opened the door for tourists. A general, systematic plan of land use emerged from the ecological reconnaissance missions undertaken on behalf of international environmental organizations, which included a Fulbright program that sent U.S. experts in wildlife ecology and range management to Africa. Critical of pastoralism, and particularly of Maasai tribal practices that maintained livestock, a symbol of wealth and social standing, as a renewable high-protein food source through a practice of bloodletting, ecologists recommended the controlled harvesting of wildlife populations for protein production on lands unsuited for pastoralism or permanent agriculture. They reasoned that such government programs would undermine the economic viability of poaching for game meat and substitute for pastoralism on marginal lands a more ecologically sound practice of land use.

Despite the emphasis on *human* ecology in these studies, ecologists in international development efforts looked at humans primarily as a disturbance that would threaten the integrity of national parks as important field sites and tourist attractions. While the Conservation Foundation promoted multiple land-use conservation strategies in the United States in places like Jackson Hole, when their attention shifted to the international scene, the interests of ecologists and American tourists took precedence over those of local inhabitants. Huxley, for instance, insisted that the annexation of Maasai tribal lands around Ngorongoro Crater into Serengeti National Park meant "turning the Masai out."[32] Recommendations to fence all national park lands in order to ensure the preservation of "primeval Africa" reinforced a conservationist vision that separated humans from nature.[33]

In an effort to obtain endorsement for wildlife conservation from African leaders of newly independent states, the IUCN, in conjunction with UNESCO and the FAO, sponsored in 1961 the Pan-African Symposium on the Conservation of Nature and Natural Resources in Modern African States, which met in Arusha, Tanganyika. Many conservationists regarded the remarks of Tanganyikan president Julius Nyerere as a crucial turning point in international conservation when he declared that "wild creatures amid the wild places they inhabit are not only important as a source of wonder and inspiration but are an integral part of our natural resources and of our future livelihood and well-being."[34] For many ecologists associated with international development, however, wildlife constituted more than a national resource. "Africa's wild life belongs not merely to the local inhabitants," Huxley insisted, "but to the world. To let it die or be destroyed would be to allow a precious element in that rich variety to be submerged forever in the drab monotonous flood of uniformity that is threatening to engulf our mass-produced technological civilization."[35] Nature was a global resource, to be put under the watchful eye of ecologists and opened to the view-seeking tourists from the First World.

Economic prosperity, inexpensive air travel, and a penchant to escape the uniformity of mass culture combined, Huxley believed, to make Africa a prime tourist playground. Ecologists regarded tourism as central to their efforts in saving wildlife from destruction. With tourism the second largest source of revenue in Kenya, African leaders could be persuaded of the local economic payoff accrued by treating wildlife as an international asset. In his

1960 UNESCO report, Huxley called for U.N. aid to survey the tourist trade in East Africa, to consider the possibility for expansion, and to estimate the capital expenditure needed to provide increased facilities and personnel. Once the survey was complete, African nations might then apply to the World Bank for funds to embark on projects that would help accommodate increased tourist traffic. Furthermore, the U.N. Technical Assistance Board could facilitate tourist promotion by supplying wildlife and national parks departments with staff to research and produce popular education films.

As Huxley, Osborn, and other conservationists realized, wildlife films could help champion their cause. Nature documentaries enhanced the image of international conservation from a number of standpoints. In reinforcing an aesthetic relationship in which the links between humans and nature were established primarily through the eye, wildlife films accentuated a nature seen through the tourist's lens.[36] The removal of animals as a source of produced goods was critical to this transformation of nature into a recreational economy. Wildlife was there to be seen, not poached or culled. For audiences with the finances and leisure to travel, nature documentaries fostered a market for the photographic safari. Compared to North America, the abundance of wildlife in Africa led one to believe that spectacular shots of African flora and fauna could be captured readily by the amateur photographer. Even if individuals lacked the resources to travel, sublime images could still arouse individual concern for the plight of Africa's wildlife that translated into protests and contributions to international conservation organizations.

The director of the Frankfurt Zoo, Bernhard Grzimek, was one master in the use of natural history film as propaganda to enlist public support for his conservation efforts. In 1953, he and his son Michael traveled through equatorial Africa on a collecting expedition for the Frankfurt Zoo in search of the rare okapi. Upon their return to Germany, Michael made a film of their adventures. *No Room for Wild Animals* won the "gold bear" award at the Berlin Film Festival and was distributed to over sixty-three countries. The film portrayed Africa as a doomed ecological paradise and protested against a British government proposal to cede three-fifths of the land in Serengeti National Park to the Maasai.[37]

The image of Africa as an animal paradise, devoid of human inhabitants, that Grzimek and others helped promote on television and film reinforced the perceived need to exclude human inhabitants from African national parks and game reserves. Perkins had been explicit about the incompatibility of wildlife and people in Africa. Such a dichotomy was critical to the aesthetic of pristine nature that he, Disney, and others had cultivated for a mass market. Although rooted in the history of American environmentalism, that aesthetic became dominant in the philosophy of international environmental organizations after the Second World War. Exported to Africa, the notion of pristine wilderness shaped both wildlife conservation on the ground and in the global visions and tourist dreams of Africa that appeared on screen.

Grzimek's objections to the British government proposal for Serengeti National Park were firmly rooted in the political implications of pristine nature. When the park was created in 1951, no limits were placed on the number of Maasai or their livestock. By 1956, an expanding human population and an increase in herds forced the Maasai to graze their cattle on the central plain and floor of the Ngorongoro crater, a large extinct volcano of immense tourist value, which Grzimek described as "the largest zoo in the world."[38] The Tanganyikan Trusteeship Council believed that under the United Nations mandate, any attempt to limit the numbers of the Maasai would be a violation of their native rights. They therefore proposed making over half the park a controlled area that would be open to the Maasai, their cattle, and wild game. In Britain, the Fauna Preservation Society, a conservation organization with close connections to the IUCN and World Wildlife Fund, voiced public opposition to the proposal. Julian Huxley expressed his concerns to the assistant secretary general of the United Nations that the Trusteeship Council had acted in "the immediate interest of the native populations" rather than "for the benefit of all the peoples of the world." Huxley called for an international body to oversee the park.[39] Michael Grzimek offered revenues from the film *No Room for Wild Animals* to the British authorities for the purchase of lands to expand the game preserve.

Although the government declined Grzimek's offer, they invited father and son to conduct a census of wildlife herds in Serengeti National Park to obtain better estimates of population sizes and migration routes. In the meantime, the Fauna Preservation Society had commissioned ecologist W. H. Pearsall to investigate the problem. Based on Pearsall's study, a Committee of Enquiry recommended that the western plains area remain a national park, the Ngorongoro crater be turned into a nature sanctuary, and the crater highlands become a multiple land-use conservation area. The government followed the committee's recommendations, but Huxley wrote disparagingly of the park's management after completing his 1960 UNESCO tour and was particularly critical of land use by the Maasai.[40] Grzimek shared Huxley's concerns about the park's ecological condition and his disdain of Maasai land use. Using a Dornier light aircraft, the Grzimeks conducted aerial surveys of the Serengeti. They reported greatly reduced herds and migration patterns that went outside park boundaries, although their estimates were discounted three years later by a survey that revealed significantly higher numbers. In 1959, Michael was killed in an airplane crash when a griffon-vulture collided with the right wing of the airplane. The elder Grzimek returned to Germany, published a best-selling book, *Serengeti Shall Not Die,* and produced a film under the same title that won a 1960 Oscar for best documentary. The film and book helped draw international attention to Africa's wildlife crisis and underscored Grzimek's belief that there was no room for people in national parks. In the early 1960s, *Serengeti Shall Not Die* became the first propaganda film in a mass public outreach campaign, initiated by John Owen, Director of Tanzania National Parks, to educate Africans about the value of wildlife conservation and the national parks.[41]

In the United States, Fairfield Osborn preached a similar message of the incompatibility between people and wildlife through film. Through the African Wildlife Fund, Osborn directed monies toward a number of educational outreach projects such as the training of African game scouts in wildlife management and the development of educational films and programs on conservation and wildlife for Tanganyikan school children. The largest amount of funds, however, were spent on the production of a film distributed in Africa and the United States in 1961 that portrayed East Africa as a land in crisis, fueled by a conflict between wildlife and the Maasai.

Wild Gold was filmed by James R. Simon, former director of the Jackson Hole Wildlife Park and lead animal photographer for Disney, and written by Irving Jacoby, who also scripted *High Over the Borders*. The film opens with a scene of pristine herds of wildlife grazing on the Ngorongoro crater. Africa, according to the narrator, is the only remaining continent where individuals can witness this untouched animal paradise. The audience is then instructed about the meanings of wildness, from the speed of the cheetah to the slow and majestic pace of the elephant, accompanied by spectacular wildlife footage. Part of an intricate web of food relationships that regulates their numbers and keeps the system in perfect balance, the animals, we are told, "are not the only inhabitants of the African interior." This spectacle of wildlife on screen is abruptly ended by the intrusion of the Maasai. In the following segment, the Maasai are portrayed as the harbingers of destruction, as their herds of livestock create food and water shortages for wild game. Although Simon had to trade maize flour and tobacco to photograph the Maasai, the film depicts them willingly accepting the advice and technical assistance of Western conservationists. Conservationists hoped that a planned system of wells, financed in part by the New York Zoological Society, would lure the Maasai away from national park lands and prevent wildlife from migrating outside park boundaries. Assimilation of the Maasai into Western culture was inevitable, the narrator confidently asserts, once they realized that "herds of useless cattle do not represent the real wealth." In *Wild Gold,* real wealth is not found in nomadic pastoralism, but in tourism. A leaping impala on screen is "wild gold waiting to be minted." North American dollars will fill the pockets of Africans, now that it is possible for the "middle income American to take his own pictures of [a] lion family at home in the Serengeti." With financial assistance from countries like the United States to help build roads and construct tourist facilities, "this wealth of wildlife," the narrator assures the audience, "can belong to all of us and our children." Once again, international conservation is woven into a story in which nature, a global resource, is to be sustained through the altruistic acts of native populations, enlightened by the scientific knowledge, financial assistance, and tourist dollars of the West.[42]

Julian Huxley also shared Osborn's conviction that natural history film was a powerful ally of the conservationist. Huxley's enthusiasm for nature film as an educational and propaganda tool dated back to the late 1920s, when, as Secretary of the Zoological Society of London, he conducted a survey of East African education and endorsed cinema as the "most important instrument for awakening the young African mind."[43] In the years to follow, he associated with leading figures in the British documentary film movement, such as John Grierson, who assisted him in making *The Private Life of the Gannet.* Huxley recruited Grierson as director of mass communications at UNESCO, although Grierson, under FBI pressure, was forced to resign because of a Canadian spy scandal in which he and the National Film Board were implicated in a Russian espionage ring.[44] Huxley's enthusiasm for film continued unabated through the 1950s. Immediately prior to his 1960 UNESCO tour, he requested Osborn's help in securing financial assistance from either the Conservation Foundation or the United Nations to fund the making of a propaganda film for the Tanganyika National Parks directed at African audiences. Huxley also enlisted the talents of his college friend and established filmmaker, Armand Denis, in the making of a "propaganda film" about wildlife conservation and the future of Africa for BBC television. Denis produced a three-part series titled *The Dying Plains of Africa* that aired on the BBC in the fall of 1961.[45]

The following year Huxley became patron of another Denis project to enlist public support for international wildlife conservation: *Animals* magazine. A British weekly, *Animals* devoted its pages to skillful wildlife photography and stories about animals from the human interest angle to natural history accounts of threatened and endangered species. The giant panda, featured in the first issue, symbolized the close alliance between *Animals* magazine and the newly established World Wildlife Fund. It also illustrated the significance of individual charismatic species to the campaign strategies of environmental organizations. The first issue of *Animals* magazine included endorsements from luminaries like Prince Bernhard of the Netherlands, the president of the World Wildlife Fund, who sat on the magazine's editorial board and helped generate additional publicity for the conservationist cause. But Huxley and primatologist Solly Zuckerman expressed concern that the anthropomorphic references found in many of the articles in *Animals*—ants "living happily" and the "triangle of affections" in ducks—"distort[ed] the untrained reader's appreciation of biological science" and undermined the magazine's value as publicity for the World Wildlife Fund. The editor-in-chief responded that if the magazine's authors had to submit to the discipline required in a scientific journal, the magazine would be useless since it would not reach a very wide audience. In the cause of conservation, science had to relax its rigorous exactitude.[46]

All these promotional efforts at wildlife conservation, from *Serengeti Shall Not Die,* to *Wild Gold,* to *The Dying Plains of Africa,* paled in comparison to the reverberations that sounded through American and British culture with the 1960 publication of Joy Adamson's *Born Free.* The story of a couple's rearing of a lioness cub in captivity and the eventual release of Elsa into the Kenya bush, *Born Free* captivated the public's interest and quickly rose to the top ten of the *New York Times* bestseller list in the spring of 1960. Two book sequels and a highly acclaimed movie version followed. By 1965, Joy Adamson's income from the *Born Free* trilogy and movie had escalated to 250,000 pounds.[47]

Born Free is a story about the human desire for, and attainment of, intimate contact with nature. For an anonymous writer in the *New Yorker,* the book fulfilled "the irrepressible human dream of entering into wild nature. Elsa, gay and loving, and the Adamsons, patient and perceptive, [were] model feline and human characters, ideal figures in a Peaceable Kingdom." This image of African paradise masked the stormy marriage of a promiscuous wife and an abusive game-warden husband, revealed to the public thirty years later.[48] In 1960, Elsa's story carried a message that audiences longed to hear. Huxley himself gave scientific legitimacy to Joy Adamson's story in an introduction he wrote for Adamson's second book, *Living Free*. Huxley had met the Adamsons and visited Elsa's camp on his 1960 UNESCO tour of East and Central Africa. In his introduction, he praised Joy Adamson's contributions to mammalian ethology, despite her lack of scientific training. "By a passionate patience and an understanding love," Huxley wrote, "Joy Adamson succeeded in eliciting something in the nature of an organised personality out of an animal's individuality." Joy Adamson had discovered something of nature—not through science, but through the more accessible qualities of patience, empathy, and understanding. Her story affirmed the belief that a tourist in Africa, with sufficient time, patience, and understanding, might also make contact with nature, provided nationalism and cultural mistrust among Africans did not get in the way.[49]

In his introduction, Huxley alluded to the allegory of African independence found within Adamson's story. Patience and understanding, Huxley suggested, could be as effective in "attempts at contact between hostile or mutually suspicious groups . . . as they were with Elsa."[50] Caught between the worlds of white civilization and African wilderness, Elsa symbolized Western images of Africans, who were said to be lost between the same two worlds. Would Elsa resort to the man-eating ways of her father when returned to the wild? Or did the patience and love Joy Adamson bestowed upon her outshine Elsa's primitive instincts and enable the lioness to successfully straddle both worlds? Similar questions about Africans echoed in the psyches of many Americans and Europeans. Joy and George Adamson had lived through the Mau Mau uprisings in Kenya between 1953 and 1955. George had killed sixteen Kikuyu when he and his game scouts were placed into military service due to increased Mau Mau activity in his district. The Mau Mau insurgency became a symbol of African savagery within British memory. Had Britain bestowed enough patience and understanding in its colonial administration to guarantee an easy transition to independence, one in which Africans would grant white settlers contact with the African landscape? Or would independence for Kenya be marked by the violence and savagery reminiscent of the Mau Mau rebellion and end in the expulsion of Westerners from this Garden of Eden? Elsa gained her freedom, but as illustrated by her return with her cubs to the Adamsons' camp a year after her release, not at the expense of friendly contact with whites. Westerners projected the same hopes on Kenya as a whole.[51]

Although *Born Free* did not address the issue of wildlife conservation in Africa, in *Living Free* and *Forever Free,* and in her lectures and public appearances, Joy Adamson became an outspoken advocate for international conservation. She channeled the royalties from her books and movie into Elsa Ltd. (Kenya), Elsa Trust (Kenya), and The Elsa Wild Animal Appeal (U.K.), trusts that funded projects ranging from the establishment of game reserves in Kenya, to educational programs, to the creation of a wolf sanctuary by Carol and Marlin Perkins in the United States. Although the Elsa Wild Animal Appeal was initially affiliated with the World Wildlife Fund, Huxley expressed uneasiness about publicly endorsing Adamson when she wrote in 1964 of plans for a television series on the mysterious activities of animals. Adamson's suggestion that Elsa controlled her cubs through thought communication raised concerns similar to those expressed by Huxley in his criticisms of *Animals* magazine. How far could one stray from science in enlisting public support through entertainment for international conservation? Patience and understanding were tolerable, even respected, traits within Huxley's vision of nature entertainment; telepathy was not.[52]

The world that conservationists aimed to protect in the 1950s and early 1960s was a landscape fundamentally changed by the Second World War and by the strategy of containment that germinated in the war's aftermath. In their eyes, overpopulation and scarce food resources created an environment conducive to the growth of economic nationalism and communism in the developing world. But if the war had taught one lesson, it was the danger of nationalism and the costs of isolationism in a global society. Interdependence, not independence, was the image that the United States and its allies hoped to impress upon the developing nations of the world. And nature was an ally in this cause.

As part of the world's heritage, wildlife was to be appreciated and enjoyed by all citizens of the globe. Nature documentaries helped raise the environmental awareness of Americans, brought the plight of African wildlife to the world's attention, and made Planet Earth a much smaller place. But there are associated costs when film is used to enlist public support for international conservation efforts, costs that both people and wild animals are living with to this day. By embracing an aesthetic of pristine wilderness, nature films reinforced a management scheme that effectively divorced humans from the natural landscape. In East Africa, the Maasai and their livestock, once active agents in a biologically productive system, were expelled when international environmental organizations sought to promote tourism as the future source of wealth. These organizations met their goals in that tourism in Kenya now accounts for one-third of the country's foreign-exchange income. But there is yet another sense in which nature films separate us from nature, while simultaneously making us impassioned supporters of the global environment. In bringing wildlife and their natural habitats into our living rooms, we are merely informed viewers, estranged from the material relationships, from the struggles and joys that local people, living with these animals, encounter on a daily basis. Today, local participation in developing plans of sustainable economic use has become important in a strategic shift in conservation away from protectionism to community-based conservation efforts. Some of these plans include game ranching on the edge of national park lands and sea-cucumber farming near marine reserves. To the extent that our affection for and relationship with wildlife is sustained by entertainment, such conservation schemes based on productive labor face a difficult challenge in gaining worldwide public support.[53]

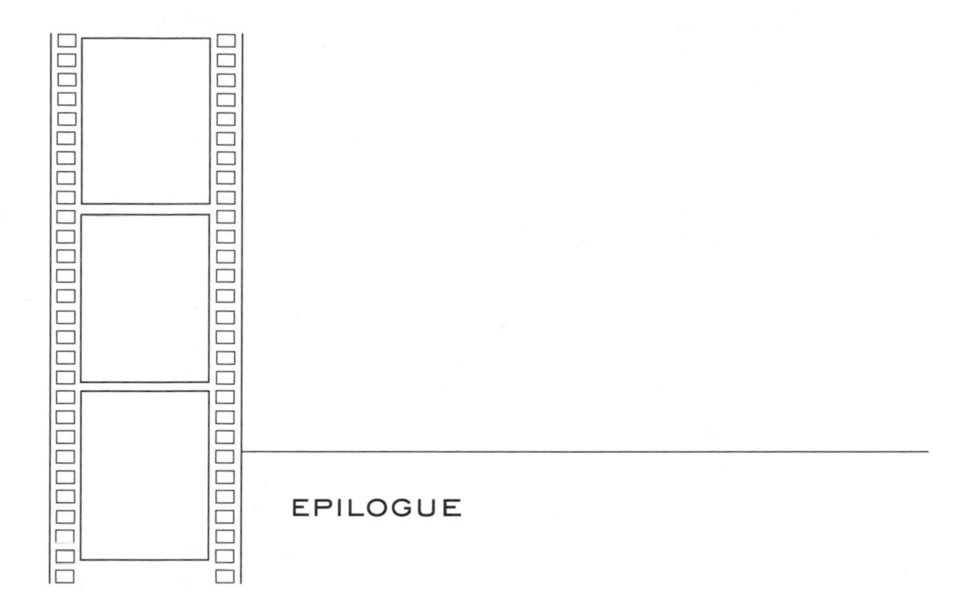

EPILOGUE

On February 9, 1996, the *Denver Post* published a front-page story that made national news. Marty Stouffer, an Aspen filmmaker and creator of the popular nature series *Wild America,* broadcast on PBS for eleven seasons, was "accused of staging scenes in his documentaries, mistreating animals, and defiling public lands." The charges stemmed from a lawsuit filed by the Aspen Center for Environmental Studies that alleged Stouffer had illegally cut a trail across its property to reach an unauthorized encampment in a national park. The trial and the $300,000 judgment awarded to the plaintiffs brought out further disclosures from former animal handlers and staff of Stouffer's production company. Allegedly, caged, tethered, and tame animals and splices of footage shot in different locations were used to construct the dramatic chase, kill, and mating scenes of animals in *Wild America.*[1]

In the early 1900s, President Theodore Roosevelt had attacked the credibility of realistic animal stories written by authors such as Jack London and William Long, accusing them of using sentiment to sell nature. In the early 1930s, William Douglas Burden and others publicly condemned Paul Hoeffler's *Africa Speaks* for its Hollywood fabrication of a lion charge in a supposedly authentic picture about the Colorado African Expedition. In the early 1950s, naturalist-photographers, critics, and the public had questioned Disney's claim that the True-Life Adventures were just that, unstaged and candid. Now once again, as the century drew to a close, national attention was drawn to the artificial (and sometimes cruel) means employed in the production and marketing of nature for popular consumption.

Editorials that followed the Stouffer incident reveal how firmly Americans wish and expect nature films to be the real thing. "Films that depict otherwise," *The Denver Post* remarked, "do a disservice to the animals and negate the very point of making wildlife documentaries, which is to help humans appreciate the natural world around them." Nature films are expected to give us direct, unadulterated access to wildlife. Immersed in nature through the camera lens, we depend upon the naturalist-photographer to give us an experience that is pure and unadorned. The conscience and integrity of the naturalist-photographer are our only assurances that we have not been deceived and nature has not been exploited. Stouffer initially claimed that the charges of faking nature leveled against him were "character assassination." PBS appealed to Stouffer's unquestioned "integrity" in assuaging public concerns about *Wild America*. Eighty-five years earlier, Theodore Roosevelt had vouched for Cherry Kearton's character to assure the authenticity of his African wildlife motion pictures. Integrity and the issue of character have always been important criteria in distinguishing the scientist from the showman and authenticity from artifice in nature on screen.[2]

Although PBS cleared Stouffer of any wrongdoing, newspaper copy and editorials consistently judged that Stouffer went astray by succumbing to commercialism. Denver Zoo officials recalled that when Stouffer approached them to photograph the zoo's two famous animal celebrities, the polar bears Klondike and Snow, he had wanted to shoot a scene that involved a fake den that would represent their "home." The zoo had refused, asserting: "We're a scientific organization, and this was theatrics." Like Martin and Osa Johnson in the 1920s, Stouffer was accused of manipulating nature—for example, using explosives to provoke endangered trumpeter swans to take flight—strictly for financial gain.[3]

The temptations were great. Unlike the limited market and fleeting success of travelogue-expedition films, nature entertainment has become a lucrative business with market staying-power. What was educational entertainment in the early decades of the twentieth century for urban upper and middle classes is now a part of popular culture and mass consumption. The immediate postwar years saw an explosive growth in nature audiences when Disney, Perkins, and others successfully appealed to, and captured, a market composed of America's baby-boomers and their parents. War-weary Americans looked to nature for wholesome entertainment for their children. When the baby-boom generation came of age, these childhood experiences of the wide world of nature became the seeds of 1960s environmental activism. The environmental movement in turn sparked even greater audience demand for natural history shows. *National Geographic*'s television specials, *The Undersea World of Jacques Cousteau, The World of Survival,* and *The Wild Wild World of Animals* are just a few of the nature programs that appeared in the late 1960s and early 1970s to meet escalating demand for knowledge and observation of wildlife and the natural world. By the mid- to late 1970s, however, television networks filled programming slots with less expensive and equally popular game shows as costs for more sophisticated technological productions of nature programs kept pace with viewers' increasingly exacting taste for more dramatic, hyperreal scenes of wildlife.

In the 1980s, the expansion of cable television created new specialized markets for cinematic nature. Cable TV's Discovery channel, which started in 1986, grew into a $500 million company in ten years largely on the basis of its popular nature and science documentaries. The number of nature films produced each year far exceeded the limited exhibition outlets of network and public television by the early 1980s. Initially, the Discovery Channel capitalized on this surplus, buying shows for much less than their cost of production. Furthermore, by dramatizing sex and violence found throughout the animal kingdom, such shows as *Fangs, Predators,* and weekly specials like *Shark Week* engaged new generations of enthusiasts with graphic, close-up scenes of animals copulating or predators killing prey.[4]

Through savvy marketing strategies that appealed to consumer tastes for violent, real-life programs, home video gave nature entertainment an even more profitable mass audience. *The Trials of Life* series, produced by Sir David Attenborough and distributed jointly by Turner Home Entertainment and Time-Life Video, had earned over $21 million five years after its release in 1991 by taking viewers on a stunning twelve-part photographic journey into the challenges faced by animals in their struggle to survive. Action nature videos, such as *Dangerous Animals,* which consist primarily of staged attack scenes like that of a mountain lion pouncing on a cross-country skier, netted Stouffer's company $20 million in sales between 1993 and 1996.[5] Time-Life Video advertised *The Trials of Life* by condensing all the attack scenes, including the scene of killer whales snatching sea-lion pups from shore and "exultantly" tossing their mangled prey into the air, into a brief promotional spot. A voice-over barked, "See up close how the law of the jungle is kill or be killed . . . Find out why they call them animals." Sir David Attenborough was so disturbed by this exploitative sensationalism that he considered taking legal action. Out of a twelve-hour series on the natural history of animal behavior, Time-Life Video singled out "the deadly but fascinating drama" of the hunt in order to sell nature videos.[6]

Some critics believe that in the commercial exploitation of dramatic nature to attract audiences, ratings, and financial revenues, nature films have "sunk like so much daytime entertainment into cheap emotions, sex and violence." In the early 1900s, similar sentiments were voiced. Reformers and naturalists expressed concerns that fabricated images of jungle wildlife adventures, such as Selig's *Hunting Big Game in Africa* or Congo Pictures' *Ingagi,* which capitalized on melodrama and sensationalism to attract crowds, had sacrificed science for showmanship, education for entertainment. They argued that in manipulating nature for profit, hucksters had polluted it with commerce, tainting the sanctity of nature as a landscape of authenticity within American culture.[7]

Then and now, however, such debates about the commercial exploitation of nature for entertainment mask perhaps a more profound anxiety. Our uneasiness with the exploitation of nature for financial gain reveals how much we wish nature to appear pristine, set apart from the hands of man. Nature films, naturalistic habitat displays, and animal theme parks like Disney's Animal Kingdom capitalize on our desire to be close to nature, yet curiously removed from it. By making animals into spectacle, rather than beings we engage with in work and play, nature films and other recreations of nature reinforce this dichotomy of humans and nature. In nature as spectacle, the animal kingdom exists solely to be observed, objectified, and enjoyed. We have our world and they have theirs. This voyeurism precludes any meaningful exchange because we remain at a physically and emotionally safe distance, far removed from the shared labor of animals and humans, whose interactions have made such vicarious experiences possible. We no longer work with animals, we predominantly watch them. And film—as a technology of art, science, and entertainment, but above all vision—overwhelmingly has come to mediate our relationship with animals and the natural world.[8]

One cannot underestimate the benefits science and environmentalism have accrued from the interest in wildlife and nature stimulated by film. For many of my generation, *Flipper, Sea Hunt,* and *The Undersea World of Jacques Cousteau,* along with family summer vacations spent in America's national parks, were decisive childhood influences in shaping our aspirations to become more deeply involved with animals and nature. By eliciting an emotional relationship with wildlife on screen (who among us did not long for a pet dolphin just like Flipper?), film strengthened my desire to make intimate contact with animals in the wild. And by encouraging a sense of wonder and intrigue, natural history film instilled a passion for biology and the natural environment that motivates me to this day.

But the glamour of science and the drama of nature on screen match neither our experiences in the field or lab nor our everyday experiences with the natural world. Nature is not all action. And science is filled with more methodical and sustained labors than adventures. Conditioned by nature on screen, we may fail to develop the patience, perseverance, and passion required to participate in the natural world with all its mundanity as well as splendor. Trained as spectators, we make little effort to accommodate ourselves to nature. When presented to us through the hidden labor of others, nature is crafted to meet our demands, expectations, and values.

Visiting Alaska several summers ago, I witnessed the tangible consequences of film's impact in shaping our relationship with wildlife. On a bus full of tourists, I was riding toward the interior of Denali National Park, where car access is limited in an effort to preserve the purity of this remnant wilderness. The driver narrated our tour across the tranquil, desolate tundra. The experience had all the makings of a natural history film. We hoped for a glimpse of spectacular Denali—the highest mountain peak in North America—but it remained hidden in the clouds. Suddenly the bus slowed, then stopped. Off to our right a wolf had been spotted. The atmosphere on the bus was electric. The quiet chatter of a few seconds earlier was replaced by a flurry of activity as passengers rushed for their cameras and camcorders to capture the moment on film. The wolf stalked a ground squirrel less than twenty yards away. As the predator killed its prey, shutters snapped. The wolf lingered, but the drama had subsided. Everyone was impatient, ready to leave in search of the next dramatic scene. The bus continued on. Dramatic nature had proven more satisfying than the serene and subtle qualities of this vast Arctic landscape. The camera had shaped our expectations, defined our experience.[9]

Four years later, on an early spring morning, I set off through the back yard of an urban housing complex for my office, where I would reflect on the subject of animals and human society. Over the sounds of commuter traffic on the highway, I heard a screech of anguish. Fifteen feet away, near a chain-link fence along common green space dotted with strollers and plastic toys, a sharp-shinned hawk held a European starling to the ground. Aware of my presence, the hawk released its prey and took flight. I stood still and waited. Again, the hawk seized the smaller bird in its talons and landed close by. I was transfixed by the raptor's piercing eye as the bird cautiously observed me.

Captivated by the hawk's power and by the starling's suffering, I thought about how this scene would have been framed in nature films. In the sentimental nature films of the 1950s, the violence would have been tempered by reference to the harmonious web of life. The individual struggle and suffering would have been resolved into one of beneficence, as each species helped keep nature in balance, testimony to the wonder of nature's grand design. In the animal snuff films of the 1990s, the choice shot would be a close-up of the hawk tearing the flesh of the starling, with an amplified audio-track of the starling's shrieks. The scene would be made more intense than my actual experience could be.

Despite the contrasts in these sentimental and violent framings of nature, a single continuity would be found. Every effort would be made to eliminate the urban and human setting from the scene. If this vicarious experience of nature—lived through the camera lens—is to feel intimate and pure, the artifice of civilization must be hidden, for any sign of artificiality would destroy the illusion of this recreated nature as God's place of grace. The separateness of humans and nature must be maintained. But my presence, the chain link fence, the children's toys, and the traffic sounds were equally a part of the hawk's natural environment. To erase the so-called artifice—humans and our constructions—is to deny our presence in the natural world. Natural history film, like the ecological restoration of wilderness in the national parks or in the simulated habitats of animal theme parks and zoos, is a product of human labor, technology, and cultural values as much as it is a product of nature itself. We only know nature because we intervene. The critical issue is not how to remain separate, but how to act with integrity in our relationships with wildlife and the natural world.

AFTERWORD: THE NEW GREEN WAVE

In the spring of 2005, emperor penguins were on the move. From their home in Antarctica, they had, within a matter of months, appeared on every continent of the globe. This was not some remarkable feat of biological migration. Rather, it was a triumph of technology, storytelling, and the multinational media industry. The film, *March of the Penguins*, released by Warner Independent Pictures, a subsidiary of the multimedia giant Warner Brothers Entertainment, made movie history. With gross earnings of $77 million in the United States, it became the second-highest grossing documentary ever to be released in North America—just behind Michael Moore's *Fahrenheit 9/11.* The Academy Award–winning documentary film by French director Luc Jacquet captured worldwide audiences and worldwide sales of $127 million with its stunning cinematography and epic story of struggle and survival in one of the harshest landscapes and climates on Earth.

March of the Penguins also became a movie favorite among American conservatives. After Massachusetts became the first state to issue marriage licenses to same-sex couples in 2004, and as debates about same-sex marriage raged across the country, Rich Lowry told fellow conservatives in the *National Review* that *March of the Penguins* was the "most endearing portrayal of monogamy . . . that you are likely to see on the Big Screen."[1] Conservative film critic and radio host Michael Medved agreed: *"March of the Penguins* is the motion picture this summer that most passionately affirms traditional norms like monogamy, sacrifice, and child rearing," he asserted emphatically.[2] Writing in the conservative Christian evangelical publication, *World Magazine*, Andrew Coffin remarked that the remarkable feat of a single egg, laid each year by a penguin female, surviving in the most inhospitable conditions on the planet was also "a strong case for intelligent design."[3]

America's religious right had found an ally in nature. And the public response caught movie industry executives off guard. "You know what?" Laura Kim, vice president of Warner Independent, told the *New York Times,* "They're just birds." But birds are never *just* birds, at least not on screen. Storytelling is the basic craft of cinema.[4] And animals on film perform, willingly or not, stories through which humans make meaning of the world.[5] "In the harshest place on earth, love finds a way," Morgan Freeman affectionately narrates the tag line of the film. This universal story of love and devotion was crafted as much by clever scriptwriters, as it was drawn from nature. But religious conservatives were quick to see in the film and in nature a moral fable for our time. Birds buttressed a faith in God and traditional family values when both appeared to be threatened. It was an episode reminiscent of an earlier era when *Nature's Half-Acre*, produced by the Disney studio as part of its True-Life Adventure series, sanctified the family and God as the central pillars of American democracy during the Cold War.

Loving penguins had captured the hearts, minds, and pocketbooks of moviegoing Americans in the summer of 2005. But on the major wildlife television networks—Animal Planet, Discovery, and National Geographic—viewers could find quite different animal stories being told. Sex, not love, rocked the television airwaves. Swollen, reddened buttocks, ten-foot penises, half-ton testicles, and *Wild Sex* had become, by the new millennium, a staple part of cable-television programming, where viewers could see "all animals, all the time." Animal sex on television was nothing like the artfully choreographed, ballet-like scene of courtship displayed in *March of the Penguins*. On shows such as *When Animals Attract*, *Wild and Weird—Wild Sex*, *The Most Extreme*, or *The New Chimpanzees*, viewers found large male elephant seals mounting half-grown pups, Tasmanian devils engaged in vicious foreplay, and female bonobos rubbing their genitals against one other. On cable television, animal porn trumped parental love in attracting viewers to a saturated and competitive market.

As media scholar Cynthia Chris observes, images of animal sex began to appear on wildlife television programs in the 1990s with a frankness and frequency never before seen in the history of the genre. In fact, such images have become so pervasive, Chris notes, that pop-culture references to animal sexuality as seen on TV have become common. The alternative band Bloodhound Gang, for example, scored a huge hit in 2000 with their song "The Bad Touch" which featured the following refrain: "You and me baby ain't nothin' but mammals / So let's do it like they do on the Discovery Channel." In the British comedy *The Full Monty*, Gerald embarrassingly admits to his friends that he gets sexually aroused watching nature programs.[6]

What might account for the increase in sexual content on wildlife television programming as the twentieth century drew to a close? Filmmaker and historian Derek Bousé offers a compelling explanation. By the mid-1990s, the Discovery Channel was rapidly and aggressively expanding into the global communications marketplace. In 1996, it joined forces with the BBC to launch Animal Planet. Ten years later, Discovery Communications International had become a media giant that extended its reach into 170 countries with over 100 networks of distinct programming and 1.5 billion subscribers.[7] In 1997, National Geographic launched its own 24-hour cable channel in Europe and Australia devoted to exploring nature, science, history, and culture. It now boasts an audience of 160 million viewers in 140 countries across the globe due, in large part, to a partnership forged with Fox Cable Networks in 2001. The proliferation of channels, however, created a problem of supply and demand. Networks like Discovery could no longer afford to rely on what in the business are known as blue-chip films to fill their content. Expensive to make, often years in production, and gorgeous to watch, blue-chip films like *March of the Penguins* or Discovery's *Planet Earth* series were far too costly and time-consuming to rely on as the main source of wildlife television programming. They served as eye candy for attracting viewers, but they could never be the main course in a highly competitive marketplace.

So the networks shifted to less expensive, quickly produced, half-hour, action-adventure programs, like the immensely popular series, *The Crocodile Hunter,* which catapulted the Australian animal handler, Steve Irwin, to international fame.[8] And just when veteran production companies, such as *Survival*, were shut down because of the expense of producing blue-chip films, a study by the Glasgow Media Group sealed the fate of wildlife television programming over the next few years.[9] The study revealed that younger audiences—eagerly sought after by advertisers—wanted more "informal," less serious, nature shows. They also wanted more sex. In fact, much to the industry's surprise, viewers preferred sex over death when it came to animal television. The result, Bousé succinctly remarks, was "more action, more youth, more humour, and more (animal) sex."[10]

Given the trends in wildlife television, no one could have predicted the success of *March of the Penguins*. Other surprises soon followed that are altering the landscape of cinematic nature. Werner Herzog's documentary *Grizzly Man*, released in the summer of 2005, paled in comparison to the box-office success of *March of the Penguins* but nevertheless received widespread critical acclaim. Praised by *Chicago Sun-Times* film critic Roger Ebert as "unlike any nature documentary I've ever seen," Herzog's film is a character study of the life, death, and madness of amateur expert and grizzly-bear activist Timothy Treadwell as well as a glimpse into Herzog's philosophical musings on the human condition in relation to an indifferent and amoral nature.[11] For a film, made from 90 hours of Treadwell's video footage shot in Alaska, interspersed with interviews and stock footage, $3.1 million seemed like a healthy return to the film's production company, Discovery Studios. By the spring of 2006, *The Hollywood Reporter* noted that "with recent nonfiction efforts as varied as *March of the Penguins, Super Size Me* and *Grizzly Man* doing bang-up box office, interest in docus from distributors and broadcasters alike is at an all-time high."[12]

If industry analysts still harbored doubts about the ability of documentaries to attract crowds, *An Inconvenient Truth*, released Memorial Day weekend in 2006, did much to dispel them. A cautionary tale about the consequences of global climate change, the film weaves together an illustrated lecture Al Gore delivered more than 1,000 times with personal stories of his own life to offer a message of urgency and hope. In February of 2007, *An Inconvenient Truth* won an Oscar for best documentary. Eight months later, the Norwegian Nobel Committee awarded the Nobel Peace Prize to Gore and the Intergovernmental Panel on Climate Change for "their efforts to build up and disseminate greater knowledge about man-made climate change, and to lay the foundations for the measures that are needed to counteract such change." Heralding Gore as "the great communicator," the Nobel Committee certainly had *An Inconvenient Truth* in mind when it awarded Gore the Nobel Peace Prize.[13]

Suddenly, it was cool to be green. Hollywood stars got involved. TV and movie actor Ed Begley Jr. hosted *Living with Ed*, the "first-of-its-kind reality green show" on the Home and Garden Television channel. As the lives of Ed and his celebrity friends make clear, even when you're rich, it's not always easy being green in Tinseltown. Leonardo di Caprio tried his own hand at scaring the public into environmental action with his 2007 film, *The 11th Hour*. Its rather abysmal box-office showing suggested the public was tired of seeing stories of environmental doom and despair. Individuals wanted to feel they could make a difference given the Bush administration's refusal to take any governmental action to combat global climate change. In April of 2007, the Sundance Channel stepped up to the challenge and launched *The Green*, television's first regularly-scheduled programming destination "dedicated entirely to the environment." Committed to providing a "focused, entertaining source of information and inspiration about the planet we call home," *The Green* has opted for inspiring stories of eco-heroes and visionaries in calling the public to environmental action. Through original programs and interstitial segments, it aims to provide "viewers with ideas and tangible opportunities for all facets of their lives, demonstrating how to work green, play green, eat green, dress green and live green."[14] Similarly, in 2005, PBS launched a new series, *Strange Days on Planet Earth*, with the Sea Studios Foundation and National Geographic. The program reached 20 million viewers and won the Best Series award at WildScreen, the international showcase festival of wildlife and environmental film. It was back for a second season in 2008. Hosted by Academy Award–nominee Edward Norton, *Strange Days on Planet Earth* aims to use "engaging storytelling" and "innovative imaging" to expose the "web of invisible connections of the Earth's life support systems" and reveal the "profound global consequences of our simple everyday actions." Viewers feeling discouraged, frustrated, or inspired, can turn to the program's Web site for concrete ideas on how they can become a part of the solution.[15]

Amidst this recent explosion of green, watching blue-chip nature films is a reminder of how disconnected many wildlife films are from the environmental issues facing planet earth. The end of nature, foretold by environmental writer Bill McKibben in 1989, has come.[16] It has come because the stories of nature are now inextricably bound to the stories of our own lives. Global climate change has finally brought that realization home. Over the last two decades, however, in the hands of global media giants, wildlife films have largely become escapist fantasy. We may relish the beauty of *Winged Migration*, Jacques Perrin's 2001 magisterial and poetic meditation on bird migration across seven continents, or wonder in amazement at the *March of the Penguins.* But we, as viewers, are never forced to confront how the behaviors of individuals, institutions, corporations, and governments are altering both animal and human lives. It is ironic that humans are largely absent in big-budget wildlife films and wildlife are absent in the film credits. But the stories of people and wildlife are intertwined, both in the production of nature films and in the survival of all living species on planet earth.

Until quite recently, broadcasters have sought to portray a "nature above politics," as Randy Malamud writes.[17] By the 1990s, conservation had become a dirty word in the business. And wildlife filmmakers, many of whom cared passionately about their subjects, became increasingly disillusioned by an industry driven more by ratings, profits, and entertainment value than any commitment to the power of the media as a force of environmental change. BBC filmmaker Stephen Mills captured the dilemma well in 1997. "So long as we sustain the myth of nature," he wrote, "our programmes find a wide and appreciative audience. So many viewers could do a lot for conservation. But as viewing figures adamantly prove, once we make a habit of telling the bad news our audience slinks away. Television, after all, is primarily an entertainment medium, and wildlife films fill an escapist, non-controversial slot."[18]

As the media capitalizes on the new eco-chic, there are signs that the green wave is reshaping the contours of reel nature. The first hint that change was in the air came with *Happy Feet*, the 2006 animated sequel to *March of the Penguins.* This time the religious right found little to praise in a movie about penguins. Michael Medved, who sung the virtues of *March of the Penguins*, dubbed *Happy Feet* "Crappy Feet" and lambasted its message "that the biggest menace for the lovable penguins is the human race—stealing the fish on which the birds depend, or ruining planet earth through pollution and global warming." Medved found other hidden political subtexts in the film to rail against as well. The penguin protagonist Mumbles, ostracized for his inclinations to break out tap dancing rather than sing his own heartsong, eventually finds acceptance among his fellow penguins for who he is. It was, Medved asserted, "an obvious plead for endorsement of gay identity."[19] Despite Medved's complaints, *Happy Feet* earned the number one spot at the box office on opening weekend, beating out *Casino Royale* and proved that audiences weren't averse to films with a strong environmental message so long as they were also fun.[20]

Even production companies like National Geographic Films, which have done their best to edit out environmental politics from nature (witness *March of the Penguins*), are recognizing that being green makes for a good public image.[21] The release of *Arctic Tale* in the summer of 2007, a joint production between National Geographic Films and Paramount Classics, suggests that nature films are catching up to the realities of global warming and the trend in green consumerism. Fifteen years in the making, *Arctic Tale* is, according to Adam Leipzig, president of National Geographic Films, an attempt to create a new genre—the "wildlife adventure."[22] It's good spin, but many of the dramatic techniques the film employs, from the creation of individual animal characters, like Nanu the polar bear cub and Sela the walrus pup, to humorous scenes of farting walruses, to the storytelling narration of Queen Latifah, are tried and true methods worked out by Disney in the 1950s with the then-new genre of True-Life Adventures. What is novel in *Arctic Tale* is an explicit acknowledgment that human actions are changing the life (hi)stories of other living creatures. Through stunning cinematography and seamless editing, we follow, up close and personal, the lives of Nanu and Sela from birth to motherhood. Aided by a sentimental musical score and the soothing, sometimes hip, and often sappy narration of Queen Latifah, we are meant to empathize with these fictional characters, and the real animals from which their composites are drawn, as they struggle to adapt to a changing world of warmer winters and receding ice. "All across this special place," the movie ends, "children of the Arctic make their way into the world. Their fates are forever tied to the shifting rhythms of the ice that defines them. . . . What will their children do if it disappears?" asks Queen Latifah. A telling pause. A glance back from Sela and her two cubs. "What will ours?" The credits role and children from diverse ethnic backgrounds inform the audience of individual choices they can make that will "help save the lives of people and animals all over the world."[23] National Geographic's decision to use children and target individual consumption in addressing the politics of global warming was a safe one. It wouldn't draw heat from the Bush administration or alienate potential corporate sponsors. In fact, Starbucks seized the opportunity to promote the film in its 6,800 retail stores and enhance the coffeehouse chain's green image. And the absence of scenes featuring polar bear kills on bloodied snow or graphic animal sex ensured that the film would not repel religious conservatives or families with young children. Film critics were less than enthusiastic, decrying the over-the-top anthropomorphizing and gooey narration.[24] Financially, it bombed. But unlike *March of the Penguins*, *Arctic Tale* makes an effort to connect the stories of animal and human lives.

Ten years ago, when the first edition of *Reel Nature* appeared, I doubt anyone foresaw that growing public concern about global warming, a fashionable trend in green consumerism, and a couple of documentary hits at the box office would spark a revival of interest in theatrical films focused on environmental subjects. In April of 2008, almost fifty years after the release of its last True-Life Adventure, *White Wilderness*, the Walt Disney Company announced it was back in the business of making nature films.[25] From a business standpoint, the prospects seem promising. Of the $631 million in gross revenues earned by 275 documentaries released between 2002 and 2006, $163.1 million came from eight wildlife documentaries. When you add films like *An Inconvenient Truth* to the mix, environmental subjects are capturing a large market share of non-fiction film. Even box-office failures, like *Arctic Tale*, could be rationalized away as improving the corporate image of producers and sponsors at a time when it's hip to be green. Paramount Classics never expected *Arctic Tale* to match the kind of returns enjoyed by *March of the Penguins*. Megan Colligan, Executive Vice President of Publicity for Paramount Classics, explained that there was more to Paramount's decision to back the film than the bottom line. "It's important for us to put our money where our mouth is," she remarked. "We feel it's important people feel that just by going to the movie, they've done something. They've taken a step in the right direction."[26]

Many of us would like to believe that by going to the movies, we're helping to save the world. Surely this is part of what is driving an increase in environmental film festivals and audience numbers across the United States. The Environmental Film Festival in Washington, D.C., the largest and perhaps best known, boasted a record attendance of over 24,000 people in the spring of 2008, its 16th year. The festival featured 115 films from over 30 countries that ranged across a spectrum of environmental subjects, from the unintended consequences of war to the inequitable flow of water, from nuclear waste sites to endangered wildlife habitats. In November of 2007, the University of Wisconsin–Madison's Center for Culture, History, and Environment, in partnership with the Nelson Institute for Environmental Studies, hosted *Tales from Planet Earth*, the first environmental film festival in Madison, Wisconsin. We hoped for 500 people opening night. We were in awe as more than 1,000 people gathered for a lecture by renowned author and global warming activist, Bill McKibben, and to watch *Everything's Cool*, Daniel Gold and Judith Helfand's "toxic comedy" on global warming. Like other curators across the country, I hoped people would walk away from the festival inspired and motivated by what they saw. And, based on audience feedback, they did. But unless that energy is mobilized into action, the power of film as a force of environmental change will remain ephemeral at best.

Wildlife filmmakers are themselves coming to this realization. For a long time, broadcasters and filmmakers alike assumed that "if people know about it, they'll care for it, and do something." Richard Brock, a 35-year veteran of the BBC Natural History Unit disagrees.[27] And in his 1992 book, *The Age of Missing Information*, Bill McKibben struck a similarly pessimistic tone: "Virtually everyone in the industrialized world has . . . seen many hours of gorgeous nature films. . . . And yet we're still not willing to do anything very drastic to save that world."[28] Awareness is not enough. If the current explosion of interest in nature and environmental film, from both the production and consumption side, is to have any lasting impact on the world, we need to find ways of linking content to action.[29]

Working Films is an example of one successful organization that is harnessing the "visceral power of the moving image" and channeling it into a force of community activism and engagement.[30] Cofounded by Judith Helfand and Robert West in 1999, Working Films is neither a production company nor a distributor. Instead, it works to bring non-fiction filmmakers together with community organizers and activists in ways that ensure a film has a measurable, long-lasting impact on the world. Take, for example, the 2002 Sundance award-winning and two time Emmy-nominated documentary, *Blue Vinyl*, codirected by Judith Helfand and Daniel Gold. An irreverent, humorous, and personal tale that makes visible the toxic lifecycle of polyvinyl chloride and its tragic impact on the lives of workers and communities in the petro-chemical corridor of Louisiana, *Blue Vinyl* has had a lasting affect on communities, institutions, and corporations far beyond most nature or environmental films. The strategy of Working Films is to involve community partners at a film's pre-production stage. Healthcare without Harm and the Healthy Building Network were both involved in the making of *Blue Vinyl*. These partnerships led to the creation of My House is Your House, a consumer education and advocacy campaign linked to *Blue Vinyl* that is transforming the consumption of PVC and its bioaccumulation. Victoria's Secret eliminated PVC packaging of its products; Kaiser-Permanente, one of the largest healthcare organizations in the United States, agreed to phase out its use of PVC from building materials to vinyl gloves; and Habitat for Humanity built it first PVC-free house, all as a direct result of Working Film's efforts to link non-fiction film to cutting-edge activism.[31]

It is a strategy that is also gaining ground among wildlife filmmakers. In 2001, at the International Wildlife Film Festival, a group of experienced professionals in the wildlife film industry, disenchanted with how little their work seemed to be affecting the state of the planet and recognizing that it was "payback time for wildlife television," formed Filmmakers for Conservation.[32] One important strategy has been to link conservation films to local people, organizations, and governments where change can be most effective. One FFC member, Sarita Siegel, for example, has worked closely with the Borneo Orangutan Survival Foundation in developing public service announcements and having her National Geographic film, *The Disenchanted Forest*, translated into Indonesian as part of an effort to establish a 364,000 hectare area in Central Kalimantan as a protected conservation area for orangutans whose habitat is gravely endangered by the widespread growth of oil palm plantations.

Other filmmakers are successfully exploiting access to new media technologies—from inexpensive DV cameras, to cheap laptop editing systems, to Web-based video distribution through sites like YouTube.com—to get activist messages out and circumvent the stronghold that media giants like Discovery and National Geographic have had in the last two decades on the production and consumption of wildlife films. Consider Rob Stewart's hard-hitting 2007 documentary *Sharkwater*, which CBC television news host George Stroumboulopoulos aptly and approvingly described as "a nature documentary with a gunboat." Originally intended to be a spectacular underwater film strictly about sharks, *Sharkwater* evolved instead into a documentary that takes audiences on a wild ride from stunning undersea cinematography to a journey aboard the *Sea Shepherd* with its crew of swashbuckling eco-pirates as they aim to subvert the operations of the international shark-fin harvesting underworld. Stewart has been effective in using Web-based marketing strategies, from trailers and talk show clips on YouTube, to MySpace and Facebook, to activist organization Web sites in creating a buzz for the film that may transform the public image of sharks from savage predators, sustained by nature programming such as Discovery's *Shark Week*, to cherished and endangered megafauna.[33]

Enterprising entrepreneurs like Stewart aren't the only ones seizing upon the potential of viral video in getting their films and messages out there. Major broadcasters are in on it too, thanks largely to YouTube's Best Eyewitness Video of 2007. In May of 2007, Jason Schlosberg posted on YouTube's Web site an eight minute amateur video shot by his safari companion Dave "Buzz" Budzinski at Kruger National Park in South Africa. *Battle at Kruger* opens at a watering hole where a herd of cape buffalo unsuspectingly moves in close to a pride of lions, eagerly watching in anticipation. The chase is on. Down goes a baby buffalo into the water. But the excitement doesn't end there. A tug of war ensues between the lions and two crocodiles. When the lions successfully drag the baby back on land, the cape buffalo adults come to the baby's rescue, tossing lions into the air left and right. Stumbling to its feet, the baby appears to be saved. It is the kind of stuff a professional wildlife photographer would die for. The video went viral. By April of 2008, 28 million viewers had watched *Battle at Kruger* and over 25,000 people had posted comments. *Time* and ABC featured it in news coverage. National Geographic, quick to seize on the marketing potential, hired Budzinksi and Schlosberg to go back to South Africa to make a documentary about the video. *Caught on Safari: The Battle at Kruger* aired on the National Geographic Channel on Mother's Day in 2008.[34]

The posted comments on the YouTube Web site are a media scholar's dream.[35] They reveal how much we are invested in finding some kind of moral lesson in nature. One religious viewer remarked they had seen the video at their "large church, to illustrate how our small groups have intervened to save folks being whipped up on by Satan; helping by prayer & encouragement." Another found it "a breath of fresh air to see the 'good guys win,' better than the usual Darwinistic sex & violence nature flick that reduces our existance (sic) to killing & sex." Mvrck23, like a number of viewers, saw it through the lens of racial hatred: "Damn this looks just like the gang war in my area. But in reverse color. The Mexicans over powered the blacks because of numbers!"[36]

Despite the differences of opinion, conflicting religious and political views, racial prejudice and nationalistic fervor, the comments about *Battle at Kruger* point to one common thread: the need to find stories in nature that give meaning to our lives. But are the stories we tell familiar tales of violence and sex, love and family, Edenic Nature and ecological apocalypse that garner audience ratings and reinforce well-worn stereotypes? Or are they stories that challenge us to see the environment in new and surprising ways, inspiring action and hope on a planet, beautiful in its diversity of life, but also troubled by unprecedented environmental change and injustice?[37] If the new green wave of cinema is to be more than a passing consumer fad, a concerted effort is needed to strengthen the connections between filmmaking and community activism already underway. New media technologies have opened up the possibilities for "new material, new voices, and new points of view."[38] Outside the powerful networks of film distribution and promotion, new relationships across art, science, and activism are being forged, helping to create media that matters in the lives of people and animals throughout the world.

NOTES

MANUSCRIPT COLLECTIONS

AAP	Arthur Allen Papers Division of Rare and Manuscript Collection, Cornell University
CRCP	Clarence Ray Carpenter Papers Pennsylvania State University
FOP-LC	Fairfield Osborn Papers Manuscript Division, Library of Congress
FOP-WCS	Fairfield Osborn Papers Wildlife Conservation Society
GKNP	Gladwyn Kingsley Noble Papers Department of Herpetology, American Museum of Natural History
HFP	Harold Fabian Papers Rockefeller Archive Center
JHP	Julian Huxley Papers Rice University

KCC	Kenneth Chorley Collection Rockefeller Archive Center
MARP	Marineland Papers Marineland, Florida
MBC	Museum of Broadcast Communication Chicago, Illinois
MPP	Marlin Perkins Papers Western Historical Manuscript Collection, University of Missouri–St. Louis
NTP	Niko Tinbergen Papers Department of Western Manuscripts, Bodleian Library, Oxford University
OMC	Olaus Murie Collection Denver Public Library
RFA	Rockefeller Foundation Archives Rockefeller Archive Center
RFC	Rockefeller Family Collection Rockefeller Archive Center
SNLC	Steven N. Leek Collection American Heritage Center
WDA	Walt Disney Archives Walt Disney Studios, Burbank, California
WDBP	William Douglas Burden Papers American Museum of Natural History
ZPC	Zoo Parade Collection Lincoln Park Zoo

PROLOGUE

1. For multidisciplinary perspectives on the making of Disneyland and other Disney theme parks, see Karal Ann Marling, ed., *Designing Disney's Theme Parks: The Architecture of Reassurance* (New York: Flammarion, 1997).
2. "Disney's Animal Kingdom," vol. 1, no. 1, http://www.themeparks.com/wdw/dak/animal01.htm.
3. On the financial costs and animal deaths, see Mireya Navarro, "New Disney Kingdom Comes with Real-Life Obstacles," *New York Times* (16 April 1998): A14–A15.
4. "Disney's Animal Kingdom."
5. Norie Quintos Danyliw, "The Kingdom Comes: Fake Animals Outshine the Real Ones in Disney's Newest Park," *U.S. News & World Report* (6 April 1998): 64.
6. "Disney's Animal Kingdom."
7. Ibid.; Navarro, "New Disney Kingdom Comes With Real-Life Obstacles," p. A14.

8. Aldo Leopold, *A Sand County Almanac With Essays on Conservation from Round River* (New York: Ballantine Books, 1970), p. 264. For entry into the scholarly debate on the meaning and usefulness of the wilderness ideal in American environmentalism, see William Cronon, "The Trouble with Wilderness; or, Getting Back to the Wrong Nature," *Environmental History* 1 (1996): 7–28 and the commentaries that follow this article.

1 HUNTING WITH THE CAMERA

1. Theodore Roosevelt, "Wild Man and Wild Beast in Africa," *National Geographic Magazine* 22 (1911): 4. The number of specimens collected was reported in a number of newspaper articles. See "Roosevelt Sails Down Nile," *New York Times* (1 March 1910), p. 2; "Roosevelt Specimens 11,397," *New York Times* (7 April 1910), p. 4. For Roosevelt's account of the expedition, see Theodore Roosevelt, *African Game Trails* (New York: Charles Scribner's Sons, 1910). For a discussion of the aesthetics of realism in the museum diorama, see Donna Haraway, *Primate Visions: Gender, Race, and Nature in the World of Modern Science* (New York: Routledge, 1989); Susan Leigh Star, "Craft vs. Commodity, Mess vs. Transcendence: How the Right Tool Became the Wrong One in the Case of Taxidermy and Natural History," in *The Right Tools for the Job: At Work in Twentieth-Century Life Sciences,* eds. Adele E. Clarke and Joan H. Fujimura (Princeton, N.J.: Princeton University Press, 1992), pp. 257–286; Karen Wonders, *Habitat Dioramas: Illusions of Wilderness in Museums of Natural History* (Uppsala: Almquist & Wiksell, 1993).
2. Frederic Lucas, "The Scope and Needs of Taxidermy," *Annual Report of the American Society of Taxidermists* 3 (1883): 52–53.
3. H. F. Hoffman, "The Roosevelt Pictures," *The Moving Picture World* 6 (30 April 1910): 683. See also "Roosevelt in Africa—Special Two-Reel Licensed Subject Released April 18," *The Moving Picture World* 6 (2 April 1910): 528–529; "See Roosevelt Hunt in Moving Pictures," *New York Times* (19 April 1910), p. 5; Cherry Kearton, *Wild Life Across the World* (London: Hodder and Stoughton, 1914). On Kearton, see C. A. W. Guggisberg, *Early Wildlife Photographers* (London: David & Charles, 1977).
4. *The Moving Picture World* (14 May 1910): 793; Hoffman, "The Roosevelt Pictures," p. 682.
5. Rev. E. Boudinot Stockton, "The Educational Picture Discussed," *The Moving Picture World* 17 (9 April 1913): 627. On the social history of early motion pictures, see Lary May, *Screening Out the Past: The Birth of Mass Culture and the Motion Picture Industry* (Chicago: University of Chicago Press, 1983); Robert Sklar, *Movie-Made America: A Cultural History of American Movies* (New York: Vintage Books, 1975). David Nasaw, *Going Out: The Rise and Fall of Public Amusements* (New York: Basic Books, 1993).
6. Stockton, "The Educational Picture Discussed," p. 626.
7. "Theodore Roosevelt: The Picture Man," *The Moving Picture World* 7 (22

October 1910): 920. On the Buffalo Jones expedition, see Guy H. Schull, *Lassoing Wild Animals in Africa* (New York: Frederick A. Stokes Co., 1911). On Marey, Muybridge, and scientific cinema, see Lisa Cartwright, *Screening the Body: Tracing Medicine's Visual Culture* (Minneapolis: University of Minnesota Press, 1995). See also Marta Braun, *Picturing Time: The Work of Etienne-Jules Marey (1830–1904)* (Chicago : University of Chicago Press, 1992).

8. W. Stephen Brush, "How Wild Animals Live," 18 *The Moving Picture World* (8 November 1913): 592; "Among the Fishes. The Stickleback," *The Moving Picture World* 18 (15 November 1913): 724.
9. *The Moving Picture World* (29 May 1909): 712. On the Selig Roosevelt film, see Kevin Brownlow, *The War, the West, and the Wilderness* (New York: Alfred A. Knopf, 1979), pp. 405–406.
10. *The Moving Picture World* 16 (14 June 1913): 1105; James S. McQuade, "Alone in the Jungle," *The Moving Picture World* 16 (7 June 1913): 1006. See also "More Selig Animals Shipped West," *The Moving Picture World* 12 (11 June 1912); McQuade, "The Leopard's Foundling," *The Moving Picture World* 20 (13 June 1914): 1516–1517; McQuade, "In Tune With the Wild," *The Moving Picture World* 21 (18 July 1914): 408–409. Quoted in Stockton, "The Educational Picture Discussed," p. 626.
11. Peter J. Schmitt, *Back to Nature: The Arcadian Myth in Urban America* (New York: Oxford University Press, 1969); Raymond Williams, "Ideas of Nature," in *Problems in Materialism and Culture* (London: Verso, 1980). On Seton, see Thomas R. Dunlap, "The Realistic Animal Story: Ernest Thompson Seton, Charles Roberts, and Darwinism," *Forest and Conservation History* 36 (1992): 56–62; H. Allen Anderson, *The Chief: Ernest Thompson Seton and the Changing West* (College Station: Texas A & M University Press, 1986); and John Henry Wadlan, *Ernest Thompson Seton* (New York: Arno, 1978).
12. John Burroughs, "Real and Sham Natural History," *Atlantic Monthly* 91 (March 1903): 298–309. On the nature faking controversy, see Ralph H. Lutts, *The Nature Fakers: Wildlife, Science & Sentiment* (Golden, Colo.: Fulcrum Publishing, 1990).
13. Edward B. Clark, "Roosevelt on the Nature Fakirs," *Everybody's Magazine* 16 (June 1907): 770–774.
14. Theodore Roosevelt, "Nature Fakers," *Everybody's Magazine* 17 (September 1907): 427–430; Lutts, *The Nature Fakers,* pp. 101–138.
15. On nature as the landscape of authenticity within American culture, see William Cronon, "The Trouble with Wilderness; or, Getting Back to the Wrong Nature," in *Uncommon Ground: Toward Reinventing Nature,* ed. William Cronon (New York: W. W. Norton, 1995), pp. 69–90. See also T. Jackson Lears, *No Place of Grace: Antimodernism and the Transformation of American Culture, 1880–1920* (New York: Pantheon, 1981).
16. On the shift from a culture of imitation to a culture of authenticity, see Jackson Lears, *Fables of Abundance: A Cultural History of Advertising in America* (New York: Basic Books, 1994); Miles Orvell, *The Real Thing: Imitation and Authen-*

ticity in American Culture, 1880–1940 (Chapel Hill: University of North Carolina Press, 1989).

17. Kearton, *Wild Life Across the World,* pp. v–vi.
18. Guy H. Scull, *Lassoing Wild Animals in Africa* (New York: Frederick A. Stokes Co., 1911), xii; On Jones, bison preservation, and masculine frontier culture, see Andrew C. Isenberg, "The Returns of the Bison: Nostalgia, Profit, and Preservation," *Environmental History* 2 (1997): 179–196.
19. "Lassoing Wild Animals in Africa," *The Moving Picture World* 8 (25 February 1911): 404; Scull, *Lassoing Wild Animals in Africa,* vi; "Hunting with the Camera," *The Moving Picture World* 8 (11 February 1911): 286. See also Kevin Brownlow, *The War, the West, and the Wilderness* (New York: Alfred A. Knopf, 1979); James B. Wolf, "Buffalo Jones and the Conquest of the East African Frontier," *Journal of American Culture* 9 (1986): 39–48.
20. "The Boone & Crockett Club," *Forest and Stream* 30 (8 March 1888): 124; "Returning from the Hunt," *New York Times* (2 March 1910), p. 8. On the Boone and Crockett club, see John F. Reiger, *American Sportsmen and the Origins of Conservation* (New York: Winchester Press, 1975).
21. Quoted in Roderick Nash, *Wilderness and the American Mind,* 3d ed. (New Haven: Yale University Press, 1982), p. 150. Roosevelt's mythologizing of the frontier and hunter are discussed in Richard Slotkin, *Gunfighter Nation: The Myth of the Frontier in Twentieth-Century America* (New York: HarperPerennial, 1992); Slotkin, "Nostalgia and Progress: Theodore Roosevelt's Myth of the Frontier," *American Quarterly* 33 (1981): 608–637. On Roosevelt and the issue of masculinity in early twentieth-century America, see Peter Filene, "Between a Rock and a Soft Place: A Century of American Manhood," *South Atlantic Quarterly* 84 (1985): 339–355; Kim Townsend, *Manhood at Harvard: William James and Others* (New York: W. W. Norton, 1996). Roosevelt's conservation policies are discussed in Samuel P. Hays, *Conservation and the Gospel of Efficiency: The Progressive Conservation Movement, 1890–1920* (1959; New York: Athenaeum, 1975); Paul Russell Cutright, *Theodore Roosevelt: The Making of a Conservationist* (Urbana, Ill.: University of Illinois Press, 1985); Stephen Fox, *John Muir and His Legacy: The American Conservation Movement* (Boston: Little, Brown, and Co., 1981).
22. "I Propose to Smoke Roosevelt Out—Dr. Long," *New York Times* (2 June 1907), pt. 5, p. 2; William Hornaday, *Wild Animal Round-Up,* p. 331; Hornaday, *Our Vanishing Wild Life: Its Extermination and Preservation* (New York: Charles Scribner's Sons, 1913), pp. 384–385; Kearton, *Wild Life Across the World,* p. 127; Roosevelt, "Wild Man and Wild Beast in Africa," pp. 4–5.
23. George Fortiss, "Paul Rainey, Sportsman," *The Outing Magazine* 58 (1911): 746. *National Cyclopaedia of American Biography* 19: 411–412. See also Paul Rainey, "Bagging Arctic Monsters with Rope, Gun, and Camera," *Cosmopolitan* 50 (1910): 91–103; "Polar Bear Fight at Close Range," *New York Times* (8 September 1910): 8.
24. William T. Hornaday, "Odious Nicknames," *New York Zoological Society Bulletin* 54 (1901): 24. On American zoos as places of moral and social reform, see Eliza-

beth Anne Hanson, "Nature Civilized: A Cultural History of American Zoos, 1870–1940" (Ph.D. diss., University of Pennsylvania, 1996). See also William Bridges, *Gathering of Animals: An Unconventional History of the New York Zoological Society* (New York: Harper & Row, 1974). On the philanthropic interests and educational mission of New York City's elite, see Thomas Bender, *New York Intellect: A History of Intellectual Life in New York City, From 1750 to the Beginnings of Our Own Time* (Baltimore: Johns Hopkins University Press, 1987).

25. Fortiss, "Paul Rainey, Sportsman," 749; "Polar Bear Fight at Close Range," *New York Times* (8 September 1910): 8.
26. "The Paul Rainey African Pictures," *The Moving Picture World* 12 (20 April 1912): 214.
27. *Paul Rainey's African Hunt* (1912; Motion Picture, Broadcasting and Recorded Sound Division, Library of Congress); "Paul Rainey Going for a 3-Year Hunt," *New York Times* (1 February 1911): 7; "Abandoning the Sport for the Dogs," *New York Times* (24 August 1911): 6; "Natural History," *The Moving Picture World* 12 (4 May 1912): 411. See also Paul Rainey, "The Royal Sport of Hounding Lions," *The Outing Magazine* 59 (1911): 131–152.
28. "The Paul Rainey African Pictures," 214–215; *The Moving Picture World* 18 (15 November 1913): 770–771; *The Moving Picture World* 12 (20 April 1912): 243; "The Paul J. Rainey African Pictures," *The Moving Picture World* 21 (11 July 1914): 245.
29. *The Moving Picture World* 19 (10 January 1914): 217; Charles Musser, in collaboration with Carol Nelson, *High-Class Moving Pictures: Lyman H. Howe and the Forgotten Era of Traveling Exhibition, 1880–1920* (Princeton: Princeton University Press, 1991), p. 10. *The Moving Picture World* 18 (22 November 1913): 907.
30. On the image of science as an ascetic pursuit in America during the 1920s, see Charles Rosenberg, "Martin Arrowsmith: The Scientist as Hero," in *No Other Gods: On Science and American Social Thought* (Baltimore, Md.: Johns Hopkins University Press, 1997).
31. Quoted in May, *Screening Out the Past,* p. 113.
32. For biographical information on Burden, see *National Cyclopaedia of American Biography,* vol. 24, pp. 224–225; "William D. Burden, Naturalist, Is Dead," *New York Times* (16 November 1978): D19; "Autobiographical Statement," Box 1, Folder 1, WDBP.
33. On Andrews's life, see his autobiography, Roy Chapman Andrews, *Under a Lucky Star: A Lifetime of Adventure* (New York: Viking Press, 1943). On the Central Asiatic Expedition, see Andrews, *The New Conquest of Central Asia: A Narrative of the Central Asiatic Expeditions in Mongolia and China, 1921–1930* (New York: American Museum of Natural History, 1932); Ronald Rainger, *An Agenda for Antiquity: Henry Fairfield Osborn & Vertebrate Paleontology at the American Museum of Natural History* (Tuscaloosa: University of Alabama Press, 1991), pp. 100–104.
34. Andrews, *Under a Lucky Star,* pp. 178, 186, 243–244; William Douglas Burden, *Look to the Wilderness* (Boston: Little Brown & Co., 1956), pp. viii–ix.

35. "Autobiographical Statement," Box 1, Folder 1, WDBP.
36. Burden's account of the expedition was published in a number of versions throughout his life: see W. D. Burden, "The Quest for the Dragon of Komodo," *Natural History* 27 (1927): 2–18; Burden, *The Dragon Lizards of Komodo* (New York: Putnam, 1927); and Burden, *Look to the Wilderness.* See also "Dragons of Legend Come to Bronx Zoo," *New York Times* (19 September 1926), sec. X, pp. 5, 11.
37. Burden, *Look to the Wilderness,* p. 193.
38. On zoos as middle landscapes, see Hanson, "Nature Civilized."
39. Burden, *Dragon Lizards of Komodo,* p. 67. Burden's romantic primitivism was shared by others of his period. See Marianna Torgovnick, *Gone Primitive: Savage Intellect, Modern Lives* (Chicago: University of Chicago Press, 1990).
40. *Burden East Indian Expedition,* American Museum of Natural History Film Archives, New York, New York. For a path-breaking analysis of race and gender in the safari expedition, see Donna Haraway, "Teddy Bear Patriarchy: Taxidermy in the Garden of Eden, New York City, 1908–1936," *Social Text* 11 (1984–1985): 20–64.
41. Burden to Emmet Reid Dunn, 29 March 1927, Box 2, Folder 4, WDBP; Burden, "Quest for the Dragon of Komodo," p. 16.
42. Dunn to Burden, 21 March 1928, and Burden to Dunn, 29 March 1927, Box 2, Folder 4, WDBP.
43. Burden, "Quest for the Dragon of Komodo," p. 10; and Burden, "Observations on the Habits and Distribution of *Varanus Komodoensis Owens,*" *American Museum Novitates* 316 (18 May 1928): 5, 8. On the addition of the scene showing Burden and Katherine in the blind, see J. R. Bray to Burden, 15 May 1928, Box 2, Folder 9, WDBP. On the transformation of visual representations in the life sciences, see Michael Lynch, "The Externalized Retina: Selection and Mathematization in the Visual Documentation of Objects in the Life Sciences," in *Representation in Scientific Practice,* ed. Michael Lynch and Steve Woolgar (Cambridge, Mass: MIT Press, 1988), pp. 153–186.
44. "They Took Off the Lid When They Wrote Their Reviews," *The Moving Picture World* 72 (14 February 1925): 62; Bray to Burden, 15 May 1928, Box 2, Folder 9, WDBP. For documentation on the showing of the Komodo dragon film, see Burden to Franklin L. Fisher, 1 March 1928, Box 2, Folder 10; Alden Root to Burden, 2 February 1927, Box 2, Folder 10; George Putnam to Burden, 14 November 1927, Box 2, Folder 8; and Alfred M. Collins to Burden, 6 January 1928, Box 2, Folder 10, WDBP. On Burden's refusal of a commercial offer, see Burden to Dunn, 29 March 1927, Box 2, Folder 4, WDBP.

2 SCIENCE VERSUS SHOWMANSHIP ON THE SILENT SCREEN

1. Helen Bullitt Lowry, "New Adam and Eve Among the Gentle Wild Beasts," *New York Times Magazine* (29 April 1923), p. 8. On the formation of MGM and the rise of the Hollywood studio system, see Richard Koszarski, *An*

Evening's Entertainment: The Age of the Silent Feature Picture 1915–1928 (New York: Charles Scribner's Sons, 1990); Thomas Schatz, *The Genius of the System: Hollywood Filmmaking in the Studio Era* (New York: Pantheon Books, 1988).

2. For a comprehensive biography of Martin and Osa Johnson, see Pascal James Imperato and Eleanor M. Imperato, *They Married Adventure: The Wandering Lives of Martin & Osa Johnson* (New Brunswick, N.J.: Rutgers University Press, 1992).
3. Penelope Bodry-Sanders, *Carl Akeley: Africa's Collector, Africa's Savior* (New York: Paragon House, 1991), p. 209.
4. Quoted in ibid., p. 141. On the African Hall, see Haraway, "Teddy Bear Patriarchy: Taxidermy in the Garden of Eden, New York City, 1908–1936," *Social Text* 11 (1984–1985): 20–64.
5. Carl Akeley, *Meandering in Africa,* American Museum of Natural History Film Archives, New York, New York.
6. Quoted in Bodry-Sanders, *Carl Akeley,* p. 210.
7. Lowry, "New Adam and Eve," p. 8.
8. Ibid.
9. On the history of advertising in America, see T. Jackson Lears, *Fables of Abundance: A Cultural History of Advertising in America* (New York: Basic Books, 1994); Roland Marchand, *Advertising the American Dream: Making Way for Modernity, 1920–1940* (Berkeley: University of California Press, 1985).
10. Kenneth M. Cameron, *Africa on Film: Beyond Black and White* (New York: Continuum, 1994), pp. 46–47; Imperato, *They Married Adventure,* pp. 92–109.
11. Quoted in Donna Haraway, *Primate Visions: Gender, Race, and Nature in the World of Modern Science* (New York: Routledge, 1989), p. 45. The complex negotiations resulting in the funding of the Johnson expedition are discussed in Imperato, *They Married Adventure,* pp. 110–119.
12. Martin Johnson, "Camera Hunts in the Jungles of Africa," *New York Times* (17 July 1927), sec. IV, p. 4. *Simba* (Martin Johnson African Expedition Corp., 1928). The revenue figures for *Simba* are from Brownlow, *The War, the West, and the Wilderness.* On the box-office price, see *Variety* 90 (25 January 1928): 12.
13. The history of ethnographic film as salvage ethnography and the taxonomy of race is discussed in Fatimah Tobing Rony, *The Third Eye: Race, Cinema, and Ethnographic Spectacle* (Durham, N.C.: Duke University Press, 1996). On the racial stereotyping of African-Americans in Hollywood cinema, see Donald Bogle, *Toms, Coons, Mulattoes, Mammies, and Bucks* (New York: Continuum 1973). See also Daniel Leab, *From Sambo to Superspade: The Black Experience in Motion Pictures* (Boston: Houghton Mifflin Co., 1975) and N. Frank Ukadike, "Western Film Images of Africa: Genealogy of an Ideological Formulation," *The Black Scholar* 21(1990): 30–48.
14. On Pickford, see Lary May, *Screening Out the Past: The Birth of Mass Culture and the Motion Picture Industry* (Chicago: University of Chicago Press, 1983), pp. 118–146.

15. On the faking of the lion-spearing sequence in *Simba,* see Bodry-Sanders, *Carl Akeley,* pp. 212–213; Cameron, *Africa on Film,* p. 49; Imperato and Imperato, *They Married Adventure,* pp. 136–137, 144.
16. Harold Coolidge to W. D. Burden, 17 February 1931, Box 11, Folder 7, WDBP. On the filming of *Congorilla,* see Imperato and Imperato, *They Married Adventure,* pp. 153–169; Martin Johnson, *Congorilla: Adventures with Pygmies and Gorillas in Africa* (New York: Brewer, Warren and Putnam, 1932).
17. Burden to Merian C. Cooper, 16 April 1928, Box 2, Folder 12; Burden to J. B. Shackelford, 9 November 1927, Box 12, Folder 2, WDBP.
18. "Chang," *Moving Picture World* 85 (2 May 1927): 848. For biographical information on Cooper and Schoedsack, see Brownlow, *The War, the West, and the Wilderness,* pp. 515–516; Steve Vertlieb, "The Man Who Saved King Kong," in *The Girl in the Hairy Paw: King Kong as Myth, Movie, and Monster,* ed. Ronald Gottesman and Harry Geduld (New York: Avon Books, 1976), pp. 29–36; Ernest B. Schoedsack, "The Making of an Epic," *American Cinematographer* 64 (1983): 41–44, 109–114; "Thrills of Making Jungle Film," *New York Times* (10 April 1927), sec. VIII, p. 7, c. 6.
19. Paul Rotha, *Documentary Film* (London: Faber & Faber, 1936), p. 78. Merian Cooper, *Grass* (New York: G. P. Putnam's Sons, 1925), p. v. See also Richard M. Barsam, *Non-Fiction Film: A Critical History,* rev. ed. (Bloomington: Indiana University Press, 1992); William Guynn, *A Cinema of Nonfiction* (Rutherford, N.J.: Fairleigh Dickinson University Press, 1990).
20. Cooper, *Grass,* pp. ix–x.
21. Marguerite Harrison, *There's Always Tomorrow: The Story of a Checkered Life* (New York: Farrar & Rinehart, 1935), p. 576. Cooper, *Grass,* pp. vi, 137–141. *Grass* (Famous-Players-Lasky Corporation, 1925). See also Ernest B. Schoedsack, "The Making of an Epic," *American Cinematographer* 64 (February 1993): 41–44, 109–114, 571–664. The importance of the filmmaker's presence in documentary realism is discussed in Bill Nichols, *Representing Reality: Issues and Concepts in Documentary* (Bloomington: Indiana University Press, 1991), pp. 180–187.
22. *Chang: A Drama of the Wilderness* (Paramount, 1927). See also Merian C. Cooper and Ernest B. Schoedsack, "'Mr. Crooked': Some Pages from a Siamese Jungle Diary," *Asia* 27 (1927): 475–481, 506–513; Cooper and Schoedsack, "The Warfare of the Jungle Folk: Campaigning Against Tigers, Elephants, and Other Wild Animals in Northern Siam," *National Geographic Magazine* LIII (1928): 233–268; Brownlow, *The War, the West, and the Wilderness,* pp. 529–541.
23. Earning estimates are from Burden to Shackelford, 9 November 1927, Box 12, Folder 2, WDBP. Quote is from "Straight from the Shoulder Reports," *Moving Picture World* 88 (24 September 1927): 253. See also "*Chang* Passes 250th Showing," *Moving Picture World* 86 (18 June 1927): 508; "*Chang* Packs L.A. Theatre," *Moving Picture World* 87 (2 July 1927): 17; "Paramount Films in 3 Los Angeles Theaters," *Moving Picture World* 87 (23 July 1927): 253.

24. "Chang," *Moving Picture World* 85 (2 May 1927): 848. Quoted in Arthur Calder-Marshall, *The Innocent Eye: The Life of Robert J. Flaherty* (New York: Harcourt, 1963), p. 97. Quoted in Rudy Behlmer, "Merian C. Cooper," *Films in Review* 17 (January 1966): 22–23. See also "*Chang*," *New York Times* (30 April 1927), p. 25, col. 3; "*Chang*," *Variety* 87 (4 May 1927): 20.
25. Burden to Cooper, 16 April 1928, Box 2, Folder 12, WDBP.
26. Burden to Monsieur Mallet, 25 April 1928, Box 2, Folder 12; Mallet to Burden, 27 April 1928, Box 2, Folder 12, WDBP.
27. *The Silent Enemy* (Paramount, 1930).
28. Shackelford to Burden, 8 December 1927, Box 12, Folder 2, WDBP. Information on financing of the film is available in Burden to C. V. Whitney, 17 May 1933, Box 3, Folder 13, and Burden to Paramount Publix Corporation, 7 May 1931, Box 11, Folder 13, WDBP. For secondary accounts on the making of *The Silent Enemy,* see Brownlow, *The War, the West, and the Wilderness,* pp. 545–560; Bunny McBride, *Molly Spotted Elk: A Penobscot in Paris* (Norman: University of Oklahoma Press, 1995), pp. 96–127; Donald B. Smith, *Long Lance: The True Story of an Impostor* (Lincoln: University of Nebraska Press, 1982).
29. Quoted in McBride, *Molly Spotted Elk,* p. 105. On Burden's need to escape New York City, see Burden to Douglas Johnson, 14 January 1927, Box 11, Folder 13, and Burden to Henry Fairfield Osborn, 26 June 1928, Box 1, Folder 5, WDBP.
30. *The Silent Enemy* (Paramount, 1930); "How The Silent Enemy Was Made," p. 6, Box 3, Folder 4, WDBP.
31. See Haraway, *Primate Visions,* pp. 57–58; Susan R. Schrepfer, *The Fight to Save the Redwoods: A History of Environmental Reform, 1917–1978* (Madison: University of Wisconsin Press, 1983), pp. 43–45.
32. "How *The Silent Enemy* Was Made," p. 4, 17, Box 3, Folder 4, WDBP. Ralph Friar and Natasha Friar, *The Only Good Indian . . . The Hollywood Gospel* (New York: Drama Book Publications, 1972), pp. 67–68.
33. Quoted in McBride, *Molly Spotted Elk,* pp. 71, 88.
34. Molly Spotted Elk to Mr. Fernandez, undated, Box 2, Folder 11, WDBP.
35. Quoted in McBride, *Molly Spotted Elk,* p. 125.
36. William L. Laurence, "The Death Chant of Red Gods and Man," in "How *The Silent Enemy* Was Made," p. 7, Box 3, Folder 4, WDBP. William Douglas Burden, *Look to the Wilderness* (Boston: Little, Brown, 1956), p. 248.
37. "Indian Hunters of Old," *New York Times* (20 May 1930), p. 32, col. 3; Grant to Burden, 6 May 1930, Box 2, Folder 13; Grant to Chanler, 6 May 1930, Box 2, Folder 13, WDBP.
38. Burden-Chanler Productions to Carl Lowman [sic], 15 May 1930, Box 2, Folder 13; Lomen to Burden, 21 May 1930, Box 2, Folder 13, WDBP.
39. Burden to Osborn, 26 June 1928, Box 1, Folder 5, WDBP. See also Burden to Finne, 6 December 1928, Box 2, Folder 12; Burden to Stone, 28 March 1930, Box 2, Folder 13, WDBP.

40. Quoted in Smith, *Long Lance: The True Story of an Impostor*, pp. 176–177.
41. See ibid., p. 154, and Chief Buffalo Child Long Lance, *Long Lance* (1928; rpt. Jackson, Miss.: Banner Books, 1995).
42. Quoted in Smith, *Long Lance,* p. 195.
43. Ibid., pp. 203–204.
44. *The Silent Enemy;* Burden, *Look to the Wilderness,* p. 247.
45. Burden to Chanler, 12 August 1930, Box 3, Folder 11, WDBP. Richard Dana Skinner, "The Silent Enemy," *The Commonweal* (11 June 1930): 165; Chanler to Hamilton, 14 July 1930, Box 3, Folder 9, WDBP. See also Bruce Bliven, "Films, Russian and American," *The New Republic* 63 (4 June 1930): 73; Mordaunt Hall, "Indian Hunters of Old," *New York Times* (20 May 1930), p. 32, col. 3; "The Silent Enemy," *Time* (26 May 1930): 58. On the distribution problems and bankruptcy, see Sidney Kent to Chanler, 24 June 1930, Box 3, Folder 9; Chanler to Kent, 25 June 1930, Box 3, Folder 9; Burden to Julian Johnson, 5 January 1931, Box 3, Folder 9, Burden to Paramount Publix Corporation, 7 May 1931, Box 11, Folder 13; Chanler to C. V. Whitney, Box 3, Folder 13, WDBP. On the struggle between urban and small-town exhibition in the 1920s, see Kathryn H. Fuller, *At the Picture Show: Small-Town Audiences and the Creation of Movie Fan Culture* (Washington, D.C.: Smithsonian Institution Press, 1996).
46. "Ingagi," *Variety* (16 April 1930): 46; Advertisement, *Chicago Tribune* (20 April 1930), pt. 7, p. 4. Gerald Peary, "Missing Links: The Jungle Origins of *King Kong,*" in *The Girl in the Hairy Paw: King Kong as Myth, Movie, and Monster,* ed. Ronald Gottesman and Harry Geduld (New York: Avon Books, 1976), pp. 37–42.
47. "Twelfth Annual Meeting of the American Society of Mammalogists," *Journal of Mammalogy* 11 (1930): 426–431.
48. "'Ingagi'—True or False?" *Nature Magazine* 16 (July 1930): 66. See, also, "The Gorilla Film 'Ingagi,'" *Science* suppl. 71 (6 June 1930): x.
49. "Hays Bars 'Ingagi' Film," *New York Times* (12 June 1930), p. 30, col. 2; Peary, "Missing Links."
50. Mae Tinée, "*Ingagi,*" *Chicago Tribune* (26 March 1930), p. 27; Burden to Arthur James, 22 September 1930, Box 3, Folder 13, WDBP.
51. Quoted in Robert J. Gordon, *Picturing Bushmen: The Denver African Expedition of 1925* (Athens, Ohio: Ohio University Press, 1997), p. 17; *Variety* 87 (1 June 1927): 24.
52. *Variety* 100 (24 September 1930): 23, 30.
53. Burden to James, 22 September 1930, Box 3, Folder 13; James to Burden, 24 September 1930, Box 3, Folder 13, WDBP. *Africa Speaks* (Motion Picture Division, Library of Congress). The Library of Congress version is missing the reel containing the lion mauling scene. A version with this scene is available through Nostalgia Family Video, P.O. Box 606, Baker City, Oregon. See also Paul L. Hoeffler, *Africa Speaks: A Story of Adventure* (Philadelphia: John C. Winston Co., 1931).

54. Julian Johnson to Burden, 6 August 1930, Box 3, Folder 1, WDBP.
55. Rotha, *Documentary Film,* p. 6. Information on film libraries and projectors comes from Gloria Waldon, *The Information Film* (New York: Columbia University Press, 1949); and Ford L. Lemler, "The University or College Film Library," in *Film and Education,* ed. Godfrey M. Elliott (New York: Philosophical Library, 1948), pp. 501–521. On the impact of World War II on the development of education film, see Charles F. Hoban, Jr., *Movies That Teach* (New York: Dryden, 1946).
56. "General Outline and Plan of Operation of Adventure Pictures, Inc.," n.d., Box 3, Folder 13, WDBP. See also Betty Shannon to Burden, 8 January 1931, Box 3, Folder 10; Burden to Elizabeth Perkins, 18 November 1932, Box 2, Folder 13, WDBP.
57. C. V. Whitney, *High Peaks* (Lexington, Ky.: University Press of Kentucky, 1977); Schatz, *The Genius of the System,* pp. 176–198.
58. Burden to Harold Coolidge, 13 February 1931, Box 11, Folder 7, WDBP. On their joint interest in education film, see Burden to August Belmont, 5 April 1935, Box 10, Folder 5, WDBP.
59. Burden to Cooper, 17 April 1931, Box 11, Folder 7, WDBP. See also Burden to Coolidge, 10 March 1931, Box 11, Folder 7, WDBP.
60. Quoted in Rudy Behlmer, "Foreword," *The Girl in the Hairy Paw: King Kong as Myth, Movie, and Monster,* ed. Ronald Gottesman and Harry Geduld, pp. 10–11 (New York: Avon Books, 1976). The relationships between *King Kong* and ethnographic cinema of the 1920s are explored in Rony, *The Third Eye,* pp. 157–191.
61. On the filming of *Trader Horn,* see Brownlow, *The War, the West, and the Wilderness,* pp. 560–566; Robert C. Cannom, *Van Dyke and the Mythical City, Hollywood* (Culver City, Calif.: Murray & Gee, Inc., 1948), pp. 187–227.

3 ZOOMING IN ON ANIMALS' PRIVATE LIVES

1. Niko Tinbergen, "Watching and Wondering," in *Studying Animal Behavior: Autobiographies of the Founders,* ed. Donald A. Dewsbury (Chicago: University of Chicago Press, 1989), p. 432.
2. Niko Tinbergen, *The Herring Gull's World: A Study of the Social Behaviour of Birds* (London: Collins, 1953), pp. 231–232.
3. Niko Tinbergen to John Sparks, 28 November 1982, MS. Eng. C3157, E36, NTP.
4. Konrad Lorenz, *On Life and Living* (New York: St. Martin's Press, 1990), pp. 59–60.
5. On the importance of film in the study of behavior, see Niko Tinbergen, *The Study of Instinct* (Oxford: Clarendon Press, 1951), p. 7 and Lorenz, *On Life and Living,* p. 55. The shift from the gun to the camera in natural history is mentioned in David Allen, *The Naturalist in Britain: A Social History,* 2nd ed. (Princeton: Princeton University Press, 1994), pp. 210–212 and Ralph H.

Lutts, *The Nature Fakers: Wildlife, Science & Sentiment* (Golden, Colo.: Fulcrum Publishing, 1990). For a provocative analysis of the history of hunting and the myth of man the hunter, see Matt Cartmill, *A View to a Death in the Morning: Hunting and Nature through History* (Cambridge, Mass.: Harvard University Press, 1993).

6. For a valuable guide to scientific research films produced prior to 1955, see Anthony R. Michaelis, *Research Films in Biology, Anthropology, Psychology, and Medicine* (New York: Academic Press, 1955).
7. Film Evaluation Sheets, MS. Eng. C3132, C34, NTP. On the blurring of art and science in the filming of behavior, see Walter Benjamin, "The Work of Art in the Age of Mechanical Reproduction," in Hannah Arendt, ed., *Illuminations* (New York: Schocken Books, 1969), pp. 217–252.
8. "Museum Wants," *Time* (May 21, 1937), p. 28. "Museum Visitors Can See as Fish Do," *New York Times* (9 June 1937), p. 27. J. von Uexkull and G. Kriszat's *Streifzüge durch die Umwelten von Tieren und Menschen* (Berlin: Julius Springer, 1934) was used to help construct the visual worlds displayed in these exhibits. See Beach to Carr, 13 April 1942, Box 2, Folder 8, Department of Animal Behavior Papers, American Museum of Natural History, New York, New York.
9. Burden to Perry Osborn, 17 December 1941, Box 1, Folder 8; Burden to Malcolm Aldrich, 27 February 1946, Box 1, Folder 8, WDBP.
10. W. Douglas Burden, "A New Vision of Learning," *The Museum News* (15 December 1943): 9–12. On the chameleon and live exhibits, see Noble to Preston Mabry, 23 September 1938, Exhibits: Hall of Animal Behavior I Folder; Noble to Roy Chapman Andrews, 14 January 1939, Departments: Experimental Biology Folder; "The Habitat Group Comes to Life," 11 February 1938, Exhibits: Miscellaneous Folder, GKNP.
11. E. Weyer to A. E. Parr, 25 January 1943, Box 1, Folder 14; Parr to Burden, 8 March 1943, Box 1, Folder 13, WDBP. For the heated exchange between Parr and Burden over this issue see the correspondence in Box 1, Folder 13, WDBP.
12. Burden to Malcolm Aldrich, 27 February 1946; Burden to Aldrich, 5 November 1946, Box 1, Folder 8, WDBP.
13. G. K. Noble, "Hunting With a Camera," GKNP.
14. For biographical information on Noble, see Clifford H. Pope, "Gladwyn Kingsley Noble," *National Cyclopaedia of American Biography* 31 (1944) 396; William K. Gregory, *Copeia* (1940): 274–275; "Gladwyn Kingsley Noble," Noble, G. K.: Biography II Folder, GKNP. On Gregory, see Ronald Rainger, *An Agenda for Antiquity: Henry Fairfield Osborn and Vertebrate Paleontology at the American Museum of Natural History, 1890–1935* (Tuscaloosa: University of Alabama Press, 1991).
15. For the series of negotiations that led to the laboratory's establishment, see Osborn to Noble, 17 March 1928; Noble to Osborn, 20 March 1928, AMNH Presidency: Henry F. Osborn I Folder; Sherwood to Noble, 18 May 1928, Department History Folder; Noble to Sherwood, 22 May 1928, Noble, G. K.:

Biography I Folder; Sherwood to Noble, 1 June 1928, Department: Budgets Folder; GKNP.

16. For a brief description of the laboratory, see Noble to Burden, 15 October 1934, AMNH Departments: Experimental Biology Folder, GKNP.
17. See Sherwood to Noble, 13 January 1933; Draft of Budget for 1934, 13 November 1933; Department: Budgets Folder; Faunce to Noble, 9 April 1935, AMNH Departments: Experimental Biology Folder; GKNP.
18. Burden to Noble, n.d., Douglas, W. Burden Folder, GKNP.
19. William Engle, "Going and Getting It for Science: Biologists Look to Lizards for Answers to Riddles of Life," *New York World-Telegram* 63 (19 June 1931).
20. For fiscal budget statements documenting these funds, see Foundations & Institutes: National Research Council Folder; Foundations & Institutes: Josiah Macy, Jr. Folder; and AMNH Departments: Experimental Biology Folder, GKNP.
21. Noble to Burden, 15 October 1934, Departments: Experimental Biology Folder, GKNP.
22. The overall scope of Noble's research is evident from foundation progress reports in the GKNP. For representative publications, see G. K. Noble and Brian Curtis, "The Social Behavior of the Jewel Fish, *Hemichromis Bimaculatus* Gill," *Bulletin of the American Museum of Natural History,* 76 (1939): 1–46; G. K. Noble, M. Wurm, and A. Schmidt, "Social Behavior of the Black-Crowned Night Heron," *Auk,* 55 (1938): 7–40; G. K. Noble, "The Experimental Animal From the Naturalist's Point of View," *American Naturalist,* 73 (1939): 113–126.
23. Noble to Fremont-Smith, 15 February 1937, "Foundations & Institutes: Josiah Macy, Jr." Folder, GKNP.
24. Julian Huxley, *Bird-Watching and Bird Behavior* (London: Dennis Dobson Ltd., 1930), pp. 16–17. For an excellent analysis of Huxley's contributions to the science of ethology and his indebtedness to the amateur tradition, see Richard W. Burkhardt, Jr., "Huxley and the Rise of Ethology," in *Julian Huxley: Biologist and Statesman of Science,* eds. C. Kenneth Waters and Albert Van Helden (Houston: Rice University Press, 1992), pp. 127–149.
25. Konrad Z. Lorenz, "My Family and Other Animals," in *Studying Animal Behavior: Autobiographies of the Founders,* p. 263.
26. The influence of the biological sciences in the naturalization of gender roles has received extensive treatment by feminist scholars. See, e.g., Ruth Bleier, *Science and Gender: A Critique of Biology and Its Theories on Women* (New York: Pergamon Press, 1984); Linda Birke, *Women, Feminism, and Biology: The Feminist Challenge* (Brighton: Harvester Press, 1986); Anne Fausto-Sterling, *Myths of Gender: Biological Theories about Women and Men* (New York: Basic Books, 1985); Donna Haraway, *Primate Visions: Gender, Race, and Nature in the World of Modern Science* (New York: Routledge, 1989).
27. Thanks go to Lester Aronson and Ethel Tobach for helping me locate *The Social Behavior of the Laughing Gull* in an old walk-in refrigerator at the American

Museum of Natural History. Noble's study was published posthumously in G. K. Noble and M. Wurm, "The Social Behavior of the Laughing Gull," *Annals of the New York Academy of Sciences* 45 (1943): 179–220.

28. For a historical perspective on the development of ethology, see Richard W. Burkhardt, Jr., "On the Emergence of Ethology as a Scientific Discipline," *Conspectus of History* 1 (1981): 62–81; Burkhardt, "The Development of an Evolutionary Ethology," in *Evolution From Molecules to Men,* ed. D. S. Bendall (Cambridge: Cambridge University Press, 1983), pp. 429–444; John Durant, "Innate Characters in Animals and Man: A Perspective on the Origins of Ethology," in *Biology, Medicine, and Society, 1840–1940,* ed. C. Webster (Cambridge: Cambridge University Press, 1981), pp. 157–192; Gregg Mitman and Richard W. Burkhardt, Jr., "Struggling for Identity: The Study of Animal Behavior in America, 1930–1950," in *The Expansion of American Biology,* eds. Keith R. Benson, Ronald Rainger, and Jane Maienschein (New Brunswick, N.J.: Rutgers University Press, 1991), pp. 164–194.
29. Frank Fremont-Smith of the Macy Foundation had Noble's translation of Lorenz's paper mimeographed and sent out to a number of psychologists. Noble himself also sent the paper out to biologists working on animal behavior. See Fremont-Smith to Noble, 6 May 1937, Foundations & Institutes: Josiah Macy, Jr. Folder; Noble to David Davis, 1 October 1937, Davis, David E.: Biological Labs, Harvard Folder; Allee to Noble, 2 January 1940, Universities & Colleges: U.S.-Chicago Folder, GKNP. Konrad Lorenz, "Der Kumpan in der Umwelt des Vogels," *Journal für Ornithologie,* 83 (1935): 137–213, 289–413.
30. Niko Tinbergen, *The Study of Instinct* (Oxford: Clarendon Press, 1951). For Noble's criticisms, see G. K. Noble to Margaret M. Nice, 24 October 1938; Noble to Nice, 31 October 1938; Noble to Nice, 18 November 1938; Noble, G. K.: Birds I Folder; Noble to Nice, 14 November 1940, N-Misc. Correspondence Folder; Noble to Konrad Lorenz, 26 July 1936, Lorenz, Konrad Folder, GKNP.
31. Daniel Lehrman, "A Critique of Konrad Lorenz's Theory of Instinctive Behavior," *Quarterly Review of Biology* 28 (1953): 337–363. On the importance of the American critique in the modification of Tinbergen's views, see Niko Tinbergen to Julian Huxley, 14 July 1958, Box 26, Folder 5, JHP; Niko Tinbergen to Konrad Lorenz, 31 January 1961, MS. Eng. C3156, E10 and Niko Tinbergen to Richard Burkhardt, 16 June 1982, MS. Eng. C3156, E2, NTP.
32. Tinbergen, "Watching and Wondering," p. 438.
33. On the RWU, see IWF, *Wissenschaftlichen Film in Deutschland* (Göttingen: Instituten für den Wissenschaftlichen Film, 1981). Karl von Frisch noted that all university teachers were required under National Socialism to submit their films to the Reichstelle für den Unterrsichfilm. See Karl von Frisch, *A Biologist Remembers,* trans. Lisbeth Gombrich (New York: Pergamon Press, 1967), p. 95. On the importance of nature protection and biology in the education system of the National Socialist state, see Änne Bäumer-Schleinkofer, *Nazi Biology and Schools,* trans. Neil Beckhaus (Frankfurt am Main: Peter Lang, 1995); Anna

Bramwell, *Ecology in the Twentieth Century: A History* (New Haven: Yale University Press, 1989); Raymond H. Dominick, *The Environmental Movement in Germany: Prophets and Pioneers, 1871–1971* (Bloomington: Indiana University Press, 1992); Gert Gröning and Joachim Wolschke-Bulmahn, "Politics, Planning and the Protection of Nature: Political Abuse of Early Ecological Ideas in Germany, 1933–1945," *Planning Perspectives* 2 (1987): 127–148.

34. Konrad Lorenz, "Vergleichende Bewegungsstudien an Anatinen," *Journal für Ornithologie* 89 (1941): 19–29.
35. On the difference in their research programs, see Richard W. Burkhardt, Jr., "The Development of an Evolutionary Ethology," in *Evolution From Molecules to Men,* ed. D. S. Bendall (Cambridge: Cambridge University Press, 1983), pp. 429–444.
36. Niko Tinbergen, "Comparative Studies of the Behavior of Gulls (Laridae): A Progress Report," *Behaviour* 15 (1959): 1–70. See also Tinbergen, *The Herring Gull's World: A Study of the Social Behaviour of Birds* (London: Collins, 1953). The importance of film to Tinbergen's study of adaptive radiation in gulls is not only mentioned in his published articles, but is also detailed in private correspondence. See Tinbergen to Julian Huxley, 10 June 1953, Box 21, Folder 3; Tinbergen to Huxley, 27 June 1958, Box 26, Folder 5; Tinbergen, "Report for the Year Ending 1st October 1958," Box 27, Folder 2; Tinbergen to Huxley, 10 October 1958, 17 October 1958, Box 27, Folder 2, JHP.
37. Tinbergen regarded the study on Kittiwakes undertaken by his student Esther Cullen as one of the most significant research contributions. See Esther Cullen, "Adaptations in the Kittiwake to Cliff-Nesting," *Ibis* 99 (1957): 275–302.
38. See Osborn to Marshall, 24 July 1939; Osborn to Marshall, 9 April 1940, Box 263, Folder 3136, RG1.1 200R, RFA. Noble died before production of any of the films began, but Osborn did complete the film on bird migration, *Birds on a Wing,* which was produced with the assistance of the acclaimed documentary filmmaker John Grierson and released for theatrical distribution by Columbia Pictures.
39. Noble had an active correspondence with other researchers involved in the production of scientific film, including Adelbert Ford. See Photography I Folder, and Photography II Folder, GKNP.
40. Osborn to Marshall, 9 April 1940, Box 263, Folder 3136, RG1.1 200R, RFA.
41. C. R. Carpenter, "Use of Motion Picture Films to Simulate Field Observations of Primates," January 1973, Box 2, Reports (2) Use of Motion Picture Films; Carpenter to Carmichael, 18 June 1964, Box Q3, Correspondence-N; CRCP. For a detailed and provocative historical analysis of Carpenter's professional research and career, see Donna Haraway, "Signs of Dominance: From a Physiology to a Cybernetics of Primate Society, C. R. Carpenter, 1930–1970," *Studies in the History of Biology* 6 (1983): 129–219.
42. Carpenter's publications on instructional television and film are extensive. See, for example, C. R. Carpenter and L. P. Greenhill, "A Scientific Approach to Informational-Instructional Film Production and Utilization," *Journal of the*

Society of Motion Picture and Television Engineers 58 (1952): 415–427; Carpenter and Greenhill, "Final Report: Instructional Film Research Program," *SDC Technical Report* 296 (1956): 1–61; C. R. Carpenter, "Further Studies of the Use of Television for University Teaching," *Audio-Visual Communication Review* 4 (1956): 200–215.

43. Noble to Adelbert Ford, 18 January 1939, Photography II Folder; Noble to Rüppell, 9 November 1938, Photography I Folder, GKNP.
44. For a list of European films on animal behavior imported by Noble, see Noble to Rüppell, 9 December 1938; Noble to Rüppell, 14 September 1938; Rüppell to Noble, 7 October 1938; Werneke to Noble, 10 October 1938; Werneke to Noble, 17 January 1939; Noble to Werneke, 2 February 1939; Photography I Folder; Noble to Werneke, 12 April 1939; Werneke to Noble, 3 September 1939, Photography II Folder, GKNP. The best collection of films produced by von Frisch, which includes all of those seen and purchased by Noble, is available through the Institut Wissenschaften Film, Göttingen, Germany.
45. Julian Huxley to G. K. Noble, 13 November 1937, Huxley, Julian Folder, GKNP.
46. Julian Huxley, *Memories* (London: George Allen and Unwin, 1970), p. 239. See also Julian Huxley to Kenneth Clark, 11 September 1935, Kenneth Clark Collection, Rice University and Julian Huxley to Hugh Casson, 26 April 1960, Box 29, Folder 4, JHP.
47. Julian Huxley, "The Courtship Habits of the Great Crested Grebe *(Podiceps cristatus);* with an addition to the Theory of Sexual Selection," *Proceedings of the Zoological Society of London* 35 (1914): 491–562. See also Julian Huxley, "Courtship Activities in the Red-throated Diver *(Colymbus stellatus Pontopp.);* Together with a Discussion of the Evolution of Courtship in Birds," *Journal of the Linnaean Society of London, Zoology* 35 (1923): 253–292. On Huxley's importance in the study of ethology, see Burkhardt, "Huxley and the Rise of Ethology."
48. Huxley, "The Courtship Habits of the Great Crested Grebe," p. 492. On the importance of amateur ornithology for Huxley's behavioral studies, see Burkhardt, "Huxley and the Rise of Ethology."
49. Julian Huxley, "Making and Using Nature Films," *The Listener* 13 (1935): 595. See also Lockley to Huxley, 10 April 1934 and 10 May 1934, Box 11, Folder 5, JHP.
50. Forsyth Hardy, ed., *Grierson on Documentary* (London: Faber and Faber, 1966), p. 201. On Korda, see Karol Kulik, *Alexander Korda: The Man Who Could Work Miracles* (London: W. H. Allen, 1975). See also Forsyth Hardy, *John Grierson: A Documentary Biography* (London: Faber and Faber, 1979).
51. On this dispute and sales figures for the film, see W. F. Sturt to Huxley, 3 June 1935, 17 June 1935; R. M. Lockley to Huxley, 22 April 1935, 7 July 1935, Box 11, Folder 8, JHP; Sturt to Huxley, 9 March 1936, Box 12, Folder 1, JHP; James B. Pinker to Huxley, 31 January 1936, Box 12, Folder 1; Pinker to Huxley, 29 June 1936, Box 12, Folder 2, JHP.

52. Hardy, *Grierson on Documentary,* p. 148.
53. *The Private Life of the Gannet* (London Film Productions, 1934). Available through the British Film Institute. The 1937 American version is available through Kit Parker Films, Monterey, Calif.
54. On the meeting between Lorenz and Huxley, see Konrad Lorenz to Julian Huxley, 26 November 1963, Box 35, Folder 6, JHP.
55. CBS *Adventure,* "Animal Communication and Behavior," 1/23/55. Available through the American Museum of Natural History Film Archives. On Osborn's association of caged sideshow zoo exhibits and totalitarianism, see Fairfield Osborn, "The New York Zoological Park," *Science* 93 (1941): 467–468.
56. Niko Tinbergen, "Extramurals last lecture," MS. Eng. C3132, C35, NTP.
57. See MS. Eng. C3143, C238, NTP. Tinbergen to Derek Goodwin, 14 October 1968, MS. Eng. C3156, E6; Tinbergen to Miss Gerson, 10 April 1971, MS. Eng. C3143, C240; Tinbergen to the 'Experts,' 8 May 1971, MS. Eng. C3143, C243, NTP.
58. Niko Tinbergen to Julian Huxley, 20 June 1967, Box 41, Folder 6, JHP. For material on the making of *Signals for Survival,* see Folders C240–C243, MS. Eng. C3143, NTP.
59. *Signals for Survival: A Study of Animal Language* (1970). Available through PCR: Films and Video in the Behavioral Sciences, Pennsylvania State University, University Park, Penn.
60. Tinbergen, "Watching and Wondering," pp. 460–461. See also Tinbergen, "Extramurals last lecture," MS. Eng. C3132, C35, NTP.
61. Tinbergen to Hugh Falkus, 2 October 1972, MS. Eng. C3142, C233; Tinbergen to Dieter, 8 July 1972, MS. Eng. C3143, C244; Tinbergen to David Attenborough, 18 November 1971, MS. Eng. C3143, C243, NTP.
62. Juliette Huxley to Niko Tinbergen, 24 September 1986, MS. Eng. C3156, E7, NTP.
63. Tinbergen to John Sparks, 22 April 1980, MS. Eng. C3157, E36; Tinbergen, "Film Seminars 1971, No. 1," MS. Eng. C3132, C36, NTP.

4 WILDLIFE CONSERVATION THROUGH A WIDE-ANGLE LENS

1. Fairfield Osborn, "The Opening of the African Plains," *Bulletin of the New York Zoological Society* 44 (1941): 67. For a history of the New York Zoological Park, see William Bridges, *Gathering of Animals: An Unconventional History of the New York Zoological Society* (New York: Harper & Row, 1974).
2. "The New York Zoological Park," *Science* 93 (16 May 1941): 467. Visitation estimates are from Bridges, *Gathering of Animals,* p. 452.
3. For a historical analysis of scientific interest in peaceful nature in response to the Second World War, see Gregg Mitman, *The State of Nature: Ecology, Community, and American Social Thought* (Chicago: University of Chicago Press, 1992), pp. 146–201.
4. "The New York Zoological Park," 467–468.

5. For biographical information on Osborn, see *National Cyclopaedia of American Biography* 62 (1984): 172–173; *Current Biography 1949:* 463–465.
6. On Hagenbeck, see Nigel T. Rothfels, "Bring 'Em Back Alive: Carl Hagenbeck and Exotic Animal and People Trades in Germany, 1848–1914," (Ph.D. diss., Harvard University, 1994); Herman Reichenbach, "A Tale of Two Zoos: The Hamburg Zoological Garden and Carl Hagenbeck's Tierpark," in *New World, New Animals: From Menagerie to Zoological Park in the Nineteenth Century,* ed. R. J. Hoage and William A. Deiss (Baltimore: Johns Hopkins University Press, 1996), pp. 51–62. On the influence of Hagenbeck in American zoo design, see Elizabeth A. Hanson, "Natural Settings: Interpretations of Nature in Early 20th Century American Zoos," paper delivered at the Davis Center Seminar, 17 October 1997.
7. Aldo Leopold, *A Sand County Almanac. With Essays on Conservation from Round River* (New York: Ballantine Books, 1966), pp. 289, 294, 285.
8. Fairfield Osborn, "The Zoological Society is Going to Wyoming," *Animal Kingdom* 48 (1945): 168.
9. Osborn, "The Zoological Society is Going to Wyoming," 168.
10. "Laurance Rockefeller Reviews Aims of Wildlife Park at Dedication," KCC, RGIV3a3.1, Box 2, Folder 21.
11. *The Jackson Hole Wildlife Park Research & Training Program, Summer 1947.* This film is available through the Media Services Department, Wildlife Conservation Society; Publicity Release, 20 December 1945, KCC, RGIV3A3.1, Box 2, Folder 21.
12. Fabian to Rockefeller, 6 August 1945, HFP, RGIV3A7.3, Box 44, Folder 437.
13. For biographical information on Murie, see *The Living Wilderness* 84 (1963): 3–21. For autobiographical accounts of Murie's life in Alaska and Jackson Hole, see Margaret and Olaus Murie, *Wapiti Wilderness* (New York: Alfred A. Knopf, 1966); Olaus J. Murie, *Journeys to the Far North* (Palo Alto: Wilderness Society and American West Publishing Co., 1973). Murie's elk study was published as Olaus J. Murie, *The Elk of North America* (Harrisburg, Penn.: Stackpole Company, 1951).
14. See Olaus Murie to Vanderbilt Webb, 26 November 1945, Box 1, "Correspondence 1945" File, OMC. Quote is from Olaus Murie to Carl Hubbs, 30 July 1946, Box 361, Folder 7, OMC.
15. *Jackson Hole Courier,* 8 November 1945; Olaus J. Murie, "Fenced Wildlife for Jackson Hole," *National Parks Magazine* (January–March 1946): 8–11.
16. For a comprehensive history of the efforts to establish Jackson Hole as a national park, see Robert W. Righter, *Crucible for Conservation: The Creation of Grand Teton National Park* (Colorado Associated University Press, 1982). See also David J. Saylor, *Jackson Hole, Wyoming: In the Shadow of the Tetons* (Norman: University of Oklahoma Press, 1970).
17. Murie to Hubbs, 30 July 1946. George M. Wright, Joseph S. Dixon, and Ben H. Thompson, *Fauna of the National Parks of the United States* (Washington, D.C.: Government Printing Office, 1933), p. 148.

18. John J. Craighead, Jay S. Sumner, and John A. Mitchell, *The Grizzly Bears of Yellowstone: Their Ecology in the Yellowstone Ecosystem, 1959–1992* (Washington, D.C.: Island Press, 1995), pp. 13–47; Paul Schullery, *The Bears of Yellowstone* (Worland, Wyo.: High Plains Publishing Co., 1992), pp. 89–108.
19. Murie to Newton Drury, 27 December 1945, Box 264, Bears of the Yellowstone (1945–1934) Folder, OMC.
20. Wright et al., *Fauna of the National Parks,* p. 54.
21. On the display of wildlife in the national parks, see Lisa Mighetto, *Wild Animals and American Environmental Ethics* (Tucson: University of Arizona Press, 1991); Alfred Runte, *National Parks: The American Experience,* 2d ed. (Lincoln: University of Nebraska Press, 1987); Richard Sellars, *Preserving Nature in the National Parks: A History* (New Haven: Yale University Press, 1997).
22. Murie to Struthers Burt, 5 December 1945, Folder 7, Box 361, OMC.
23. Wright et al., *Fauna of the National Parks,* pp. 54, 80; George M. Wright and Ben H. Thompson, *Fauna of the National Parks of the United States: Wildlife Management in the National Parks* (Washington, D.C.: United States Government Printing Office, 1935), p. 21.
24. Leopold, *A Sand County Almanac,* p. 262. On the shift to an ecological approach to wildlife management, see Thomas Dunlap, *Saving America's Wildlife: Ecology and the American Mind, 1850–1990* (Princeton, N.J.: Princeton University Press, 1988); Curt Meine, *Aldo Leopold: His Life and Work* (Madison: University of Wisconsin Press, 1988).
25. "The Game Display in Jackson Hole," 26 March 1946, Box 361, Folder 7, OMC.
26. Olaus J. Murie, "Wild Country as a National Asset," *The Living Wilderness* 45 (1953): 12.
27. Murie to Conrad L. Wirth, 13 March 1958, Box 265, U.S. National Park Service Folder, OMC.
28. Murie to Conrad L. Wirth, 21 October 1957, U.S. National Park Service Folder, Box 265, OMC. Murie to C. R. Carpenter, 18 March 1947, Folder 7, Box 361, OMC. Fairfield Osborn, "Purpose of the Jackson Hole Wildlife Park," *National Parks Magazine* (July–September 1946): 35; Keynote Address by Fairfield Osborn before the National Conference on Land-Use Policy, Omaha, Nebraska, May 7–8, 1948, p. 4; Box 1; Speeches & Writings, 1946–49, May File, FOP-LC.
29. See newspaper clipping, "Work of Preserving the Elk Herds in the Far Western Country," Box 2, "Elk History" Folder, SNLC. Leek's story is detailed in miscellaneous correspondence and newspaper clippings within this collection. See also Fern K. Nelson, *This Was Jackson's Hole: Incidents & Profiles from the Settlement of Jackson Hole* (Glendo, Wyo.: High Plains Press, 1994), pp. 40–47. For an early account of the problems facing the Jackson Hole elk herd, see National Conference on Outdoor Recreation, *The Conservation of the Elk of Jackson Hole, Wyoming* (Washington, D.C., 1927).
30. "Work of Preserving the Elk Herds in the Far Western Country."

31. S. N. Leek, "Denizens of the Wild," *In the Open* 5 (1915): 15. Leopold, *A Sand County Almanac,* p. 262. See also S. N. Leek, "White Patch," *The Grand Teton* (19 January 1932), Box 2, Miscellaneous Items folder, SNLC; Leek, "The Life of An Elk," *Outdoor Life* 42 (1918): 357–360; Leek, "The Starving Elk of Wyoming," *Outdoor Life* 24 (1910): 121–134.
32. "Work of Preserving the Elk Herds in the Far Western Country." "Leek-Wallace Camp for Boys," Box 4, Pamphlets Folder, SNLC. "The First National Conference on Out-of-Door Recreation Emphasizes the Delights of Nature," *Nature Magazine* (July 1924): 8–10. On the establishment of Leek's camp, see *Wyoming: From Territorial Days to the Present,* vol. 2, ed. Frances Birkhead Beard (Chicago: American Historical Society, 1933), pp. 421–422.
33. On Mather and the importance of the automobile in his vision of the national parks, see Runte, *National Parks;* Linda Flint McClelland, *Building the National Parks* (Baltimore: Johns Hopkins University Press, 1998). On autocamping, see Warren James Belasco, *Americans on the Road: From Autocamp to Motel, 1910–1945,* pap. ed. (Baltimore: Johns Hopkins University Press, 1997).
34. Arthur Newton Pack, "Hunting Nature on Wheels," *Nature Magazine* 13 (June 1929): 392. *Nature Magazine* 3 (April 1924): 194; *Nature Magazine* 9 (March 1927): 197. *Nature Magazine* 7 (April 1926): 258. *Nature Magazine* 12 (September 1928): 199; Pack, "Hunting Nature on Wheels," p. 392.
35. *Nature Magazine* 13 (June 1929): 353 for the Bell & Howell ad featuring Finley. For biographical information, see Worth Mathewson, *William L. Finley: Pioneer Wildlife Photographer* (Corvallis, Ore.: Oregon State University, 1986); William L. and Irene Finley, "Dinty, a Pet Porcupine," *Nature Magazine* 3 (March 1924): 133–137, 170.
36. *Nature Magazine* 7 (January 1926): 66. *Queer Creatures of the Cactus Country,* Nature Magazine Collection, American Museum of Natural History Film Archives, New York, New York.
37. See *Getting Our Goat, When Mountains Call, Babes in the Woods, Ramparts of the North, The Big Game Parade,* Nature Magazine Collection. For coverage of the expedition in *Nature Magazine,* see Arthur Newton Pack, "Camera Hunting on the Continental Divide," *Nature Magazine* 11 (January 1928): 9–15; Pack, "Bandits of the Border," *Nature Magazine* 12 (July 1928): 21–25; Pack, "Camera Hunting on the Continental Divide, or 'Getting Your Goat,'" *Nature Magazine* 11 (February 1928): 88–94.
38. *Riding the Rim Rock,* Nature Magazine Collection. William L. Finley, "Riding the Rim Rocks," *Nature Magazine* 12 (September 1928): 154.
39. *The Forests,* Nature Magazine Collection. See also Eleanor B. Pack, "Camera Hunting on the Continental Divide: We Prove How Busy the Beaver Is," *Nature Magazine* 11 (March 1928): 149–152.
40. For an overview of conservation under Roosevelt's administration, see A. L. Riesch Owen, *Conservation Under F.D.R.* (New York: Praeger, 1983).
41. Quoted in Robert L. Snyder, *Pare Lorentz and the Documentary Film* (Norman: University of Oklahoma Press, 1968), p. 41.

42. *The River* (Farm Security Administration, 1937). See also Pare Lorentz, *The River* (New York: Stackpole Sons, 1938); Lorentz, *FDR's Moviemaker: Memoirs & Scripts* (Reno, Las Vegas: University of Nevada Press, 1992).
43. Fairfield Osborn, *Our Plundered Planet* (Boston: Little, Brown and Co., 1948), p. 193. Fairfield Osborn to John Marshall, 24 July 1939, RG1.1 200R, Box 263, Folder 3136, RFA.
44. For the objections of the Izaac Walton League, see *The Living Wilderness* 17 (1946): 28.
45. *Journal of Mammalogy,* 28 (1947): 431.
46. For a history of the field studies conducted at the Jackson Hole Biological Research Station, see Gregg Mitman, "When Nature *Is* the Zoo: Vision and Power in the Art and Science of Natural History," *Osiris* 11 (1996): 117–143. On the distribution of the film and its importance in the public relations campaign, see Kenneth Chorley to Mr. Duncan, 27 December 1948, KCC, RGIV3A3.1, Box 2, Folder 17; Minutes of a Special Meeting of the Members and Board of Trustees of Jackson Hole Wildlife Park, 3 September 1949, RFC, Box 89, Folder 825, Cultural Interest Series, RG2; RES to Osborn, 12 October 1948; "Conservation News Letter," September 1948, Box 2, Speeches 1948, Oct.–Dec. Correspondence and Notes File, FOP-LC.
47. *The Jackson Hole Wildlife Park Research and Training Program, Summer 1947.*
48. Ibid.
49. The Conservation Foundation, "A Report of Progress," November 1, 1948, RG2 1949, 200, Box 448, Folder 3013, RFA.
50. For figures on gross revenues of the Conservation Foundation's films, see John C. Gibbs to Staff, 15 September 1952, 1952 Motion Pictures Folder, FOP-WCS.
51. Channing Cope to Fairfield Osborn, 17 November 1950, "Speeches, 1950, May Correspondence & Notes" Folder, Box 3, FOP-LC. Audience figures are from V. C. Arspiger to Fairfield Osborn, 16 September 1952, "1952 Motion Pictures" File, FOP-WCS. For a list of Osborn's speaking engagements, see "Schedules of Speaking Engagements" Folder, Box 1, FOP-LC.
52. *This Vital Earth* (New York Zoological Society & Conservation Foundation, 1948). Available through the Media Services Department, Wildlife Conservation Society.
53. For attendance figures to Grand Teton National Park, see Connie Wirth to Olaus Murie, 14 February 1958, U.S. National Park Service, 1958 Folder, Box 265, OMC. For Yellowstone figures, see John J. Craighead, *A Biological and Economic Appraisal of the Jackson Hole Elk Herd* (New York: New York Zoological Society and the Conservation Foundation, 1952), p. 30. See also Runte, *National Parks,* p. 171.
54. John J. Craighead and Frank C. Craighead, *Hawks, Owls and Wildlife* (Washington, D.C.: Wildlife Management Institute, 1956), p. 5. See also Frank C. Craighead and John J. Craighead, "The Ecology of Raptor Predation," *Trans-*

action of the 15th North American Wildlife Conference (1950); 209–223. For their other Jackson Hole studies, see John J. Craighead, *A Biological and Economic Appraisal of the Jackson Hole Elk Herd* (New York: New York Zoological Society and Conservation Foundation, 1952); Frank C. Craighead, *A Biological and Economic Evaluation of Coyote Predation* (New York: New York Zoological Society and Conservation Foundation, 1951); and Frank C. and John J. Craighead, "Nesting Canada Geese on the Upper Snake River," *Journal of Wildlife Management* 13 (1949): 51–64.

55. Craighead, *A Biological and Economic Appraisal of the Jackson Hole Elk Herd,* p. 25.
56. Frank and John Craighead, "Wildlife Adventuring in Jackson Hole," *National Geographic Magazine* 109 (1955): 13, 28.
57. William Beebe, "In Jackson Hole," *New York Herald Tribune* (28 September 1947), sec. 7, p. 10. Sally Carrighar, *One Day at Beetle Rock* (New York: Alfred A. Knopf, 1944); Carrighar, *One Day at Teton Marsh* (New York: Alfred A. Knopf, 1947); Rachel Carson, *Under the Sea Wind* (Boston: Houghton Mifflin, 1941). On the emergence of this ecological genre of nature writing, see Dunlap, *Saving America's Wildlife,* pp. 85–88; Mighetto, *Wild Animals and American Environmental Ethics,* pp. 104–106.
58. Fairfield Osborn, "Keynote Address," *American Institute of Biological Sciences Bulletin* (February 1961): 21; Osborn, *Our Plundered Planet,* p. 35. See also A. Starker Leopold and F. Fraser Darling, *Wildlife in Alaska* (New York: Ronald Press Co., 1953) for an early study sponsored by the Conservation Foundation that attempted to outline a plan for integrated land use.
59. Struthers Burt to Olaus Murie, 30 November 1945, Box 361, Folder 7, OMC; Olaus Murie to Struthers Burt, 5 December 1945, Box 361, Folder 7, OMC.
60. Murie to Wirth, 6 February 1958; Murie to Wirth, 21 October 1957; Wirth to Murie, 14 February 1958; Box 265, U.S. National Park Service, 1958 Folder, OMC. Osborn, "Purpose of the Jackson Hole Wildlife Park," *National Parks Magazine* (July-September 1946), pp. 35–36.

5 DISNEY'S TRUE-LIFE ADVENTURES

1. *Motion Picture Herald* (21 May 1949): 4618. The Crown Theater opening is discussed in N. Paul Kenworthy, Jr., "Walt Disney's True Life Adventures," n.d., p. 5, WDA.
2. *Seal Island* (Walt Disney Productions, 1948). On *The March of Time* series, see Richard M. Barsam, *Non-Fiction Film: A Critical History,* rev. ed. (Bloomington: Indiana University Press, 1992); Raymond Fielding, *The March of Time, 1935–1951* (New York: Oxford University Press, 1978). Kenworthy, "Walt Disney's True-Life Adventures," p. 22, WDA also mentions the symbolic importance of the animation sequence at the beginning of every True-Life Adventure.
3. *Seal Island* (Walt Disney Productions, 1948). In a flyer that accompanied the

nontheatrical release of *Seal Island* to schools and other educational organizations, the movie was suggested for teaching "science, social studies, and moral and spiritual values." I obtained a copy from Walt Disney Educational Materials Company, 800 Sonora Avenue, Glendale, California. The history of environmentalism in Cold War American culture is a neglected topic. See, however, Andrew Jamison and Ron Eyerman, *Seeds of the Sixties* (Berkeley: University of California Press, 1994). For a survey of postwar American responses to the atomic bomb, see Paul Boyer, *By the Bomb's Early Light: American Thought and Culture at the Dawn of the Atomic Age* (New York: Pantheon Books, 1985); Spencer R. Weart, *Nuclear Fear: A History of Images* (Cambridge, Mass.: Harvard University Press, 1988). On the importance of religion in Cold War America, see James Gilbert, *Redeeming Culture: American Religion in an Age of Science* (Chicago: University of Chicago Press, 1997); Mark Silk, *Spiritual Politics: Religion and America Since World War II* (New York: Simon & Schuster, 1988).

4. *Variety* (5 July 1950): 10. Kenworthy, "Walt Disney's True-Life Adventures," p. 21, WDA. The importance of the family in the ideology of Cold War American culture is most systematically explored in Elaine Tyler May, *Homeward Bound: American Families in the Cold War Era* (New York: Basic Books, 1988).
5. The economic difficulties faced by Walt Disney Studios during and immediately after the Second World War are discussed in Marc Eliot, *Walt Disney: Hollywood's Dark Prince* (Secaucus, N.J.: Carol Publishing, 1993); Richard Schickel, *The Disney Version: The Life, Times, Art, and Commerce of Walt Disney* (New York: Simon & Schuster, 1968); Bob Thomas, *Walt Disney: An American Original* (New York: Simon & Schuster, 1976).
6. Quoted in Schickel, *The Disney Version,* pp. 201, 267. See also Thomas, *Walt Disney;* Christopher Finch, *Walt Disney's America* (New York: Abbeville Press, 1978); Ralph H. Lutts, "The Trouble with Bambi: Walt Disney's *Bambi* and the American Vision of Nature," *Forest and Conservation History* 36 (1992): 160–171; Matt Cartmill, *A View to a Death in the Morning: Hunting and Nature through History* (Cambridge, Mass.: Harvard University Press, 1993), pp. 161–188.
7. Thomas, *Walt Disney,* pp. 161–187.
8. Ibid., p. 206. Eliot, *Walt Disney,* pp. 198–205. See also "Disney Biography. Bob Thomas with Jim Algar" and "Elma Milotte Interview," WDA.
9. Kenworthy, "Walt Disney's True-Life Adventures," p. 5, WDA; Eliot, *Walt Disney,* pp. 198–213.
10. Review of *The Olympic Elk, Natural History* 61 (April 1952): 190.
11. Kenworthy, "Walt Disney's True-Life Adventures," pp. 6–9, WDA. Disney did begin marketing books with the full-length feature True-Life Adventures. See Jane Werner et al., *Walt Disney's Living Desert: A True-Life Adventure* (New York: Simon & Schuster, 1954); Werner et al., *Walt Disney's Vanishing Prairie: A True-Life Adventure* (New York: Simon & Schuster, 1955).
12. Eliot, *Walt Disney,* pp. 212–213; Schickel, *The Disney Version,* pp. 308–309; Thomas, *Walt Disney,* pp. 238–239.

13. Herman Quick, "Disney Finds the Fur Seals—and Wins Another 'Oscar,'" *Nature Magazine* 42 (June–July 1949): 259–262, 292. The importance of the frontier setting in Disney's nature films is also discussed in Alexander Wilson, *The Culture of Nature: North American Landscape from Disney to the Exxon Valdez* (Cambridge, Mass.: Blackwell, 1992), pp. 117–131.
14. Walt Disney Studios, *The Story of Walt Disney's True-Life Adventure Series* (Burbank, Calif.: Walt Disney Productions, 1952).
15. *Beaver Valley* (Walt Disney Productions, 1950); *The Olympic Elk* (Walt Disney Productions, 1951); *The Vanishing Prairie* (Buena Vista Productions, 1954).
16. Quoted in Roderick Nash, *Wilderness and the American Mind,* 3d ed. (New Haven: Yale University Press, 1982), p. 288. For accounts of the Brooks Range expedition, see Olaus J. Murie, *Journeys to the Far North* (Palo Alto, Calif.: Wilderness Society and America West Publishing Co., 1973); Margaret E. Murie, *Two in the Far North* (New York: Alfred A. Knopf, 1962). *Letter from the Brooks Range* is available from the Media Services department of the Wildlife Conservation Society.
17. Olaus J. Murie, "Wild Country as a National Asset. II. Wild Country Round the World," *The Living Wilderness* 45 (1953): 12. William O. Douglas, *My Wilderness: The Pacific West* (Garden City, N.Y.: Doubleday, 1960), p. 101. See also Joseph Wood Krutch, *The Voice of the Desert: A Naturalist's Interpretation* (New York: William Sloane Associates, 1954), esp. p. 77. William H. Whyte, *The Organization Man* (New York: Simon & Schuster, 1956) and David Riesman, *The Lonely Crowd: A Study of the Changing American Character* (New Haven: Yale University Press, 1950) are two classic texts that underscore the anxieties about mass society and corporate culture in 1950s America. See for example Lary May, ed., *Recasting America: Culture and Politics in the Age of the Cold War* (Chicago: University of Chicago Press, 1989); Stephen J. Whitfield, *The Culture of the Cold War* (Baltimore, Md.: Johns Hopkins University Press, 1991).
18. Crisler's experience is detailed in Jean Muir, "Camera in the Wilderness," *True: The Man's Magazine* (May 1954): 43, 104–109. For other sketches of Disney's naturalist-photographers, see Fred C. Ells, "How Disney Does It," *The Movie Makers* (November 1951): 355, 376–377; Fred C. Ells, "How Disney Does It: 2," *The Movie Makers* (December 1951): 398–399, 408; Joseph V. Mascelli, "Filming True-Life Adventures," *International Photographer* (January 1953): 5–7, 20; Arthur Rowan, "Husband and Wife Camera Team," *American Cinematographer* (July 1951): 263–254; Anon., "Filming 'The African Lion,'" *American Cinematographer* (September 1955): 534–535, 542, 544; Frederick Foster, "Walt Disney's Naturalist-Cinematographers," *American Cinematographer* (February 1954): 74–75, 104–105, 109; Anon., "Time-lapse and Telephotos Probe Nature's Secrets," *American Cinematographer* (October 1956): 598–599, 622–624. The story of Knowles is retold in Nash, *Wilderness and the American Mind,* pp. 141–143.
19. Olaus Murie to Walt Disney, 29 July 1949, Box 264, Correspondence 1965–1949 File, OMC.

20. Lois Crisler, *Arctic Wild* (New York: Harper & Bros., 1958), pp. 11, 96, 118, 158.
21. Christopher Finch, *The Art of Walt Disney* (New York: Harry N. Abrams, 1973), p. 422.
22. Lloyde Beebe interview, n.d., WDA. Disney's populist persuasion is a recurrent theme in Steven Watts, *The Magic Kingdom: Walt Disney and the American Way of Life* (Boston: Houghton Mifflin, 1997).
23. J. P. McEvoy, "McEvoy in Disneyland," *Reader's Digest* 66 (1955): 20–21.
24. Knowing nature through labor is an idea introduced in Richard White's *The Organic Machine: The Remaking of the Columbia River* (New York: Hill & Wang, 1995).
25. Algar interview, 1968, WDA.
26. Julian Huxley to Walt Disney, 5 November 1964, Box 37, Folder 5, JHP.
27. Walt Disney Studios, *The Story of Walt Disney's True-Life Adventure Series;* Crisler, *Arctic Wild,* p. 34.
28. Walt Disney Studios, *The Story of Walt Disney's True-Life Adventure Series.*
29. Walt Disney to Olaus Murie, 4 December 1953, Box 264, "Correspondence, Jan. 8, 1963–" File, OMC. Murie to Disney, 18 September 1950, OMC.
30. Jack Couffer, *Songs of Wild Laughter* (London: Constable and Co., 1963), p. 152.
31. Ann Shelby Blum, *Picturing Nature: American Nineteenth-Century Zoological Illustration* (Princeton: Princeton University Press, 1993), p. 113.
32. Ernest Thompson Seton, *Wild Animals I Have Known* (New York: Charles Scribner's Sons, 1917), pp. 9–10.
33. Walt Disney, "What I've Learned from Animals," *American Magazine* 155 (1953): 106.
34. Sally Carrighar, *One Day at Teton Marsh* (New York: Alfred A. Knopf, 1947). James Algar referred to Carrighar as "one of the finest nature writers in the land." See Jim Algar, "Lecture to Talent Development Discussion Group," 29 November 1973, WDA. Although Carrighar's books downplayed the anthropomorphism found in the True-Life Adventures, both the choice of settings and the way in which she used a particular habitat to weave together the ecological relationships of individual animal lives was strikingly similar to the narrative conventions found in the True-Life Adventures. See also Sally Carrighar, *Icebound Summer* (New York: Alfred A. Knopf, 1953).
35. James Algar, "Film Music and Its Use in Beaver Valley," *Film and TV Music* (November–December 1950): 17–19. See also Ross B. Care, "Threads of Melody: The Evolution of a Major Film Score—Walt Disney's *Bambi,*" *Quarterly Journal of the Library of Congress* 40 (1983): 76–98.
36. Algar interview by Richard Hubler, 7 May 1968, WDA. Couffer, *Songs of Wild Laughter,* p. 15.
37. Bosley Crowther, "Disney's Nature," *New York Times* (22 August 1954), sec. 2, p. 1.
38. See "Interview with Elma Milotte," 5 November 1985, and Kenworthy, "Walt Disney's True-Life Adventures," p. 18, WDA.

39. *Bear Country* (Walt Disney Productions, 1953). Review of *Bear Country, Natural History* (March 1953): 139.
40. Arthur Rowan, "Husband and Wife Camera Team," *American Cinematographer* (July 1951): 276.
41. "Water Birds," *Natural History* (September 1952): 330.
42. For a brief account of the Audubon Screen Tours, see Frank Graham, Jr., *The Audubon Ark: A History of the National Audubon Society* (New York: Alfred A. Knopf, 1990), pp. 178–180.
43. Wanda Elvin to Arthur A. Allen, 29 January 1950; Allen to Elvin, 2 February 1950; Elvin to Allen, 6 February 1950; Allen to Elvin, 11 February 1950; Elvin to Allen, 3 May 1950; Allen to Elvin, 4 May 1950; Box 88, AAP.
44. Olaus Murie, "The World We Live In," *The Living Wilderness* 37 (1951): 17; Ernest S. Griffith, "Walt Disney's Secret," *The Living Wilderness* 50 (1955): 14; "Walt Disney Receives Audubon Medal," *Audubon Magazine* 58 (1956): 25; "The Olympic Elk," *Natural History* 61 (1952): 190.
45. Olaus Murie, "Wilderness Is for Those Who Appreciate," *The Living Wilderness* 5 (1940): 5. The tension between Murie's elitist views and Robert Marshall's democratic vision of wilderness is discussed in Robert Gottlieb, *Forcing the Spring: The Transformation of the American Environmental Movement* (Washington, D.C.: Island Press, 1993), pp. 15–19.
46. In an essay praised by Olaus Murie, Joseph Wood Krutch railed against the ways in which mass production and mass propaganda imposed upon Americans "the tyranny of the average." See Joseph Wood Krutch, "Is the Common Man Too Common," in *Is the Common Man Too Common? An Informal Survey of Our Cultural Resources and What We Are Doing about Them* (Norman: University of Oklahoma Press, 1954), pp. 3–19.
47. Murie to Disney, 18 September 1950, OMC
48. On the Mineral King project, see Schickel, *The Disney Version,* pp. 16–17; Thomas, *Walt Disney,* 345–347.
49. Cecile Starr, "Professional Movies for Home Showing," *House Beautiful* 97 (1955): 147–148. See also Anna W. M. Wolf, "In These Walt Disney Films There's Enchantment for Your Children," *Woman's Home Companion* 31 (May 1954): 36–39, for a similar endorsement of why every home needed a 16-mm film projector.
50. Disney's explicit disdain for Coney Island and the carny atmosphere in the amusement parks of his youth is discussed in John Findlay, *Magic Lands: Western Cityscapes and American Culture After 1940* (Berkeley: University of California Press, 1992), pp. 64–67; Margaret J. King, "Disneyland and Walt Disney World: Traditional Values in Futuristic Form," *Journal of Popular Culture* 15 (1981): 116–140; Schickel, *The Disney Version,* p. 310.
51. *Davy Crockett: King of the Wild Frontier* (Buena Vista Productions, 1955). On the popularity of Davy Crockett, see Thomas, *Walt Disney;* Schickel, *The Disney Version;* Eliot, *Walt Disney.*
52. Walt Disney, "What I've Learned from Animals," 109; Thomas, *Walt Disney,*

246. On the importance of the family in Cold War American culture, see Donald Katz, *Home Fires: An Intimate Portrait of One Middle-Class Family in Postwar America* (New York: HarperCollins, 1992); Robert L. Griswold, *Fatherhood in America: A History* (New York: Basic Books, 1993), pp. 185–218; May, *Homeward Bound;* May, *Barren in the Promised Land: Childless Americans and the Pursuit of Happiness* (New York: Basic Books, 1995), pp. 127–150.

53. Clifford E. Clark, Jr., "Ranch-House Suburbia: Ideals and Realities," in *Recasting America,* pp. 171–190. See also Clark, *The American Family Home* (Chapel Hill: University of North Carolina Press, 1986). Kenneth T. Jackson, *Crabgrass Frontier: The Suburbanization of the United States* (New York: Oxford University Press, 1985).

54. This legal challenge is discussed briefly in Thomas R. Dunlap, *DDT: Scientists, Citizens, and Public Policy* (Princeton: Princeton University Press, 1981), pp. 87–89 and Rachel Carson, *Silent Spring* (New York: Fawcett Crest, 1962), p. 144. The ways in which naturalists and environmentally concerned citizens equally engaged themselves with suburban nature and wilderness in the 1950s suggests that Samuel P. Hays's critique of William Cronon's essay, "The Trouble with Wilderness," is more astute than Cronon admits. See Samuel P. Hays, "Comment: The Trouble with Bill Cronon's Wilderness," *Environmental History* 1 (1996): 29–32. A history of American environmentalism and suburban nature awaits its author.

55. *Nature's Half-Acre* (Walt Disney Productions, 1951); Disney, "What I've Learned from Animals," p. 107.

56. For a history of creationism in the twentieth century, see Ronald L. Numbers, *The Creationists* (New York: Alfred A. Knopf, 1992).

57. Darrin Scot, "World's Biggest Little Studio," *American Cinematographer* (August 1961), p. 4. For an internal history of the Moody Institute of Science, see Gene A. Getz, *MBI: The Story of Moody Bible Institute* (Chicago: Moody Press, 1969). On the Moody Institute films, see Gilbert, *Redeeming Culture,* pp. 121–145.

58. Walt Disney Studios, *The Story of Walt Disney's True-Life Adventure Series.*

59. Quoted in Getz, *MBI,* p. 319. Roberts's background is mentioned in Ells, "How Disney Does It: 2," p. 399.

60. Kenworthy, "Walt Disney's True-Life Adventures," p. 21. *Nature's Half-Acre* (Walt Disney Productions, 1951).

61. *Nature's Half-Acre* (Walt Disney Productions, 1951). For a history of the balance-of-nature concept, see Frank N. Egerton, "Changing Concepts of the Balance of Nature," *Quarterly Review of Biology* 48 (1973): 320–350; Donald Worster, *Nature's Economy: A History of Ecological Ideas,* 2d ed. (Cambridge: Cambridge University Press, 1994).

62. Quoted in Mark Silk, *Spiritual Politics: Religion and America Since World War II* (New York: Simon & Schuster, 1988), pp. 41, 98–99.

63. Quoted in Eliot, *Walt Disney,* p. 171. Watts, *The Magic Kingdom,* pp. 283–302 characterizes Disney's Cold War vision as "libertarian populism."

64. For a discussion of the survival-of-the-fittest motif in the True-Life Adventures

and its relationship to American ideals of individualism and competition, see Watts, *The Magic Kingdom,* p. 305.

65. Bosley Crowther, "Review of *The Living Desert,*" *New York Times* (10 November 1953), p. 38, col. 1. Olaus Murie to Conrad L. Wirth, 6 February 1958, Box 265, U.S. National Park Service, 1958 Folder, OMC. On the importance of balance-of-nature concepts in postwar nature writing, see Thomas Dunlap, *Saving America's Wildlife: Ecology and the American Mind, 1850–1990* (Princeton: Princeton University Press, 1988), pp. 84–92; Lisa Mighetto, *Wild Animals and American Environmental Ethics* (Tucson: University of Arizona Press, 1991), pp. 104–106.
66. Griffith, "Walt Disney's Secret," p. 15. On the censorship of the buffalo birth, see "Disney Protests Any Cut in Film," *Baltimore Sun* (August 13, 1954): 17, 30.
67. On the pervasive myth of the frontier and the ways in which it has shaped the contours of American environmentalism, see William Cronon, "The Trouble with Wilderness; or, Getting Back to the Wrong Nature," in *Uncommon Ground: Toward Reinventing Nature,* ed. William Cronon (New York: W. W. Norton & Co., 1995), pp. 69–90; Kerwin L. Klein, "Frontier Products: Tourism, Consumerism, and the Southwestern Public Lands, 1890–1990," *Pacific Historical Review* (1993): 39–71.

6 DOMESTICATING NATURE ON THE TELEVISION SET

1. Marlin Perkins, *My Wild Kingdom* (New York: E.P. Dutton, 1982), p. 113. For popular press coverage of the python feeding show, see for example "Forced Feeding of a Zoo Python," *St. Louis Dispatch* (19 June 1938): 11. On the showmanship qualities of Vierheller and Perkins, see "By the Lake," *Time* (7 July 1947): 16–22.
2. For figures on television ownership, see Cobbett S. Steinberg, *TV Facts* (New York: Facts on File, Inc., 1980), p. 142. On ratings for *Zoo Parade,* see Ralph Knight, "Chicago's Sunday Jungle," *Saturday Evening Post* (25 October 1952): 32–33, 64, 68, 70; Charles and Jean Komaiko, "The Animal Kingdom of Marlin Perkins," *Pageant* (April 1952): 98–101.
3. Lynn Spigel, *Make Room for TV: Television and the Family Ideal* (Chicago: University of Chicago Press, 1992), p. 50. Both Spigel and Cecelia Tichi, *Electronic Hearth: Creating an American Television Culture* (New York: Oxford University Press, 1991) discuss the efforts by television manufacturers to make the television appear as a natural family member in 1950s American culture. For census figures on births and televisions produced, see Landon Y. Jones, *Great Expectations: America and the Baby Boom Generation* (New York: Coward, McCann & Geoghegan, 1980), p. 336, and Leo Bogart, *The Age of Television* (New York: Frederick Ungar Publishing Co., 1956), p. 11.
4. Statistics on ratings and audience estimates of *Wild Kingdom* are from Marlene Cimons, "It's Not Easy to Deceive a Grebe," *TV Guide* (26 October 1974): 29. On the 1980 survey, see Steven R. Kellert, Catherine A. McConnell,

Sharon Hamby, and Victoria Dompka, "Wildlife and Film: A Relationship in Search of Understanding," *The BKSTS Journal* (January 1986): 38–43.

5. Quoted in Bogart, *The Age of Television,* p. 249. On the ideal of domesticity in postwar America, see Elaine Tyler May, *Homeward Bound: American Families in the Cold War Era* (New York: Basic Books, 1988). On television's creation of family togetherness, see Spigel, *Make Room for TV,* pp. 37–45; Tichi, *Electronic Hearth,* pp. 42–61.
6. For an entry into the debates in the 1950s about the impact of television on children, see Wilbur Schramm, Jack Lyle, and Edwin B. Parker, *Television in the Lives of Our Children* (Stanford: Stanford University Press, 1961); Carmen Luke, *Constructing the Child Viewer: A History of the American Discourse on Television and Children, 1950–1980* (New York: Praeger, 1990), pp. 61–114; Spigel, *Make Room for Television,* pp. 50–60. On juvenile delinquency in 1950s America, see James Gilbert, *A Cycle of Outrage: America's Reaction to the Juvenile Delinquent in the 1950s* (New York: Oxford University Press, 1986). The increasing emotional value placed upon children in twentieth-century American society is analyzed in Viviana A. Zelizer, *Pricing the Priceless Child: The Changing Social Value of Children* (New York: Basic Books, 1985). On the parallels between the construction of childhood and natural innocence, see Jacqueline Rose, *The Case of Peter Pan or The Impossibility of Children's Fiction* (1984; Philadelphia: University of Pennsylvania Press, 1993), pp. 42–65.
7. Lee Brashear, "*Zoo Parade*: The TV Show Named Year's Best for Children," *St. Louis Globe Democrat* (25 November 1951), sec. F, p. 2. On the struggles of educational television in the 1950s, see R. D. Heldenfels, *Television's Greatest Year: 1954* (New York: Continuum, 1994), pp. 175–186. On the relationship of the child to educational television, see Schramm et al., *Television in the Lives of Our Children,* pp. 57–60.
8. VC ZP 37, "Nature of Animals," 18 November 1951, ZPC. On the use of *Zoo Parade* in the classroom, see Jeanne Gaede, "*Zoo Parade* Schedule," Box 10, Folder 248, MPP. For a list of *Zoo Parade* awards, see "All about the New *Zoo Parade,*" Box 10, Folder 258, MPP.
9. VC ZP 37, "Nature of Animals," 18 November 1951, ZPC.
10. Burr Tillstrom, Guest "John Crosby" column, *New York Herald Tribune* (25 November 1956), TV section, p. 1. Brashear, "*Zoo Parade,*" p. 1. For the Animal Crackers cartoon, see Patricia Forsyth, "That Awful Perkins Boy," *Missouri Life* (September–October 1978): 24.
11. For surveys on the popularity of television shows by topics and viewer's age during the 1950s, see Schramm et al., *Television in the Lives of Our Children,* pp. 37–42. On Muggs, see Erik Barnouw, *Tube of Plenty: The Evolution of American Television,* 2d rev. ed. (New York: Oxford University Press, 1990), pp. 146–148.
12. VC ZP 21, "Giant Constrictors," 7 January 1951, ZPC.
13. Ralph Knight, "Chicago's Sunday Jungle," *The Saturday Evening Post* (25 October 1952): 32. Perkins, *My Wild Kingdom,* p. 116.

14. For visual evidence of Lear Grimmer's demeanor around the animals on *Zoo Parade,* see VC ZP 4, "Why Man Needs Animals," 5 June 1950 and VC ZP 20, "Animals of South America," 8 April 1951, ZPC.
15. VC ZP 19, "Outdoors with the Lions," 12 August 1951, ZPC. For a short biography of Dillinger, see Marlin Perkins, *Marlin Perkins' Zoo Parade* (Chicago: Rand McNally & Co., 1954), pp. 16–20.
16. On the production costs of *Zoo Parade,* see Charles and Jean Komiako, "The Animal Kingdom of Marlin Perkins," p. 99; Brashear, "*Zoo Parade,*" 4F.
17. "*Wild Kingdom*" promotional brochure, 6 March 1958, Box 11, Folder 270, MPP. "*Zoo Parade*: A Sunday Coast to Coast Television Show in Forty-One Cities That Offers a Unique Sunday Comic Section Feature," Box 10, Folder 256, MPP. VC ZP 55, "Zoo Mothers," 13 May 1951, ZPC.
18. VC ZP 61, "Easter Parade & Easter Story," 5 April 1953, ZPC.
19. VC ZP 55, "Zoo Mothers," 13 May 1951, VC ZP 55. On Spock, see Nancy Pottishman Weiss, "Mother, the Invention of Necessity: Dr. Benjamin Spock's *Baby and Child Care,*" *American Quarterly* 29 (1977): 519–546.
20. On fatherhood in the 1950s, see Robert L. Griswold, *Fatherhood in America: A History* (New York: Basic Books, 1993). On the naturalization of family life on 1950s television, see Nina C. Leibman, *Living Room Lectures: The Fifties Family in Film and Television* (Austin: University of Texas Press, 1995); Spigel, *Make Room for TV,* pp. 177–180; Mary Beth Haralovich, "Sitcoms and Suburbs: Positioning the 1950s Homemaker," *Quarterly Review of Film and Video* 11 (1989): 61–83.
21. VC ZP 34, "Main Street at the Zoo," 24 June 1951, ZPC.
22. With an estimated 3 million viewers, *Adventure* had an audience approximately one quarter the size of *Zoo Parade*. See Edward Weyer, "A Year of Adventure," *Natural History* (June 1954): 288. George W. Woolery, *Children's Television: The First Thirty-Five Years, 1946–1981. Part II: Live, Film, and Tape Series* (Metuchen, N.J.: Scarecrow Press, 1985), pp. 11–12.
23. CBS/*Adventure* Series, 12 July 1953 [33], "Mouth Breeding Fish from the Amazon," American Museum of Natural History Film Archives, New York, N.Y.; Eugenie Clark, *Lady with a Spear* (New York: Harper & Bros., 1953).
24. CBS/*Adventure* Series, 3 January 1954, "Bees-Russian Film," [54]. On the concerns expressed by American biologists on the use of social insects as a biological model for totalitarianism during and after the Second World War, see Gregg Mitman, *The State of Nature: Ecology, Community, and American Social Thought, 1900–1950* (Chicago: University of Chicago Press, 1992), pp. 146–168.
25. Peter Parakeet, "They're Sad at Lincoln Park Zoo," *TV Forecast* (30 August 1952): 7.
26. For an insightful analysis on the symbolism of childhood innocence in the 1950s, see Tom Engelhardt, *The End of Victory Culture: Cold War America and the Disillusioning of a Generation* (New York: Basic Books, 1995), pp. 133–154. On the history of *Mad Magazine*, see Maria Reidelbach, *Completely MAD: A*

History of the Comic Book and Magazine (Boston: Little, Brown, 1991). On *Panic,* see Mike Benton, *The Comic Book in America: An Illustrated History* (Dallas: Taylor Publishing, 1989).

27. "Zoo Charade," *Panic* (June-July 1955), Box 31, Folder 882, MPP.
28. Perkins's syndicated newspaper columns are located in Box 10, Folder 253, MPP. For a sample of the Dell comic book, see *Marlin Perkins' Zoo Parade,* n. 662 (New York: Dell Publishing Co., 1955), Box 10, Folder 257, MPP. On the line of toy products, see Robert R. Max to Marlin Perkins, 22 March 1956, Box 3, Folder 54, MPP.
29. Shana Alexander, "A Gradual Clean Sweep," *Life* 43 (25 November 1957): 81. See also Woolery, *Children's Television,* pp. 23–26, 282–287.
30. See, e.g., VC ZP 34, "Main Street at the Zoo," 24 June 1951; VC ZP 134, "Early American Animals," 22 March 1953; VC ZP 136, "Nesting Habits," 11 March 1954, ZPC.
31. "Recently on the TV Program 'Lassie,'" *Nature Magazine* 49 (April 1956): 175.
32. On 1956 audience estimates for *Zoo Parade,* see "*Wild Kingdom* promotional brochure," 6 March 1958, Box 11, Folder 270, MPP. For an analysis of the shift of commercial sponsorship and the increase to action-adventure series on children's programs, see Joseph Turow, *Entertainment, Education, and the Hard Sell: Three Decades of Network Children's Television* (New York: Praeger, 1981), pp. 15–49. On the general shift away from live television, see William Boddy, *Fifties Television: The Industry and Its Critics* (Urbana: University of Illinois Press, 1990), pp. 187–213.
33. Larry Wolters, "Perkins to Give Show on Safari," *Chicago Daily Tribune* (5 September 1957): part 4, p. 4. For a sample of these early shows, see, e.g., VC ZP 20, "Animals of South America," 8 March 1951; VC ZP 31, "Animals of the Southwest," 14 January 1951; VC ZP 70, "Animals of Africa," 26 October 1952.
34. "Mutual of Omaha's *Wild Kingdom,* Program Guidelines," Box 12, Folder 320, MPP. For the *Zoo Parade* episodes that included Disney footage, see VC ZP 101, "Animals of the Hot Region," 27 December 1953; VC ZP 125, "Unguarded Moments," 25 October 1953. On Disney's later uncooperativeness, see Erwin L. Verity to Marlin Perkins, 1 May 1963, Box 3, Folder 57, MPP.
35. Quoted in Tichi, *Electronic Hearth,* p. 15. Bogart, *The Age of Television,* p. 238.
36. "Mutual of Omaha's *Wild Kingdom,*" p. 2, Box 12, Folder 327, MPP. For these later episodes of *Zoo Parade,* see "Jackson Hole Country, Parts 1 & 2," TV2257.1; "African Safari, Part 1," TV 2256.1; "African Safari (Part 2 of 2)," TV 2255.4; MBC. On the logistics of shooting on location, see Ben Park to Don Meier, "*Zoo Parade* Safari Progress Report," Box 10, Folder 243, MPP; "*Zoo Parade* Safari—Itinerary and Shooting Schedule," Box 10, Folder 246, MPP.
37. "World Safari," Box 12, Folder 323, MPP.
38. Dick Creyke to Don Meier, "Proposed revamping of 'Zoo Parade' Format," Box 3, Folder 54, MPP.

39. "*Wild Kingdom* promotional brochure," 6 March 1958, Box 11, Folder 270, MPP. On the chance meeting with Skutt, see Perkins, *My Wild Kingdom,* pp. 153–155. For a partial listing of the show's awards, see "Citations and Awards Presented to Mutual of Omaha's *Wild Kingdom,*" Box 13, Folder 342, MPP.
40. "Concepts to Extend *Wild Kingdom* or Replace the *Wild Kingdom* Series, 1976," Box 11, Folder 301, MPP. The snake-biting incident was recounted many times in stories about Perkins. See, e.g., Brashear, "*Zoo Parade,*" p. 4F; Janet Kern, "Never Go Near a Seal's Harem," *TV Guide* (St. Louis edition), Box 33, Folder 928, MPP; Perkins, *My Wild Kingdom,* pp. 118–119. On the importance of errors in adding to the "immediacy" of live television, see Heldenfels, *Television's Greatest Year,* pp. 59–63.
41. Terrence O'Flaherty, "Television's Wildest Show," *San Francisco Chronicle* (22 October 1966): 31. For a description and pictures of the anaconda incident, see "Annual Report, Mutual of Omaha and Affiliates, 1978 Financial Statement," pp. 15–16, Box 11, Folder 350, MPP. For a fan's recollection, see Flora Ditton to Marlin Perkins, 7 April 1972, Box 3, Folder 57, MPP.
42. "Concepts to Extend *Wild Kingdom,*" p. 7. This document contains a valuable look at the evolution of *Wild Kingdom* from the perspective of its production staff. For descriptions of Fowler in the popular press, see Patrick Walsh, "Television's Dr. Dolittle Returns to the Air," *TV Guide* (17 February 1968): 25–26; Donald Kirkley, "Look and Listen," *Baltimore Sun* (8 January 1963): 10; Lawrence Laurent, "A Man Who Loves Nature," *Washington Post* (14 February 1967): B9.
43. "Program Guidelines, Document #2," Box 12, Folder 320, MPP.
44. Cimons, "It's Not Easy to Deceive a Grebe": 31. For a complete synopsis of all the episodes of *Wild Kingdom,* see Box 12, Folder 327, MPP.
45. Flora Terzian Ditton to Marlin Perkins, 7 April 1972, Box 3, Folder 57, MPP. On Skutt's interest in the show for its wholesome values, see "Annual Report of Mutual of Omaha and Affiliates," p. 14.
46. *Zoo Parade,* "Jackson Hole Country, Part 2," MBC.
47. Richard Starnes, "Why TV's Wolf Boy Scares Me," *Outdoor Life* (August 1978): 6–10. On the environmental activism of Carol Perkins, see Patricia Forsyth, "That Awful Perkins Boy," *Missouri Life* (September–October 1978): 23–27; Maureen McNerney, "Mrs. Marlin Perkins Shares Concern About *Wild Kingdom,*" *Courier-Journal* (25 October 1973): B2; Carol and Marlin Perkins, "I am Only One," *Topics* (Summer 1970): 5, 14, 18, 22; "Interview of Carol Perkins by Christie Fritch and Natalie Drew," MPP.
48. On international tourism in the postwar years, see Maxine Feifer, *Going Places: The Ways of the Tourist from Imperial Rome to the Present Day* (London: MacMillan, 1985), pp. 219–258. See also Earle Dean MacCannell, *The Tourist, a New Theory of the Leisure Class* (London: MacMillan, 1976).
49. Raymond Williams, "Ideas of Nature," in *Problems in Materialism and Culture* (London: Verso, 1980); William Cronon, "The Trouble with Wilderness; or,

Getting Back to the Wrong Nature," in *Uncommon Ground: Toward Reinventing Nature,* ed. William Cronon (New York: W. W. Norton & Co., 1995), pp. 69–90; Richard White, "'Are You an Environmentalist or Do You Work for a Living?' Work and Nature," in *Uncommon Ground,* pp. 171–185. On the expansion of outdoor recreation as a consumer commodity after the Second World War, see L. Sue Greer, "The United States Forest Service and the Postwar Commodification of Outdoor Recreation," in *For Fun and Profit: The Transformation of Leisure into Consumption,* ed. Richard Butsch (Philadelphia: Temple University Press, 1990), pp. 152–172.

50. *Zoo Parade,* "African Safari. Part 1," TV 2256.1, MBC.
51. Marlin Perkins, as interviewed by Steve Cooper, "The *Wild Kingdom* of Marlin Perkins," *The Country Gentleman* (Spring 1978): 102–104.
52. Don Meier Productions, "Concepts to Extend *Wild Kingdom,*" p. 8, Box 11, Folder 301, MPP; "Basic Orientation Memo," *Mutual of Omaha's Wild Kingdom,* Box 12, Folder 320, MPP.

7 A RINGSIDE SEAT IN THE MAKING OF A PET STAR

1. "The World's Greatest Beaches: Australia's Magnificent Ten," *Condé Nast Traveler* (April 1993): 88–92, 102–104; Natalie Angier, "Dolphin Courtship: Brutal, Cunning and Complex," *New York Times* 141 (18 Feb. 1992), sec. C, p. 1.
2. "Camera at Marineland Records Delivery of the First Porpoise Ever Born in Captivity," *Life* 8 (26 February 1940): 28–29; "Letters to the Editors," *Life* 8 (18 March 1940): 2, 4. See also Arthur F. McBride to A. H. Schmidt, 5 March 1940, Laboratories & Field Stations: Marine Studios; Marineland I Folder, GKNP. Ralph Nading Hill, *Window in the Sea* (New York: Rinehart & Co., 1956) incorrectly identifies this as the first live birth at Marine Studios.
3. Burden discussed the idea behind Marine Studios in a number of published popular articles. See, e.g., J. Byran III, "Ocean under Glass," *Saturday Evening Post* 211 (1939): 18–19, 59–63; Morris Fradin, "The World's First Artificial Ocean," *Travel* 72 (1938): 27–54; and Arthur McBride, "The Undersea World thru 200 Portholes," *Biology Briefs* 1 (1938): 33–36. The opening of Marine Studios even made its way into the listing of scientific events in *Science* 88 (1938): 50. See also Gregg Mitman, "Cinematic Nature: Hollywood Technology, Popular Culture, and the American Museum of Natural History," *Isis* 84 (1993): 637–661.
4. W. E. Burden, "Education Possibilities of Marine Studios," Box 2, Folder 1, General Correspondence, 1938–1973, General Education Publications, MARP; and Burden, untitled manuscript, p. 8, Box 2, Folder 2, General Education Publications—PreWar, MARP; "Prospectus of Marine Studio," p. 7, Laboratories & Field Stations: Marine Studios II Folder, GKNP.
5. Noble to Ivy Lee, 17 June 1938, Laboratories & Field Stations: Marine Studios Folder, GKNP.

6. Pete Smith to William Douglas Burden, 5 December 1938, Box 9, Folder 3, WDBP; Memorandum Concerning Contract Between Marine Studios, Inc. and Metro-Goldwyn-Mayer, 19 October 1938, Box 9, Folder 2, WDBP. See also "Filming of Underwater Shots at Marine Studios is Begun," *St. Augustine Record* (18 November 1938); "'Marine Circus' is Termed 'A Short With A Porpoise,'" *St. Augustine Record* (26 April 1939), clippings scrapbook, MARP.
7. Morris Fraden, "The World's First Artificial Ocean," *Travel* 72 (1938): 27–29, 54; J. Bryan III, "Ocean Under Glass," *Saturday Evening Post* 211 (14 January 1939): 18–19, 59–63; Federal Writer's Project, *Florida: A Guide to the Southernmost State* (New York: Oxford University Press, 1939). "Grantland Rice Sportlight to Be Filmed at Marineland; New Invention Revealed," *St. Augustine Record* (2 June 1940), clippings scrapbook, MARP.
8. Burden to Pete Smith, 28 November 1938, Box 9, Folder 3, WDBP. On Burden's attitude toward publicity of the motion picture short, see Burden to Ilia Tolstoy, 9 December, 1938; Burden to Tolstoy, 20 November 1939, Box 8, Folder 11, WDBP. The trained beluga whale at Aquariel Gardens is mentioned in Forrest G. Wood, *Marine Mammals and Man: The Navy's Porpoises and Sea Lions* (New York: Robert B. Luce, Inc., 1973), pp. 35–36.
9. On the Scholander anecdote, see McBride to Burden, Box 1, Folder 3, MARP. Laurence Irving, P. F. Scholander, and S. W. Grinnell, "The Respiration of the Porpoise, *Tursiops Truncatus*," *Journal of Cellular and Comparative Physiology* 17 (1941): 145–168. Geiling to Whitney, 12 October 1937, Box 3, Cooperation in Research, Pre-War Folder, MARP.
10. "Chicago Scientists Study Porpoises at Marine Studio," *St. Augustine Record* (21 July 1939), clippings scrapbook, MARP. See also Lillian Eichelberger, E. S. Fetcher, Jr., E. M. K. Geiling, and B. J. Voss, "The Composition of Dolphin Milk," *Journal of Biological Chemistry* 133 (1940): 171–176.
11. On the Board of Scientific Directors, see Geiling to Bacon, 10 April 1940, Box 4, Folder 2, MARP; Burden to Noble, 15 April 1940, AMNH Laboratories & Field Stations: Marine Studios Folder, GKNP.
12. McBride to Burden, 9 September 1947, p. 7, Box 7, Folder 15, WDBP.
13. The anaesthetic experiments and copper treatment solution to epibdella are discussed in Hill, *Window in the Sea*. See also correspondence in Laboratories & Field Station: Marine Studios, I Folder, GKNP and Box 3, Folder 18, MARP.
14. See Tolstoy to Board of Directors, 31 July 1939, Research Lab Planing, Physical Plant & Costs Folder, OLD MFL Files; McBride to Geiling, 8 November 1939, Box 3, Folder 11; McBride to Fetcher, 25 July 1940, Box 3, Folder 11; Burden to Laurence Irving, 36 (sic) December 1939, Box 4, Folder 2; Irving to Burden, 6 January 1940, Box 4, Folder 2; MARP for concerns about conducting research on dolphins in public view.
15. E. M. K. Geiling, "A Laboratory at Marine Studios, Florida," Box 7, Folder 5, WDBP; Irving to Forrest G. Wood, 26 November 1951, Old MFL Files, Research Lab: Planning Stage, 1949–52 Folder, MARP; McBride to Burden, 9 September 1947, Box 7, Folder 15, WDBP. For Cousteau's popular account

of the invention of SCUBA, see Jacques Cousteau, *The Silent World* (London: Hamish Hamilton, 1953). On the importance of World War II in the transformation of oceanography, see Chandra Mukerji, *A Fragile Power: Scientists and the State* (Princeton, N.J.: Princeton University Press, 1989); Captain E. John Long, ed., *Ocean Sciences* (Annapolis, Md.: United States Naval Institute, 1964).

16. Rachel L. Carson, *The Sea Around Us* (1951). Reprint, with introduction by Ann H. Zwinger, afterword by Jeffrey S. Levinton (New York: Oxford University Press, 1989), p. vii. *20,000 Leagues Under the Sea* (Buena Vista Productions, 1954).
17. Burden to Fairfield Osborn, 29 January 1954, FOP-WCS. Forrest Wood to Burden, 4 August 1955, Box 1, Folder 3, MARP.
18. Eastman to Burden, 16 April 1946, Box 7, Folder 12, WDBP. Hill, *Window in the Sea,* pp. 183–184.
19. On Frohn, see Hill, *Window in the Sea,* pp. 177–182; Bill Ballantine, "The Playful Porpoise," *Holiday* 22 (1957): 22, 24–29; Richard O'Barry, *Behind the Dolphin Smile* (Chapel Hill, N.C.: Algonquin Books, 1988), pp. 85–92. "'Flippy,' The Docile Dolphin: A Strange Triumph of Animal Training," *Illustrated London News* (3 March 1951): 341; "The Accomplished Dolphin of a U.S. Aquarium: Flippy Displaying His Repertoire of Tricks," *Illustrated London News* (21 June 1952): 1059.
20. J. Bryan III, "Ocean Under Glass," *Saturday Evening Post* 211 (14 January 1939): 61. Brochure in Box 2, Folder 1, MARP.
21. Richard F. Dempewolff, "Flippy Rings the Bell," *Popular Mechanics* 98 (December 1952): 73. John W. Dillin, "Flippy the Educated Porpoise," *Natural History* 61 (1952): 152–159.
22. Arthur F. McBride and D. O. Hebb, "Behavior of the Captive Bottle-Nose Dolphin, *Tursiops Truncatus,*" *Journal of Comparative and Physiological Psychology* 41 (1948): 120. Barbara Lawrence and William E. Schevill, "*Tursiops* as an Experimental Subject," *Journal of Mammalogy* 35 (1954): 230. Schevill to Wood, 14 October 1955, Old MFL Files, Research Lab: William Schevill Folder, MARP. On the surfboard in tow, see R. J. Eastman to Burden, 6 March 1947, Box 7, Folder 12, WDBP.
23. Wood to Ken Norris, 4 April 1956, Box 1, Folder 2, MARP. On Breland, see *American Men of Science: The Social and Behavioral Sciences,* 10th ed. (1962), p. 125; Keller Breland and Marian Breland, *Animal Behavior* (New York: Macmillan Co., 1966); Daniel W. Bjork, *B. F. Skinner: A Life* (New York: Basic Books, 1993), pp. 119–127.
24. McBride to Geiling, 8 November 1939; Geiling to McBride, 29 November 1939, Box 3, Folder 2, MARP. Arthur F. McBride, "Meet Mister Porpoise," *Natural History* (January 1940): 16–29.
25. "Quiz Kids of the Ocean," *Science News Letter* (5 September 1953): 154–155. On the CBS plans, see Burden to Eastman, 27 May 1946, Box 7, Folder 12, WDBP. *Marineland Carnival* (1962), Motion Picture Collection, Library of Congress. *Revenge of the Creature* (Universal Pictures Co., Inc., 1955).

26. McBride and Hebb, "Behavior of the Captive Bottle-Nosed Dolphin," p. 114. For Townsend's descriptions of dolphin behavior, see C. H. Townsend, "Endurance of the Porpoise in Captivity," *Science* 43 (1916): 534–535; Roy Chapman Andrews, *Whale Hunting with Gun and Camera* (New York: D. Appleton & Co., 1916), pp. 278–290.
27. On Flippy, see Wood to Rolleston, 15 July 1953, Box 1, Folder 5, MARP. On the death of Herman, see Hill, *Window in the Sea,* pp. 151–162. McBride's injection of paraldehyde is mentioned in McBride to Burden, 28 February 1947, Box 8, Folder 2, WDBP.
28. Wood to Norris, 24 November 1954, Box 1, Folder 1; Schmidt to McBride, 1 December 1939, Old MFL Files, Cetacea Correspondence Folder, MARP. McBride and Hebb, "Behavior of the Captive Bottle-Nosed Dolphin," p. 122.
29. Breland and Breland, *Animal Behavior,* p. 87. Greer Williams, "Practical Joker of the High Seas," *Saturday Evening Post* 223 (October 7, 1950): 31–33, 169–172.
30. Edward Weyer to Frank Essapian, 10 July 1953, Old MFL Files, Cetacea Correspondence Folder, MARP. Frank S. Essapian, "The Birth and Growth of a Porpoise," *Natural History* (November 1953): 392–399. *Mysteries of the Deep* (Walt Disney Studios, 1959). My thanks to Donald Dewsbury for supplying me with the historical footage of dolphin sex play taken at Marine Studios.
31. On the Cape Hatteras dolphin fishery, see Andrews, *Whale Hunting With Gun and Camera,* pp. 280–281.
32. Wood to Burden, 20 June 1955, Box 1, Folder 3, MARP. Burden to Tolstoy, February 1947, Box 7, Folder 1, WDBP.
33. William N. Tavolga, *Review of Marine Bio-Acoustics, State of the Art: 1963.* U.S. Naval Training Device Center, Port Washington, N.Y.. Technical Report: NAVTRADEVCEN 1212–1. Winthrop N. Kellogg, *Porpoises and Sonar* (Chicago: University of Chicago Press, 1961). Jacques-Yves Cousteau and Louis Malle, *Le Monde du Silence* (1956).
34. McBride to Wilfred Bronson, 10 November 1938, Old MFL Files, Cetacea Correspondence Folder; Schevill to Burden, 26 May 1949, Old MFL Files, Research Lab: William E. Schevill Folder; Wood to Rolleston, 3 March 1952, Box 1, Folder 5; Wood to Frank Beach, 9 December 1952, Box 3, Folder 14, MARP.
35. Arthur McBride, "Evidence for Echolocation by Cetaceans," *Deep-Sea Research* 3 (1956): 153–154. William Schevill arranged to have the excerpt on echolocation from McBride's notes published posthumously. McBride to Hebb, 8 July 1947, Old MFL Files, Cetacea Correspondence Folder, MARP.
36. William E. Schevill and Barbara Lawrence, "Food-Finding by a Captive Porpoise *(Tursiops Truncatus),*" *Breviora* 53 (1956): 1–15. W. N. Kellogg, "Reaction of the Porpoise to Ultrasonic Frequencies," *Science* 116 (1952): 250–252; W. N. Kellogg, Robert Kohler, and H. N. Morris, "Porpoise Sounds as Sonar Signals," *Science* 117 (1953): 239–243; W. N. Kellogg, "Echo Ranging in the Porpoise," *Science* 128 (1958): 982–988. Wesley S. Griswold, "The Case of the Blindfolded Dolphin," *Popular Science Monthly* 177 (August 1960): 70–73, 184;

K. S. Norris, J. H. Prescott, P. V. Asa-Dorian, and Paul Perkins, "An Experimental Demonstration of Echolocation Behavior in the Porpoise *Tursiops truncatus* (Montagu)," *Biological Bulletin* 120 (1961): 163–176.

37. John Lilly, "Some Considerations Regarding Basic Mechanisms of Positive and Negative Types of Motivations," *American Journal of Psychiatry* 115 (1958): 501. On Sputnik and the space race, see Walter A. McDougall, . . . *The Heavens and the Earth: A Political History of the Space Age* (New York: Basic Books, 1985). For sample media coverage, see "Intelligence: This Fish is a Smarty," *Newsweek* 56 (4 July 1960): 59; Calvin Tomkins, "Conversation at Sea Level," *New Yorker* 36 (3 September 1960): 24–26; "Delphinology: Someone to Talk to," *Newsweek* 59 (23 April 1962): 58; "Space Communication: Five Peeps and a Tweet," *Newsweek* 60 (22 October 1962): 83; "Can the Dolphins Learn to Talk?" *Life* 51 (28 July 1961): 61–68.
38. Hill, *Window in the Sea,* p. 89. Cousteau and Malle, *The Silent World.* "Catalina Under Sea—With Underwater Camera," CBS/*Adventure* Series, American Museum of Natural History Film Archives.
39. Wood to Jerzy Rose, 3 June 1955, Old MFL Files, Visiting Investigators–Johns Hopkins Folder, MARP. The incident is discussed in Wood to Schevill, 31 October 1955, Old MFL Files, Research Lab–William Schevill Folder, MARP; John C. Lilly, *Man and Dolphin* (Garden City, N.Y.: Doubleday & Co., 1961) and Wood, *Marine Mammals and Man.*
40. During his first tragic visit to Marine Studios, Lilly obtained a binoural recorder from the Ampex Corporation to determine whether the animals were emitting sounds even when there were no visible signs of vocalization. Preliminary observations suggested that dolphins were beaming sound in a directional, investigative fashion, and thus added further evidence of their echolocation navigational abilities.
41. Lilly, *Man and Dolphin.*
42. Margaret C. Tavolga and William N. Tavolga, "Review of *Man and Dolphin* by John C. Lilly," *Natural History* 71 (1962): 5–7. Wood to James Atz, 5 October 1961, Box 2, Folder 4, MARP.
43. Wood to Atz, 5 October 1961; Lilly, *Man and Dolphin,* pp. 219–220.
44. On the Navy's marine mammal research program, see Wood, *Marine Mammals and Man;* Coles Phinizy, "You Can't Con a Porpoise," *Sports Illustrated* 24 (21 March 1966): 50–52, 66, 68. On Tuffy and SeaLab, see Scott Carpenter, "30 Days in Sealab," *Life* 59 (15 October 1965): 100A–106.
45. *Flipper* (Metro-Goldwyn-Mayer, Inc., 1963). On the making of *Flipper,* see O'Barry, *Behind the Dolphin Smile.* "Gamboling dolphin in a new role: My pal Flipper," *Life* 54 (7 June 1963): 61–64, 66, 69.
46. Ivan Tors, *My Life in the Wild* (Boston: Houghton Mifflin Co., 1979); Wood to Lilly, 28 October 1959, Old MFL Files, VI. Individuals: John Lilly, MARP; Susan G. Davis, *Spectacular Nature: Corporate Culture and the Sea World Experience* (Berkeley: University of California Press, 1997); p. 287 incorrectly identifies Marineland as the location where *Flipper* was filmed.

47. For an archive of this fan mail, see the website http://www.best.com/fimaryflr/P1-intro-old.html.
48. Shannon Brownlee, "A Political Casserole of Tuna and Greens," *U.S. News & World Report* (11 August 1997): 52–53.

8 GLOBAL VISIONS, TOURIST DREAMS

1. On the Apollo spacecraft image, see Denis Cosgrove, "Contested Global Visions: *One-World, Whole-Earth,* and the Apollo Space Photographs," *Annals of the Association of American Geographers* 84 (1994): 270–294.
2. On Roosevelt's Good Neighbor Policy, see Irwin F. Gellman, *Good Neighbor Diplomacy: United States Policies in Latin America, 1933–1945* (Baltimore: Johns Hopkins University Press, 1979); David Green, *The Containment of Latin America: A History of the Myths and Realities of the Good Neighbor Policy* (Chicago: Quadrangle Books, 1971); Walter LaFeber, *Inevitable Revolutions: The United States in Central America,* exp. ed. (New York: W. W. Norton & Co., 1984), pp. 59–83.
3. Allen Woll, *The Latin Image in American Film,* rev. ed. (Los Angeles: UCLA Latin American Center, 1980), p. 55. On the OCIAA, see Gellman, *Good Neighbor Diplomacy,* pp. 142–155; Fred Fejes, *Imperialism, Media, and the Good Neighbor: New Deal Foreign Policy and United States Shortwave Broadcasting to Latin America* (Norwood, N.J.: Ablex Publishing Corp., 1986); OCIAA, *History of the Office of the Coordinator of Inter-American Affairs* (Washington: U.S. Government Printing Office, 1947). On Disney films and the OCIAA, see Julianne Burton-Carvajal, "'Surprise Package': Looking Southward with Disney," in *Disney Discourse: Producing the Magic Kingdom,* ed. Eric Smoodin (New York: Routledge, 1994), pp. 131–147; Jose Piedra, "Pato Donald's Gender Ducking," in Smoodin, *Disney Discourse,* pp. 148–168; Lisa Cartwright and Brian Goldfarb, "Cultural Contagion: On Disney's Health Education Films for Latin America," in Smoodin, *Disney Discourse,* pp. 169–180.
4. Robert C. Murphy, "Inter-American Conservation," *Bird-Lore,* 42 (1940): 226. On the Western Hemisphere Convention, see Simon Lyster, *International Wildlife Law: An Analysis of International Treaties Concerned with the Conservation of Wildlife* (Cambridge: Grotius Publication, 1985), pp. 97–111; Sherman Strong Hayden, *The International Protection of Wild Life: An Examination of Treaties and Other Agreements for the Preservation of Birds and Mammals* (New York: AMS Press, 1942).
5. Osborn to Macgowan, 28 April 1941, RG1.1 200R, Box 263, Folder 3136, RFA.
6. Fairfield Osborn to John Marshall, 9 April 1940, RG1.1 200R, Box 263, Folder 3136, RFA. On the announcement of funding, see Osborn to Marshall, 14 August 1941, RG1.1 200R, Box 263, Folder 3136, RFA.
7. Both *High Over the Borders* and *Birds on the Wing* are available from the Media Services department of the Wildlife Conservation Society. My thanks to Tom

Veltre for his generosity in providing me with copies of these films. On the hummingbird footage, see the speech given by Fairfield Osborn to the Institute of Women's Professional Relations, 9 February 1940, RG1.1 200R, Box 263, Folder 3136, RFA, p. 2.

8. For an excellent account of the *Canada Carries On* series and Grierson's role in the National Film Board, see Gary Evans, *John Grierson and the National Film Board: The Politics of Wartime Propaganda* (Toronto: University of Toronto Press, 1984).
9. Snider to Osborn, 7 December 1948, Box 2, "Speeches & Writings. Carnegie Institute Lecture, Feb. 8, 1949" Folder, FOP-LC. Robert G. Snider, "The Conservation Foundation Starts Two Important Projects," *Animal Kingdom* 52 (1949): 128. For a discussion of UNSCCUR, see John McCormick, *Reclaiming Paradise: The Global Environmental Movement* (Bloomington: Indiana University Press, 1989), pp. 36–38. The international dimension of Osborn's environmentalism is also pointed out in Andrew Jamison and Ron Eyerman, *Seeds of the Sixties* (Berkeley: University of California Press, 1994), pp. 74–82.
10. Osborn, *The Limits of the Earth,* p. 29, 74. For additional comments on the Point Four program, see "Information Service Release," 3/16/49, Box 2, Speeches 1949, Jan.-Mar., Correspondence and Notes, FOP-LC.
11. "Grant Disclosure, 3/18/49," RG1.2 200D, Box 108, Folder 1666, RFA; "Grant Disclosure, 6/19/53," RG1.2 200D, Box 108, Folder 1666, RFA. On the Rockefeller Foundation's agricultural projects in Latin America, see Joseph Cotter, "The Rockefeller Foundation's Mexican Agricultural Project: A Cross-Cultural Encounter, 1943–1949," in *Missionaries of Science: The Rockefeller Foundation and Latin America,* ed. Marcos Cueto (Bloomington: Indiana University Press, 1994), pp. 97–125; Deborah Fitzgerald, "Exporting American Agriculture: The Rockefeller Foundation in Mexico, 1943–1953," *Social Studies of Science* 16 (1986): 457–483; Bruce H. Jennings, *Foundations of International Agricultural Research: Science and Politics in Mexican Agriculture* (Boulder, Colo.: Westview Press, 1988).
12. Snider, "The Conservation Foundation Starts Two Important Research Projects," p. 129.
13. *Yours Is the Land* is available from the Media Services department of the Wildlife Conservation Society.
14. These audience estimates are based on figures cited in a letter from V. C. Arnspiger to Fairfield Osborn, 16 September 1952, "1952 Motion Pictures" Folder, FOP-WCS. John C. Gibbs to Osborn, 21 April 1952, 1952 Motion Pictures Folder, FOP-WCS; Osborn to Warren Weaver, 30 June 1952, RG1.2 200D, Box 181, Folder 1671, RFA.
15. Snider to Osborn, 7 December 1948.
16. Katharine Drake, "Africa's Amazing Treetop Hotel," *Reader's Digest* (December 1959): 144.
17. *New York Times* (11 October 1938) 14:3. Reviewers at the time recognized the departure of *Below the Sahara* from the traditional African safari film. See, for example, the review in "The Screen," *Natural History* (June 1953): 284–285.

18. Julian Huxley to Armand Denis, 2 June 1960, Box 29, Folder 6, JHP. See also *New York Times* (10 October 1938), p. 14.
19. Armand Denis, *On Safari: The Story of My Life* (London: Collins, 1963), pp. 254, 234, 222–223.
20. *Below the Sahara* (RKO Pictures, 1953). One reviewer wrote of the "Walt Disney True-Life-ism" nature of the film. See Bosley Crowther, Review of *Below the Sahara, New York Times* (2 September 1953) 20:1.
21. Alan Moorehead, "A Reporter in Africa: The Birds and the Beasts Were There—I," *New Yorker* 33 (May 25, 1957):45–46; II 34 (June 1, 1957): 62; I: 66. See also Alan Moorehead, *No Room in the Ark* (New York: Harper & Bros., 1957). On Moorehead's life, see Moorehead, *A Late Education: Episodes in a Life* (London: Hamish Hamilton, 1970).
22. For a sampling of this literature, see Katharine Drake, "Last Chance for Africa's Wild Animals!" *Reader's Digest* (November 1960): 182–188; George and Jinx Rodger, "Where Elephants Have Right of Way," *National Geographic Magazine* (September 1960): 363–389; "Big Game—Casualty of Change in Africa," *U.S. News & World Report* (25 September 1961): 74–78; Smith Hempstone, "Twilight of the Great Beasts," *Saturday Evening Post* 23 (19 November 1960): 21–23, 73–74, 76–77; Elspeth Huxley, "Wildlife—Another African Tragedy," *New York Times Magazine* (11 October 1959): 22–23, 42, 44; George Treichel, "The Vanishing Herds," *Atlantic* 203 (April 1959): 85–87.
23. Julian Huxley, *Memories II* (New York: Harper & Row, 1973), p. 51. For historical analyses of Huxley's career as biologist and statesman of science, see C. Kenneth Waters & Albert Van Helden, *Julian Huxley: Biologist and Statesman of Science* (Houston: Rice University Press, 1992).
24. Fairfield Osborn, "A Field Study of African Parks and Wildlife," *Animal Kingdom* 59 (1956): 170.
25. See George Treichel, "Africa and Our Wildlife Heritage, Part I" *Animal Kingdom* 61 (1958): 162–168; Treichel, "Africa and Our Wildlife Heritage, Part II," *Animal Kingdom* 62 (1959): 22–28; F. Fraser Darling, "An Ecological Reconnaissance of the Mara Plains in Kenya Colony," *Wildlife Monographs* n. 5 (August 1960): 1–41; Darling, *Wild Life in an African Territory* (London: Oxford University Press, 1960).
26. For a listing of projects funded by the African Wild Life Fund, see George Wall Merck, "Report on the African Wild Life Fund," *Animal Kingdom* 63 (1960): 54–55 and "The Crisis Facing Wild Animals in Africa," *Animal Kingdom* 64 (1961).
27. Treichel, "Africa and Our Wildlife Heritage, Part II," pp. 23–24.
28. F. Fraser Darling, "Wildlife Husbandry in Africa," *Scientific American* 203 (1960): 123.
29. Treichel, "African and Our Wildlife Heritage," p. 167.
30. Hempstone, "Twilight of the Great Beasts," p. 77. Fairfield Osborn, "Animals and the Congo," *Animal Kingdom* 63 (August 1960): 129.

31. Julian Huxley, *The Conservation of Wild Life and Natural Habitats in Central and East Africa* (Paris: UNESCO, 1961), p. 15.
32. Juliette Huxley, *Wild Lives of Africa* (London: Collins, 1963), p. 243. Huxley's remarks about the ecological destructiveness of Maasai pastoralism as well as the view of Africa as an ecological time-bomb has recently been seriously questioned. See, e.g., Richard Bell, "Conservation with a Human Face: Conflict and Reconciliation in African Land Use Planning," in *Conservation in Africa: People, Policies and Practice,* ed. David Anderson and Richard Grove (New York: Cambridge University Press, 1987), pp. 79–102; Katherine Homewood and W. A. Rodgers, "Pastoralism, Conservation and the Overgrazing Controversy," in Anderson and Grove, *Conservation in Africa,* pp. 111–128; David Collett, "Pastoralists and Wildlife: Image and Reality in Kenya Maasailand," in Anderson and Grove, *Conservation in Africa,* pp. 129–148.
33. For perspectives that touch upon the powerful role that ecology can play in development schemes that exclude the voice of indigenous populations, see Jonathan S. Adams and Thomas O. McShane, *The Myth of Wild Africa: Conservation Without Illusion* (New York: W. W. Norton, 1992); Susanna Hecht and Alexander Cockburn, *The Fate of the Forest: Developers, Destroyers, and Defenders of the Amazon* (New York: Verso, 1989).
34. Gerald G. Waterson, ed., *Conservation of Nature and Natural Resources in Modern African States* (Morges, Switzerland: UNESCO, 1963), p. 13.
35. Huxley, *The Conservation of Wild Life,* p. 24.
36. Richard White suggests that environmentalists have predominately emphasized the eye over the hand in knowing nature. See Richard White, *The Organic Machine: The Remaking of the Columbia River* (New York: Hill & Wang, 1995).
37. Bernhard and Michael Grzimek, *Serengeti Shall Not Die* (London: Fontana Books, 1964).
38. Ibid., p. 46.
39. Julian Huxley to A. D. Owen, December 1956, Box 24, Folder 6, JHP. For a critical overview of the history of Serengeti National Park, see Adams & McShane, *The Myth of Wild Africa,* pp. 37–58. On the Fauna Preservation Society, see Richard Fitter and Sir Peter Scott, *The Penitent Butchers: The Fauna Preservation Society, 1903–1978* (London: Collins, 1978).
40. Huxley, *Conservation of Wild Life,* pp. 60–68.
41. On the importance of film in Owen's educational campaign, see Huxley to J. S. Owen, 2 June 1960, Box 29, Folder 6; Owen to A. Gille, 28 September 1961, Box 32, Folder 3, JHP.
42. *Wild Gold* is available from the Media Service department at the Wildlife Conservation Society.
43. Julian Huxley, *Africa View* (1931; reprint, London: Chatto & Windus, 1936), p. 60.
44. See, for example, Rotha to Huxley, 21 February 1941; Huxley to Rotha, 13 February 1941, Box 15, Folder 1; Huxley to Rotha, 6 August 1941, Box 15, Folder 4; Grierson to Huxley, 5 March 1942, Box 16, Folder 1; Grierson to

Huxley, 14 September 1947, Box 17, Folder 7, JHP. On the spy scandal incident, see Evans, *John Grierson and the National Film Board,* pp. 223–268.

45. Huxley to Osborn, 25 May 1960; Osborn to Huxley, 31 May 1960; J. S. Owen to Huxley, 21 May 1960; Huxley to J. S. Owen, 25 May 1960, Box 29, Folder 5; Huxley to Denis, 2 June 1960; Denis to Huxley, 17 June 1960, Box 29, Folder 6, JHP.
46. Huxley to Zuckerman, 23 February 1963, Box 34, Folder 2, JHP. Chancellor to Huxley, 5 March 1963, Box 34, Folder 3, JHP. On the history of the WWF, see Peter Scott, ed., *The Launching of a New Ark: First Report of the World Wildlife Fund* (London: Collins, 1965).
47. Joy Adamson, *Born Free: A Lioness of Two Worlds* (New York: Pantheon, 1960); Adamson, *Living Free: The Story of Elsa and Her Cubs* (London: Collins & Harvill Press, 1961); Adamson, *Forever Free* (New York: Harcourt, Brace, & World, Inc., 1962). The filmed version was also called *Born Free* (Columbia Pictures Release, 1965). Sales figures appear in Adrian House, *The Great Safari: The Lives of George and Joy Adamson* (New York: William Morrow & Co., 1993), p. 289.
48. *New Yorker,* 36 (June 4, 1960): 158. On the Adamson's marriage, see House, *The Great Safari.*
49. Adamson, *Living Free,* p. ix.
50. Ibid., p. x.
51. On the Adamsons' experiences in the Mau Mau rebellion, see House, *The Great Safari,* pp. 197–216. For the place of the Mau Mau in British memory, see John Lonsdale, "Mau Maus of the Mind: Making Mau Mau and Remaking Kenya," *Journal of African History* 31 (1990): 393–421.
52. Adamson to Huxley, 27 November 1964, Box 37, Folder 5; Huxley to Scott, 16 December 1964; Huxley to Adamson, 24 December 1964, Box 37, Folder 6, JHP. On Adamson's conservation efforts, see House, *The Great Safari.*
53. See, e.g., David Western, *In The Dust of Kilimanjaro* (Washington, D.C.: Island Press, 1997); David Western, R. Michael Wright, and Shirley C. Strum, eds., *Natural Connections: Perspectives in Community-based Conservation* (Washington, D.C.: Island Press, 1994); Adams & McShane, *The Myth of Wild Africa;* Ramachandra Guha, "The Authoritarian Biologist and the Arrogance of Anti-Humanism: Wildlife Conservation in the Third World," *The Ecologist* 21 (1997): 14–20. For an introduction to some of the criticisms leveled against Western and community-based conservation, see Michael McRae, "Survival Test for Kenya's Wildlife," *Science* 280 (24 April 1998): 510–512.

EPILOGUE

1. Mike McPhee and Jim Carrier, "Wild Photos Called Staged," *Denver Post* (9 February 1996): 1A.
2. "'Wild America' Case Raises Broader Ethical Issues," *The Denver Post* (16 February 1996): 6B; McPhee and Carrier, "Wild Photos Called Staged," p. 1A.

3. Jim Carrier, "Stouffer sought Klondike, Snow," *Denver Post* (16 February 1996): 1A.
4. On the market for contemporary nature films, see Robert La Franco, "Actors Without Agents," *Forbes* (8 April 1996): 82–83; Mark R. Orner, "Nature Documentary Explorations: A Survey History and Myth Typology of the Nature Documentary Film and Television Genre from the 1880s through the 1990s" (Ph.D. diss., University of Massachusetts–Amherst, 1996).
5. Revenue figures are from La Franco, "Actors Without Agents," pp. 82–83.
6. On the Attenborough anecdote, see Louis McElvogue, "Jaws, Claws, and Cash: Show Biz Jungle of Wildlife," *New York Times* (29 September 1997): E4. *The Trials of Life: Hunting and Escaping* (Turner Home Entertainment, 1991).
7. Mike McPhee and Jim Carrier, "Unnatural Selection? Critics Say Wildlife Shows Misleading Viewers," *Denver Post* (11 February 1996): 1A. On concerns about the commodification of nature, see Susan G. Davis, *Spectacular Nature: Corporate Culture and the Sea World Experience* (Berkeley: University of California Press, 1997); Dean MacCannell, "Nature Incorporated," in *Empty Meeting Grounds: The Tourist Papers* (London: Routledge, 1992), pp. 114–117; Jennifer Price, "Looking for Nature at the Mall: A Field Guide to the Nature Company," in *Uncommon Ground: Toward Reinventing Nature,* edited by William Cronon, pp. 186–203 (New York: W. W. Norton & Co., 1995).
8. On the consequences of the marginalization of animals from an economic and productive centrality in society and their transformation into spectacle, see John Berger, "Why Look at Animals?" in *About Looking* (New York: Pantheon Books, 1980).
9. On the role of communication media in shaping experience, see Neil Postman, *Amusing Ourselves to Death: Public Discourse in the Age of Show Business* (New York: Penguin Books, 1985). For a specific look at the impact of deregulated television in fostering a sense of impatience and shortsightedness with respect to conservation, see Thomas Veltre, "The Slums of the Global Village," *BBC Wildlife* (May 1990): 328–329. For additional perspectives on the impact of natural history film on attitudes toward nature, see Ralph H. Lutts, *The Nature Fakers: Wildlife, Science & Sentiment* (Golden, Colo.: Fulcrum Publishing, 1990), pp. 189–204; Bill McKibben, "Curbing Nature's Paparazzi," *Harper's Magazine* (November 1997): 19–24; Stephen Mills, "Pocket Tigers: The Sad Unseen Reality Behind the Wildlife Film," *Times Literary Supplement* (21 February 1997): 6; Charles Siebert, "The Artifice of the Natural," *Harper's Magazine* (February 1993): 43–51); Peter Steinhart, "Electronic Intimacies," *Audubon* 90 (November 1988): 10–13.

AFTERWORD

1. Quoted in Jonathan Miller, "March of the Conservatives: Penguin Film as Political Fodder," *New York Times,* 13 September 2005, F2.
2. Ibid.
3. Andrew Coffin, "The March of the Penguins," *World Magazine* 20 (6 August 2005).
4. Miller, "March of the Conservatives," F2. For an analysis of archetypal narrative strategies embedded in wildlife films, see Derek Bousé's informative book, *Wildlife Films* (Philadelphia: University of Pennsylvania Press, 2000). Over the last decade, scholarship on nature and environmental film has started to gain critical mass, albeit through quite different approaches informed by animal studies, ecocriticism, environmental history, and media studies to name a few. See, e.g., Andy Masaki Bellows and Marina McDougall, eds., *Science Is Fiction: The Films of Jean Painlevé* (Cambridge: MIT Press, 2000); David Ingram, *Green Screen: Environmentalism and Hollywood Cinema* (Exeter: University of Exeter Press, 2000); Akira Mizuta Lippit, *Electric Animal: Toward a Rhetoric of Wildlife* (Minneapolis: University of Minnesota Press, 2000); Jonathan Burt, *Animals in Film* (London: Reaktion, 2002); Finis Dunaway, *Natural Visions: The Power of Images in American Environmental Reform* (Chicago: University of Chicago Press, 2005); Cynthia Chris, *Watching Wildlife* (Minneapolis: University of Minnesota Press, 2006); and Scott MacDonald, "Up Close and Political: Three Short Ruminations on Ideology in the Nature Film," *Film Quarterly* 29 (2006): 4–21.
5. In the past decade, interest in approaches to animal agency and performativity have been critical in moving scholarship in animal studies beyond representation. The chapter on dolphins in *Reel Nature* was an early attempt by me to take up the issue of animal agency in wildlife film, and my own initial foray into uniting materialist and cultural approaches in environmental history. The issue of performance is further explored in the introduction to *Thinking with Animals: New Perspectives on Anthropomorphism,* ed. Lorraine Daston and Gregg Mitman (New York: Columbia University Press, 2005), pp. 1–14. For a sampling of recent critical work from very different disciplinary perspectives that addresses the problem of the animal in its otherness, see Steve Baker, "Sloughing the Animal," in *Zoontologies: The Question of the Animal,* ed. Cary Wolfe (Minnesota: University of Minnesota Press, 2003), pp. 147–164; Donna Haraway, *The Companion Species Manifesto: Dogs, People, and Significant Otherness* (Chicago: Prickly Paradigm Press, 2003); Brett Walker, *The Lost Wolves of Japan* (Seattle: University of Washington Press, 2005); Sarah Whatmore, "Embodying the Wild: Tales of Becoming Elephant," in *Hybrid Geographies: Nature, Culture, Spaces* (London: Sage, 2002), pp. 35–57.

6. For an excellent analysis of animal sex on television, see Chris, *Watching Wildlife*, pp. 122–166.
7. See Discovery Communications International, http://corporate.discovery.com/news/press/06q2/111606.html (accessed 26 May 2008).
8. Chris, *Wildlife Films*, pp. 79–121, offers a very informative summary of the changing market and content of wildlife television programming in the 1990s.
9. On the closing of *Survival*, see Robert Mendick, "*Survival* series near extinction," *The Independent*, 8 April 2001.
10. Derek Bousé, Review of "Cynthia Chris, *Watching Wildlife* (University of Minnesota Press, 2006)," *International Journal of Communication* 1 (2007): 96–110, on p. 103. Bousé's review gives a wonderful and rare insider's perspective on the changes that occurred in the production of natural history television in the late 1990s and continue to resonate to the present day.
11. Roger Ebert, review of *Grizzly Man*, *Chicago Sun-Times,* 12 August 2005.
12. "MIPDOC basking in new spotlight," *The Hollywood Reporter*, 2 April 2006.
13. Professor Ole Danbolt Mjøs, Nobel Peace Prize Presentation Speech, 10 December 2007, http://nobelprize.org/nobel_prizes/peace/laureates/2007/presentation-speech.html (accessed 26 May 2008).
14. "Sundance Channel's *The Green* Presented by Robert Redford Returns for Second Season," 28 February 2008, http://www.sundancechannel.com/blogs/thegreen/390323951 (accessed 26 May 2008).
15. See National Geographic's *Strange Days on Planet Earth*, http://www.pbs.org/strangedays (accessed 26 May 2008).
16. Bill McKibben, *The End of Nature* (New York: Random House, 1989).
17. Randy Malamud, "A New Breed of Environmental Film," *Chronicle of Higher Education*, 25 April 2008.
18. Stephen Mills, "Paper Tigers," *Times Literary Supplement*, 21 February 1997. For an insider's expressions of disillusionment, see Christopher Palmer, "Successful Environmental Filmmaking," http://www.centerforsocialmedia.org/eff_palmer.htm (accessed 26 May 2008). This observation is also based on conversations with filmmaker Sarita Siegel, fall 2007.
19. Michael Medved, "Don't Be Misled by 'Crappy Feet'," 17 November 2006, http://michaelmedved.townhall.com/blog/g/5094f586-fed7-4cf4-872c-d20b94c78024 (accessed 26 May 2008).
20. For revenue earnings of *Happy Feet*, see http://pro.imdb.com/title/tt0366548/boxoffice (accessed 26 May 2008).
21. Although the threat of global warming may have been implicit in the appeal of *March of the Penguins,* that wasn't the political message critics picked up on. For an inside look into the editorial control National Geographic exercises over independent filmmakers, see Sarita Siegel, "Reflections on Anthropomorphism in *The Disenchanted Forest*, in *Thinking*

with Animals: New Perspectives on Anthropomorphism, ed. Lorraine Daston and Gregg Mitman (New York: Columbia University Press, 2005), pp. 196–222.

22. Desson Thomson, "March of the Cuddly-Wuddly Documentaries," *Washington Post*, 29 July 2007.
23. *Arctic Tale* (National Geographic Films and Paramount Classics, 2007).
24. See, e.g., Jeanette Catsoulis, "A Lesson in Global Warming from Two Cold, Cute, Critters," *New York Times*, 25 July 2007; Roger Ebert, "Arctic Tale," *Chicago Sun-Times*, 3 August 2007; Peter Bradshaw, "Arctic Tale," *The Guardian*, 8 February 2008.
25. Brooks Barnes, "Disney Looks to Nature, and Creates a Film Division to Capture It," *New York Times*, 22 April 2008.
26. Thomason, "March of the Cuddly-Wuddly Documentaries." One has to wonder, though, how many millions of dollars companies are willing to lose in the name of being green. In an interview on SeattlePI.com, filmmaker Adam Ravetch estimated the budget of *Arctic Tale* to be around $10 million. See William Arnold, "Filmmaker tells the story behind the wildlife epic, 'Arctic Tale,'' http://seattlepi.nwsource.com/movies/325432_arcticintv30.html (accessed 26 May 2008). The film earned less than $900,000 in its first three months in theaters, see, http://pro.imdb.com/title/tt0488508/boxoffice (accessed 26 May 2008).
27. The Brock Initiative, http://www.brockinitiative.org/about.htm (accessed 26 May 2008).
28. Bill McKibben, *The Age of Missing Information* (New York: Random House, 1992), pp. 73–74.
29. Despite the market for nature and environmental film and the flourishing body of scholarship, we still have no good empirical studies that investigate the impact, intentional or unintentional, of film on particular species, communities, or landscapes. The chapters on dolphins and Africa in *Reel Nature* were a preliminary attempt to do so, as Dan Phillippon graciously pointed out in his review, "Nature on Screen," *The Review of Communication* 2.3 (2002): 273–288. Finis Dunaway's *Natural Visions*, as well as his article, "Gas Masks, Pogo, and the Ecological Indian: Earth Day and the Visual Politics of American Environmentalism," *American Quarterly* (2008): 67–99, is the best analysis we have to date of how the camera has shaped American environmental politics and, in some cases, particular places. Film has material effects on the world. George Stoney's 1978 film, *How the Myth Was Made*, which revisits the location of Robert Flaherty's 1934 classic, *Man of Aran*, offers telling insight into how a film can transform lives and landscapes. It is a subject that awaits an author.
30. Kathy de Nobriga, "Working Films Profile," http://www.workingfilms.org/article.php?id=109 (accessed 26 May 2008).
31. On *Blue Vinyl's* impact, see Patricia Thomson, "The Catalytic Role of Documentary Outreach," http://www.fundfilm.org/for_grant/for_

grant_article7_all.htm. See also the activist extras on the DVD version of *Blue Vinyl*, as well as http://www.myhouseisyourhouse.org (accessed 26 May 2008).

32. The Brock Initiative, http://www.brockinitiative.org/about.htm (accessed 26 May 2008).
33. *Sharkwater*, http://www.sharkwater.com (accessed 26 May 2008).
34. Jeffrey Kluger, "When Animals Attack—and Defend," *Time*, 7 June 2007, http://www.time.com/time/health/article/0,8599,1630667,00.html (accessed 26 May 2008). Adrienne Mand Lewin and Liz Tascio, "Germaphobes, be warned: Viral video strategies are catching," March 2008, http://publications.mediapost.com/?fuseaction=Articles.showArticleHomePage&art_aid=77025 (accessed 26 May 2008).
35. YouTube and other video Web sites are creating new possibilities for historical and media scholarship on wildlife and environmental film. Historical footage, public service announcements, and movies, which were once buried in archival and private collections, are now widely available. In a recent class I taught, for example, we used over two decades of Smokey the Bear PSA announcements on YouTube to explore changing American environmental attitudes and politics. The possibilities are endless.
36. *The Battle at Kruger*, http://www.youtube.com/watch?v=LU8DDYz68kM (accessed 29 April 2008).
37. When filmmaker Judith Helfand and I taught "Green Screen: Environmental Film in History and Action," we began the course with William Cronon's, "A Place for Stories: Nature, History, and Narrative," *Journal of American History* 78 (1992): 1347–1376. It is also a fitting place to end.
38. The quote is from Tim Robbins, http://www.mediathatmattersfest.org/about/ (accessed 25 May 2008).

CREDITS

Figure 1. Reproduced from the Collections of the Library of Congress.

Figure 3. Detail from photograph, courtesy of the Department of Library Services, American Museum of Natural History, New York.

Figure 4. Detail from photograph, neg. no. 311991, giant lizards *(Varanus komodoensis)*, habitat group in AMNH. Reprinted courtesy of the Department of Library Services, American Museum of Natural History, New York.

Figure 5. Courtesy of the Film Stills Archive, Museum of Modern Art, New York. Reprinted with permission.

Figure 9. Details from photographs, courtesy of the Department of Library Services, American Museum of Natural History, New York.

Figures 14–15. Copyright © Wildlife Conservation Society, headquartered at the Bronx Zoo.

Figures 16–17. Copyright © 1958 by Lois Crisler. Reprinted by permission of HarperCollins Publishers, Inc.

Figure 19. Courtesy of the Lincoln Park Zoo, Chicago Parks District.

Figure 20. Courtesy of the Western Historical Manuscript Collection, University of Missouri–St. Louis.

Figure 21. Reprinted by permission of William M. Gaines, Agent, Inc.

Figure 22. Courtesy of the Western Historical Manuscript Collection, University of Missouri–St. Louis.

Figures 23–25. Courtesy of Marineland of Florida.

Figure 26. Courtesy of Elgin Ciampi.

Figures 27–28. From *Serengeti Shall Not Die* by Bernhard and Michael Grzimek, translated by E. L. and D. Rewald. Translation copyright © 1959 by Ullstein A. G., Berlin. Copyright © 1960 by E. P. Dutton & Co., Inc., and Hamish Hamilton, Ltd. Used by permission of Dutton, a division of Penguin Putnam Inc.

Figure 29. Copyright © Wildlife Conservation Society, headquartered at the Bronx Zoo.

Figure 30. Reprinted with the permission of Elsa Trust, which continues the Adamsons' conservation work through the education center at Elsamere.

INDEX

WEYERHAEUSER
ENVIRONMENTAL
BOOKS

The Natural History of Puget Sound Country by Arthur R. Kruckeberg

Forest Dreams, Forest Nightmares: The Paradox of Old Growth in the Inland West by Nancy Langston

Landscapes of Promise: The Oregon Story, 1800–1940 by William G. Robbins

The Dawn of Conservation Diplomacy: U.S.-Canadian Wildlife Protection Treaties in the Progressive Era by Kurkpatrick Dorsey

Irrigated Eden: The Making of an Agricultural Landscape in the American West by Mark Fiege

Making Salmon: An Environmental History of the Northwest Fisheries Crisis by Joseph E. Taylor III

George Perkins Marsh: Prophet of Conservation by David Lowenthal

Driven Wild: How the Fight against Automobiles Launched the Modern Wilderness Movement by Paul S. Sutter

The Rhine: An Eco-Biography, 1815–2000 by Mark Cioc

Where Land and Water Meet: A Western Landscape Transformed by Nancy Langston

The Nature of Gold: An Environmental History of the Alaska/Yukon Gold Rush by Kathryn Morse

Faith in Nature: Environmentalism as Religious Quest by Thomas R. Dunlap

Landscapes of Conflict: The Oregon Story, 1940–2000 by William G. Robbins

The Lost Wolves of Japan by Brett L. Walker

Wilderness Forever: Howard Zahniser and the Path to the Wilderness Act by Mark Harvey

On the Road Again: Montana's Changing Landscape by William Wyckoff

Public Power, Private Dams: The Hells Canyon High Dam Controversy by Karl Boyd Brooks

Windshield Wilderness: Cars, Roads, and Nature in Washington's National Parks by David Louter

Native Seattle: Histories from the Crossing-Over Place by Coll Thrush

The Country in the City: The Greening of the San Francisco Bay Area by Richard A. Walker

Drawing Lines in the Forest: Creating Wilderness Areas in the Pacific Northwest by Kevin R. Marsh

Plowed Under: Agriculture and Environment in the Palouse by Andrew P. Duffin

Making Mountains: New York City and the Catskills by David Stradling

The Fishermen's Frontier: People and Salmon in Southeast Alaska by David F. Arnold

Shaping the Shoreline: Fisheries and Tourism on the Monterey Coast by Connie Y. Chiang

Dreaming of Sheep in Navajo Country by Marsha Weisiger

CYCLE OF FIRE BY STEPHEN J. PYNE

Fire: A Brief History

World Fire: The Culture of Fire on Earth

Vestal Fire: An Environmental History, Told through Fire, of Europe and Europe's Encounter with the World

Fire in America: A Cultural History of Wildland and Rural Fire

Burning Bush: A Fire History of Australia

The Ice: A Journey to Antarctica